AF556804

# INTERFACIAL SYNTHESIS

# INTERFACIAL SYNTHESIS

## VOLUME III

## Recent Advances

*edited by*

**CHARLES E. CARRAHER, JR.**
*Department of Chemistry*
*Wright State University*
*Dayton, Ohio*

**JACK PRESTON**
*Monsanto Textiles Company*
*Pensacola, Florida*

MARCEL DEKKER, INC. New York and Basel

**Library of Congress Cataloging in Publication Data**
Main entry under title:

Interfacial synthesis.

Volume 3 edited by C.E. Carraher and J. Preston.
Includes bibliographical references and indexes.
Contents: v. 1. Fundamentals -- v. 2. Polymer applications and technology -- v. 3. Recent advances.
1. Polymers and polymerization. 2. Chemistry, Organic--Synthesis. I. Millich, Frank. II. Carraher, Charles E. III. Preston, Jack, [date]
QD281.P6I48 574'.28 75-27750
ISBN 0-8247-6372-6 (v. 1) AACR2

These papers originally appeared in *Journal of Macromolecular Science - Chemistry*, Volume A15, Number 5, edited by George E. Ham.



MARCEL DEKKER, INC.
270 Madison Avenue, New York, New York 10016

Current printing (last digit):
10 9 8 7 6 5 4 3 2 1

PRINTED IN THE UNITED STATES OF AMERICA

# Preface

The present volume is an ongoing effort to update advances in the broad area of interfacial synthesis. A further aim of this volume is to present further aspects dealing with related fundamental facets of this subject. To this end, the editors have brought together in this book the several papers which were read at a symposium, organized and chaired by Carraher and Preston, entitled "Interfacial Synthesis." The symposium, sponsored by the Division of Organic Coatings and Plastics Chemistry and cosponsored by the Division of Polymer Chemistry, was held on March 24-25, 1980, at the Spring National Meeting of the American Chemical Society in Houston.

This volume is divided into three sections. The first section, subtitled "New Frontiers and Recent Applications," is intended to illustrate the diversity in the use of the interfacial technique both industrially and on a laboratory scale. The opening chapter, "Comments on the Status and Future of Interfacial Polycondensation," is by Paul W. Morgan, the author of the first book, *Condensation Polymers: By Interfacial and Solution Methods* (Wiley, 1965), dealing with this field and through whose efforts the technique was popularized to the point where it is today one of the major polycondensation techniques. Chapters dealing with the application of the interfacial concept include such topics as lipids and liposomes, microcapsulation and reverse osmosis desalination. The synthesis of polydyes, copolyamides and block copolymers is detailed.

Probably the topic of most current research interest, because of theoretical considerations and the likelihood of immediate industrial applications, is that of phase transfer catalysis (PTC) and related catalysis systems. It was our good fortune to be able to obtain papers by pioneers in the field of PTC to be included in the Symposium and in this volume. In-depth, current, and critical reviews by Mathias and Imai are featured in this second section, subtitled "Phase Transfer Catalysis." The description of new catalysis systems is detailed.

The final section, "New Polycondensation Systems," details newly developed interfacial systems including gas/liquid systems.

In summary, the several chapters of the present volume may be viewed as dealing with the broader themes of interfacial syntheses initiated in *Interfacial Synthesis,* Volume I - *Fundamentals* and Volume II - *Polymer Applications and Technology* edited by Millich and Carraher (Marcel Dekker, Inc., 1977).

The editors wish to sincerely thank those who have made this volume possible, especially the authors and reviewers. We also wish to thank Charles E. Carraher III and Shawn Carraher for their help with the indexing.

Charles E. Carraher, Jr.
Jack Preston

# Contents

# Contributors

RUSSELL G. ARNOLD *Bureau of Veterinary Medicine, Food & Drug Administration, Rockville, Maryland 20857*

AJIT K. BANTHIA *Chemistry Department and Polymer Materials and Interfaces Laboratory, Virginia Polytechnic and State University, Blacksburg, Mirginia 24061*

J. E. CADOTTE *Filmtec Corporation, 15305 Minnetonka Boulevard, Minnetonka, Minnesota 55343*

CHARLES E. CARRAHER, JR. *Department of Chemistry, Wright State University, Dayton, Ohio 45435*

ETHEL Z. CASASSA *Chemical Engineering Department, Carnegie-Mellon University, Pittsburgh, Pennsylvania 15213*

DEAN-YANG CHAO *Chemical Engineering Department, Carnegie-Mellon University, Pittsburgh, Pennsylvania 15213*

W. DEITS *GTE Laboratories, Inc., 40 Sylvan Road, Waltham, Massachusetts 02154*

VOLKER ENKELMANN *Institut für Makromolekulare Chemie der Universität Freiburg, D-7800 Freiburg 1 Br., West Germany*

WILLIAM A. FELD *Department of Chemistry, Wright State University, Dayton, Ohio 45435*

JEAN M. J. FRÉCHET *Department of Chemistry, University of Ottawa, Ottawa, Ontario, K1N-9B4, Canada*

R. G. FREELIN *Chemistry Department and Polymer Materials and Interfaces Laboratory, Virginia Polytechnic Institute and State University, Blacksburg, Virginia 24061*

KIYOSHIGE FUKUDA *Department of Chemistry, Saitama University, Urawa, 338, Japan*

S. GROSSMAN *The Upjohn Company, North Haven, Connecticut 06473*

MOHAMMED HABIB *Department of Chemistry, University of Alabama, University, Alabama 35486*

MARK HENSON *Chemical Engineering Department, Carnegie-Mellon University, Pittsburgh, Pennsylvania 15213*

SAMUEL J. HUANG *Department of Chemistry and Institute of Materials Science, University of Connecticut, Storrs, Connecticut 06268*

H. H. HUB *Institut für Organische Chemie, Johannes Gutenberg-Universität, D-6500 Mainz, West Germany*

B. HUPFER *Institut für Organische Chemie, Johannes Gutenberg-Universität, D-6500 Mainz, West Germany*

YOSHIO IMAI *Department of Polymer Chemistry, Yamagata University, Yamagata, 992, Japan*

SHOEHEI IMANISHI *Department of Chemistry, Sophia University, Tokyo, 102, Japan*

HARALD KAPP *Institut für Makromolekulare Chemie der Universität Freiburg, D-7800 Freiburg i. Br., West Germany*

R. S. KING *Filmtec Corporation, 15305 Minnetonka Boulevard, Minnetonka, Minnesota 55343*

H. KOCH *Institut für Organische Chemie, Johannes Gutenberg-Universität, D-6500 Mainz, West Germany*

JOSEPH KOZAKIEWICZ *Department of Chemistry and Institute of Materials Science, University of Connecticut, Storrs, Connecticut 06268*

EDWIN A. LEWIS* *Department of Chemistry, University of Alabama, University, Alabama 35486*

GÜNTER LIESER *Institüt fur Makromolekulare Chemie der Universität Freiburg, D-7800 Freiburg i. Br., West Germany*

---

*Current affiliation: *Thermochemical Institute, Brigham Young University, Provost, Utah 84602*

DAVID LUNSFORD *Chemistry Department and Polymer Materials and Interfaces Laboratory, Virginia Polytechnic Institute and State University, Blacksburg, Virginia 24061*

R. J. MAJERLE *Filmtec Corporation, 15305 Minnetonka Boulevard, Minnetonka, Minnesota 55343*

LON J. MATHIAS* *Department of Chemistry, Auburn University, Auburn, Alabama 36830*

J. E. MCGRATH *Chemistry Department and Polymer Materials and Inter faces Laboratory, Virginia Polytechnic Institute and State University, Blacksburg, Virginia 24061*

H. MICHAEL MOLLOY *Department of Chemistry, Wright State University, Dayton, Ohio 45435*

PAUL W. MORGAN *822 Roslyn Avenue, West Chester, Pennsylvania 19380*

HIROO NAKAHARA *Department of Chemistry, Saitama University, Urawa, 338, Japan*

NAOYA OGATA *Department of Chemistry, Sophia University, Tokyo, 102, Japan*

TAKAO ONOZAKI *Department of Chemistry, Sophia University, Tokyo, 102, Japan*

REBECCA J. PAESSUN *Department of Chemistry, Wright State University, Dayton, Ohio 45435*

R. J. PETERSEN *Filmtec Corporation, 15305 Minnetonka Boulevard, Minnetonka, Minnesota 55343*

CHARLES U. PITTMAN, JR. *Department of Chemistry, University of Alabama, University, Alabama 35486*

J. S. RIFFLE *Chemistry Department and Polymer Materials and Interfaces Laboratory, Virginia Polytechnic Institute and State University, Blacksburg, Virginia 24061*

H. RINGSDORF *Institute of Organic Chemistry, University of Mainz, D-6500 Mainz, West Germany*

KOHEI SANUI *Department of Chemistry, Sophia University, Tokyo, 102, Japan*

---

*Current affiliation: *Department of Polymer Science, University of Southern Mississippi, Southern Station, Mississippi 39401*

JACK A. SCHROEDER *Department of Chemistry, Wright State University, Dayton, Ohio 45435*

H. SCHUPP *Institute of Organic Chemistry, University of Mainz, D-6500 Mainz, West Germany*

MACY SCHWARZ* *Department of Chemistry, Wright State University, Dayton, Ohio 45435*

RICHARD A. SCHWARZ* *Department of Chemistry, Wright State University, Dayton, Ohio 45435*

M. PAUL SERVE' *Department of Chemistry, Wright State University, Dayton, Ohio 45435*

RAYMOND B. SEYMOUR *Department of Polymer Science, University of Southern Mississippi, Hattiesburg, Mississippi 39401*

YOSHIO SHIBASAKI *Department of Chemistry, Saitama University, Urawa, 338, Japan*

G. ALLAN STAHL** *B. F. Goodrich R & D Center, 9921 Brecksville Road Brecksville, Ohio 44141*

BERND TIEKE *Institut für Makromolekulare Chemie der Universität Freiburg, D-7800 Freiburg i. Br., West Germany*

LARRY P. TORRE† *Department of Chemistry, Wright State University, Dayton, Ohio 45435*

O. VOGL *Polymer Science and Engineering Department, University of Massachusetts, Amherst, Massachusetts 01003*

DEAN C. WEBSTER *Chemistry Department and Polymer Materials and Interfaces Laboratory, Virginia Polytechnic Institute and State University, Blacksburg, Virginia 24061*

GERHARD WEGNER *Institut für Makromolekulare Chemie der Universität Freiburg, D-7800 Freiburg i. Br., West Germany*

---

*Current affiliation: *Cosden Oil, Big Springs, Texas 79720*

**Current affiliation: *Research and Development, Phillips Petroleum Company, Bartlesville, Oklahoma 74003*

†Current affiliation: *Department of Materials Science and Engineering, Washington State University, Pullman, Washington 99164*

# INTERFACIAL SYNTHESIS

# NEW FRONTIERS AND RECENT APPLICATIONS

# Comments on the Status and Future of Interfacial Polycondensation

Paul W. Morgan

822 Roslyn Avenue
West Chester, Pa. 19380

ABSTRACT

Interfacial polycondensation came into prominence in the 1950's with the work of Schnell on polycarbonates, Conix on polyesters, and Wittbecker, Morgan and co-workers on polyurethanes, polyamides and other polymer classes. There were a few earlier patent references. The method has several variations and is a highly effective procedure for the rapid preparation of many polymers on a small scale. It is particularly appropriate for polymers which are thermally unstable or unmeltable, for use with volatile or unstable intermediates and for polymers with reactive functional groups. Literally thousands of polymers have been made by this process. There are a number of short-comings to the process, most of which can be overcome by adjustment of the polymerization conditions. Polycarbonates and aromatic polyamides are reported to be prepared commercially by stirred interfacial processes. Another application is the encapsulation of pesticides, inks, and other materials. As to the future, the processes are a basic part of the chemist's tool kit and will continue to be used in the laboratory for rapid syntheses and for those polymers which are not obtained as readily by other means. The basic tests of commercial use of these processes, as with any other process, will be attainment of results not available by other routes, economic considerations, and safety and environmental factors.

## INTRODUCTION

Interfacial polycondensation came into prominence as a major polymer-forming technique about 25 years ago with the work of

Schell (1) on polycarbonates at Bayer, Conix (2,3) on polyesters and polysulfonates at Gevaert, and Wittbecker, Morgan, and coworkers (4,5,6) on polyurethanes, polyamides, and other polymer classes at Du Pont. Scattered earlier reports were available in the patent literature as indicated in Table 1. (See introduction to reference 7 for greater detail on early literature.)

Interest in the polycarbonates as a new class of polymers for plastics was widespread. This interest was enhanced by the development of solution and melt methods for their preparation (8,9).

## Polymer Film Formation by Interfacial Polycondensation

Perhaps a major contribution to the excitement and interest in interfacial polycondensation was the description of its broad applicability in 1958-1959 (4) and that fascinating experiment, the Nylon Rope Trick (6). In this experiment a film of aliphatic poly-

TABLE 1

Early Reports of Interfacial Condensation Polymerization

| Authors | Type of Polymer | Date |
|---|---|---|
| I. G. Farbenindustrie | urethanes | 1944,1945 |
| Orthner, Schlack et al. | urethanes | 1954,1955 |
| Jones, McFarlane | amides, urethanes | 1952 |
| Kirby | amides | 1953 |
| Schnell, Bottenbruch, Krimm | carbonates | 1954 |
| Magat, Strachan | amides | 1955 |
| Wittbecker | urethanes | 1956 |
| Conix | aromatic esters | 1957 |
| Conix | aromatic sulfonates | 1958 |
| Wittbecker, Morgan et al. | amides and others | 1958 |
| Hill, Kwolek, Morgan | aromatic amides | 1959 |
| Morgan, Kwolek | nylon rope experiment | 1959 |

amide is formed at the liquid-liquid interface and removed continuously as a small flexible rope.

This experiment has probably been done and viewed by more people than any other in polymer chemistry which is carried out solely for the experiment itself. By this condition we exclude such acts as the application of paints, cyanoacrylate glues, and epoxy cements. The Nylon Rope Trick has been done in technical lectures, school classrooms and polymer preparation courses, science fairs, World Fairs, and as part of sales pitches.

Many thousands of reprints have been distributed. The Eastman Kodak Co. pictured the Nylon Rope Trick in advertising for the intermediates (10) and Edmund Scientific Co., Rutherford, N.J., had kits available at one time.

The unstirred polymerization system never yielded an acceptable fine nylon fiber. The fibers were not compact enough for good strength. However, the film or rope of collapsed film can be shredded in a blender, or other device, to yield fibrous particles which will form tough paper or which can be used to bind other staple fibers into a paper (11).

The film-forming process has been used to form membranes which were examined for ion-exchange properties (12). It has been adapted for the encapsulation of inks, catalysts and insecticides (13a). Such encapsulated insecticides are being produced on a commercial scale (13b). Whitfield, Miller and Wasley (14) developed a process for coating wool fibers in fabric form with a thin film of interfacially formed polymer. The purpose was to prevent fabric shrinkage during ordinary laundering. The process reached pilot plant scale but eventually gave way to other procedures. Trubek Laboratories based an advertisement for diacid chlorides on the possible use of this process (15).

Chang (16) at McGill University and others have used interfacial polycondensation to form artificial cell-like capsules containing enzymes and other materials. These cells have possible applications in artificial kidneys, catalysis and in ion-exchange processes.

## Interfacial Polycondensation with Stirring

Polymerization with stirring offered the greatest versatility in polymer making. Polymers could be prepared in a few minutes at room temperature which took hours or days to form by melt or solid-state polymerizations. Furthermore, non-melting polymers and thermally unstable polymers could be made. Thermally unstable intermediates could be used. Thousands of new polymers were prepared and examined over the first several years.

In the Pioneering Research Laboratory of the Du Pont Textile Fibers Department the search for high-temperature stable, non-melting polymers became a goal. This eventually led to the aromatic polyamides now manufactured and sold under the tradename, Nomex® aramid fibers and papers (17).

The first general interfacial procedure employed a solution of diacid chloride dissolved in a water-immiscible solvent combined with an aqueous solution of the complementary intermediate and an inorganic base. Sometimes authors report that they have used a "standard method", perhaps meaning the above process. The writer has been guilty of this statement on at least one occasion. Such a statement overlooks the fact that there are several major variations and there are so many important variables that no one set of polymerization conditions can be considered standard or broadly applicable for optimum results.

Let us review several variations of interfacial polycondensation and in doing so illustrate the point that by experimentation one can optimize the molecular weight and yield and thereby obtain better polymer samples for characterization and evaluation.

## Phase Transfer Agents

During the last ten years there has been much excitement about phase transfer agents in chemical synthesis. These agents are compounds, such as quaternary ammonium salts, which when added to a two-phase system promote the reaction by transferring one of the intermediates across the interfacial boundary into the presence of the second intermediate. At least two books have appeared on the subject (18).

Upon casual examination of the literature, it was surprising to find that in general the use of these agents is treated as being a very recent discovery, whereas in fact Schnell and co-workers (1, 8) were using phase transfer agents for polycarbonate preparation by interfacial polycondensation in the early 1950's. Since many of the recent papers have been in the field of small molecule synthesis, one gets the impression that organic chemists are not reading the polymer literature. One hopes the opposite is not true.

The many syntheses to which phase transfer catalysts are being applied open up routes to new or improved condensation polymers. Polymer chemists should take advantage of this expanding technology and make additional contributions of their own. Several papers at this Symposium point the way.

Schnell and his co-workers used tertiary amines and quaternary ammonium compounds to accelerate and improve the synthesis of both polycarbonates (1,8,19) and the monomer, diphenyl carbonate (20), from phenols and phosgene. The additives not only increase the polymerization rate but their use leads to higher molecular weight in the polymers.

Schnell and Bottenbruch proposed that the tertiary amines formed adducts with the acid chloride in the organic phase and thus provided a more reactive intermediate. This is possible but it is more likely they are phase transfer agents for the phenoxide. A few careful experiments would settle this question.

Conix (2) used quaternary ammonium, arsonium, phosphonium, and tertiary sulfonium compounds to accelerate and enhance the preparation of aromatic polyesters, polycarbonates from bischloroformates (Table 2) and polysulfonates from bisphenols and disulfonyl chlorides (22). The data indicate arsonium compounds are the most

TABLE 2

Effect of Quaternary Compounds as Accelerators in the Preparation of the Polycarbonate from 2,2-Bis(4-hydroxyphenyl)propane and Its Bischloroformate by Interfacial Polycondensation[a]

| Accelerator | Variation of intrinsic viscosity of the product with amount of accelerator and reaction time, dl./g.[b] 0.05%[c] 5 min.[d] | 0.05%[c] 30 min.[d] | 2.0%[c] 5 min.[d] | 2.0%[c] 30 min.[d] |
|---|---|---|---|---|
| Triethylbenzylammonium chloride | - | 0.26 | 0.10 | 0.70 |
| Methyltriphenyl-arsonium iodide | - | 0.68 | 0.96 | 1.65 |
| Methyltriphenyl-phosphonium iodide | 0.40 | 0.80 | 0.90 | 1.24 |
| Benzyltriphenyl-phosphonium chloride | 0.41 | 0.90 | 0.90 | 1.36 |
| p-Xylylene-α,α'-bis(di-hydroxyethyl)sulfonium bromide | - | 0.28 | 0.38 | 0.76 |

[a] From Conix (21).

[b] [η] measured in sym-tetrachloroethane at 25°C.

[c] By weight based on the theoretical weight of polymer

[d] Counted from point of complete addition of acid chloride (5 min.).

effective accelerators (Table 2; 21, 23). However, the additives are not strictly comparable in terms of substituents. Increased amounts of an accelerator gave higher viscosity numbers for the product (increased molecular weight) at the same reaction times.

Morgan (5,24) called phase transfer agents accelerators and explained their function in accord with current thought; i.e. that they act to transfer the aqueous reactant (phenoxide in the case of phenols) into the organic phase and make it available for reaction there. Experiments with acid chlorides and phenols plus quaternary ammonium salts in single organic solvents had shown no catalytic effect on the condensation reaction. The formation of aromatic polyester films in an unstirred interfacial polycondensation was shown to take place in the organic phase and the addition of tetraethylammonium chloride to the aqueous phase greatly accelerated the rate of film formation (25,26).

Lower alkyl substituted quaternary ammonium compounds, such as tetramethyl and tetraethyl ammonium salts, have been reported to be ineffective as phase transfer agents for some interfacial condensation reactions. Morgan (27) found that tetraethylammonium chloride was quite effective in the preparation of aromatic polyesters. Relatively large amounts of the salt were required for short reaction times of ten minutes or less and a maximum degree of polymerization (Figure 1). With higher concentrations of salt the degree of polymerization dropped. This could be due in part to impurities in the salt.

Brzozowski, Dubczynski and Petrus (28) have recently found tetrabutylammonium hydroxide effective in polycondensation of 2,2-bis(4-hydroxyphenyl)-1,1-dichloroethane with isophthaloyl chloride and have examined the mechanism.

The interfacial polycondensation of phenolphthalein and terephthaloyl chloride proceeds fairly well without a phase transfer agent. However, with tetraethylammonium chloride much higher inherent viscosities are obtained (29) and the small scale, stirred reactions have been completed in less than 2.5 minutes. Polyesters based on phenolphthalein have been developed for plastics applications at the Russian Academy of Sciences.

## Polymerization in Water-Miscible Solvents

An important variation of interfacial polycondensation is the use of wholly or partially water-miscible solvents (30-32). The procedure is applicable to several classes of polymers such as polyesters, polyamides, polyurethanes, and polysulfonamides, and in particular to systems wherein the diacid chloride component has relatively high hydrolytic stability. Hydrolysis is further reduced by use of weakly alkaline acid acceptors.

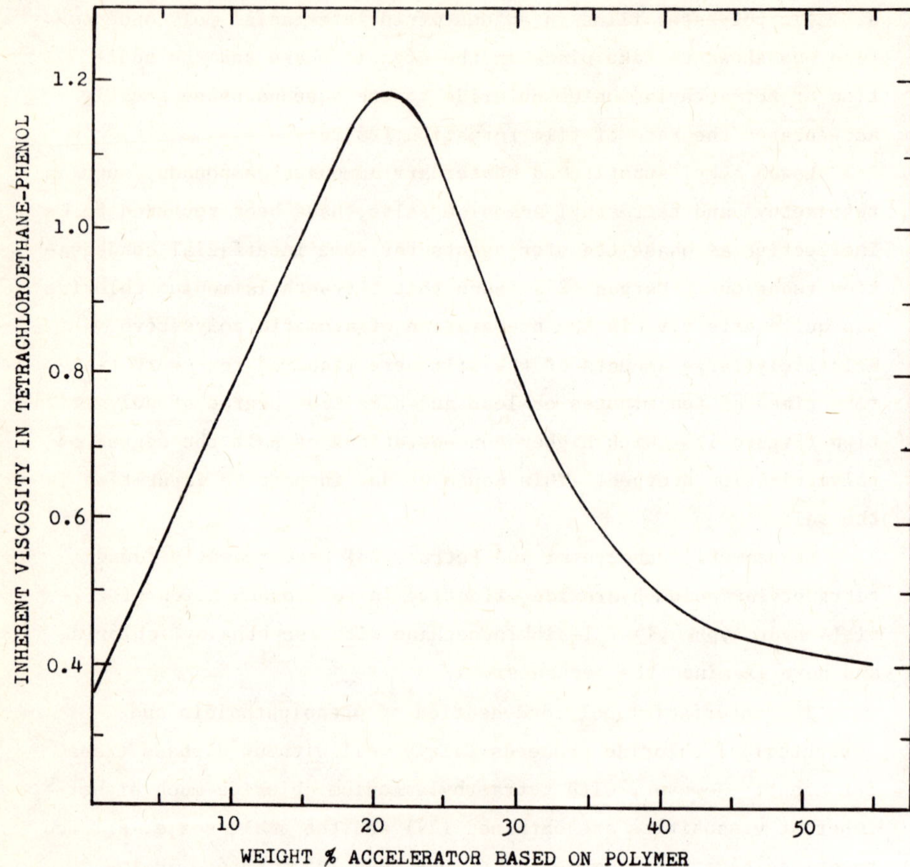


Fig. 1. Variation in inherent viscosity of the polyester from 2,2-bis(4-hydroxyphenyl)propane and terephthaloyl chloride with amount of tetraethylammonium chloride in an interfacial polycondensation system. Procedure: 0.02 mole reactants in 1,1,2-trichloroethane-water (60/130 by vol.) and 0.04 mole sodium hydroxide with stirring for 10 min. (27).

Some specific polymers which have been made in high molecular weight are:

Poly(hexamethylene sebacamide)

Poly(terephthaloyl <u>trans</u>-2,5-dimethylpiperazine)

Poly(m-phenylene isophthalamide)

Poly(1,4-phenylenedioxycarbonylpiperazine)

Poly(hexamethylene 1,3-phenylenedisulfonamide)

Poly(terephthaloyl oxalamidrazone) (33)

Examples of useful organic solvents are tetrahydrofuran, acetone, cyclohexanone, and 2,4-dimethyltetramethylene sulfone. Advantages are accelerated reaction rates, enhanced polymer swelling and ease of polymer recovery. Although the solvents by themselves may be highly water-miscible, this miscibility is greatly reduced by the presence of salt, base, intermediates and polymer.

As with other polycondensation systems, there is considerable sensitivity to variables such as solvent ratio, concentration of intermediates and manner of mixing and therefore the optimum combinations must be sought for highest molecular weight and yield. This is illustrated by the effect of tetrahydrofuran-water ratio (Figure 2)

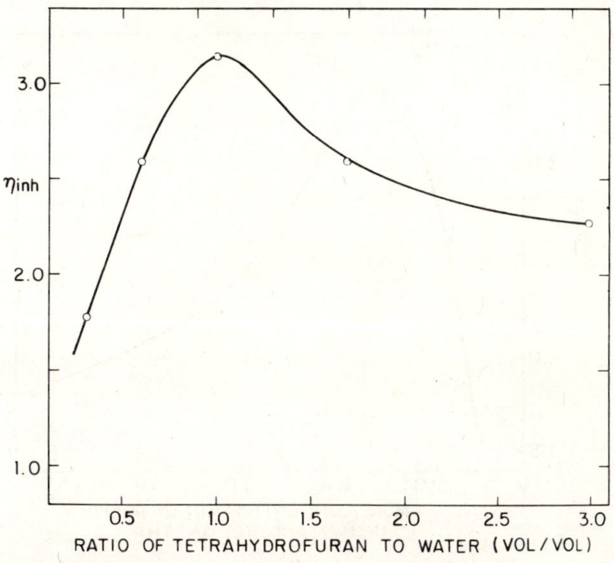


Fig. 2. Inherent viscosity versus the ratio of tetrahydrofuran to water in the preparation of poly(terephthaloyl trans-2,5-dimethylpiperazine). Total liquid volume, 200 ml.; 0.02 mole of terephthaloyl chloride, 0.02 mole of diamine, and 0.04 mole of sodium carbonate; stirred for 5 min. Yields 84-90%. Inherent viscosity in tetrachloroethane-phenol (40-60 by wt.).

and intermediates concentration (Figure 3) for the preparation of poly(terephthaloyl trans-2,5-dimethylpiperazine) (32).

Interfacial polycondensation in water-miscible solvents is being used in Japan on a large scale to prepare aromatic polyamides. The process has been much studied in Russia by Sokolov and co-workers (34).

Chemists at Glanzstoff A. G. studied the preparation of poly(terephthaloyl oxalamidrazone) in tetrahydrofuran-water quite extensively (33). A variation of the procedure, also used by Sokolov (35), is to decrease or modify the phase miscibility by the addition of salts such as sodium and lithium chlorides and sulfates.

The above polymer was produced on a pilot plant scale by Akzo Research and Engineering N.V. and in chelated form promoted

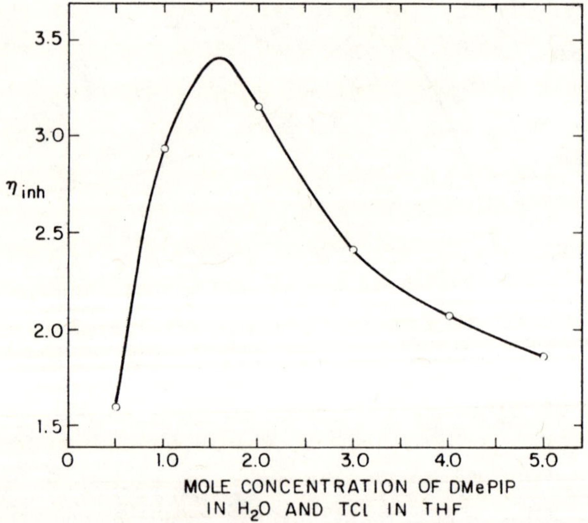


Fig. 3. Inherent viscosity versus mole concentration of intermediates in preparation of poly(terephthaloyl trans-2,5-dimethylpiperazine). 100 ml. of tetrahydrofuran and of water; 2 moles of sodium carbonate for each mole of diamine; stirred 5 min. Yields 86-94%. Inherent viscosity in tetrachloroethane-phenol (40-60 by wt.).

as a highly flame resistant fiber (36). For this program the polymer was prepared in hexane-water.

Aromatic polyesters based on 2,2-bis(4-hydroxyphenyl)propane and iso- and terephthaloyl chlorides are produced in Russia by the water-miscible solvent interfacial process for use in plastics (35,37). The procedure has been called emulsion polycondensation by Sokolov.

## Gas-Liquid Interfacial Polycondensation

The final topic is a variation of interfacial synthesis which we have recently applied to polyoxamide preparation (38). Polyoxamides have been traditionally difficult to prepare because of thermal instability and/or low solubility.

In 1960 Sokolov and his co-workers (39) introduced a new gas-liquid process which yielded good molecular weight ranges for polyoxamides from a variety of diamines but the polymer yields were about 60% at best. The process employs an aqueous solution of the diamine and oxalyl chloride vapor in an inert gaseous diluent. In the course of their extensive studies they prepared the polyoxamides from hexamethylenediamine, 1,4-phenylenediamine, and benzidine with intrinsic viscosity/percent yield as follows: 1.1/58, 1.44/66, 1.62/56. Sokolov and Astakhova (40) examined the use of inert organic liquids as the solvent for hexamethylenediamine in the gas-liquid process, but obtained lower intrinsic viscosities and yield than when water was used.

We have found that much higher yields of polymer result when both water and a water-immiscible organic solvent are used and polymers with high viscosities can be obtained.

The reactions were carried out in a small resin kettle equipped with a 2000-rpm disk stirrer. Equal volumes of water and organic solvent phases were used with the diamine and any alkali dissolved in the mixed phases. Oxalyl chloride at 5-20 mole % excess (based on diamine) was warmed in a generator and carried by means of a flow of dry nitrogen such that a gas mixture of 25% by volume of acid chloride passed over the surface of the stirred liquids at a rate of 2 liter/hr.

Several variables were studied in a limited way with 2,2'-dimethyl-4,4'-diaminobiphenyl as the diamine. Variables were the concentration of diamine, the proportion of inorganic base, the organic solvent, and the reaction temperature. Figure 4 shows the effect of diamine concentration on yield and inherent viscosity. Table 3 gives the inherent viscosities (0.5 g./100 ml. of solution in 98% sulfuric acid at 30°C.) and yields for a number of aromatic polyoxamides.

In addition to the goal of a method of synthesis, an objective was to determine whether these presumably stiff-chain polyamides would yield liquid crystalline (anisotropic) solutions. The oxamides from 2,2'-dimethyl-4,4'-diaminobiphenyl and chloro-1,4-phenylenediamine were soluble enough and of sufficient molecular weight to yield anisotropic solutions.

While the application of the gas-liquid interfacial process has been newly applied to advantage for the preparation of polyoxamides, the process is old in that it was used early by Schnell (1,8) and others in the preparation of polycarbonates.

## Future of Interfacial Polycondensation

Interfacial polycondensation and its modifications are valuable tools for the polymer researcher. They provide advantages of simplicity and speed for obtaining samples of polymer in high molecular weight and in an easily isolated form. Because of their complexity, they also offer a challenge to kineticists and others who would better understand the effects of the many variables.

The polymer chemist should and will choose among the various synthetic methods available today and not be bound by a particular route. In part the choice will be determined by the polymer and the

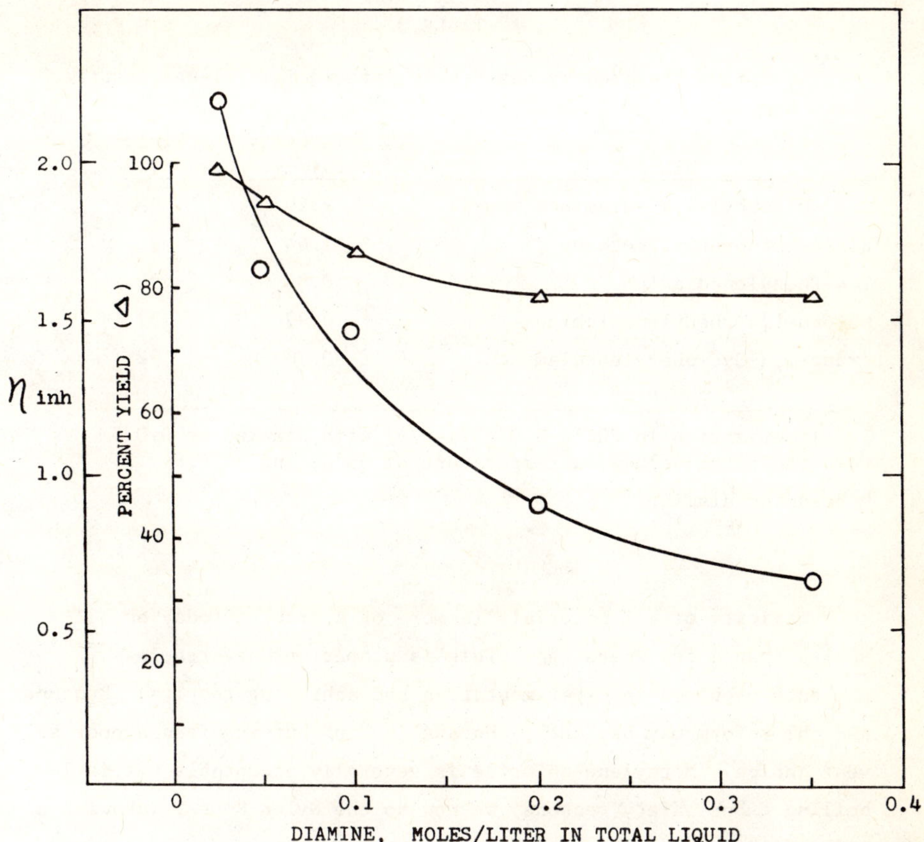


Fig. 4. Inherent viscosity and yield versus diamine concentration for the preparation of poly(2,2'-dimethyl-4,4'-biphenylene oxamide) in an oxalyl chloride vapor-chloroform-water system. Temperature, 55°C.; pH held between 6.8 and 7.2 by automatic titration.

applicability of the processes. On the other hand the type of sample and its expected use may be determining.

The industrial chemist may have other considerations to take into account. The politics of selling a product to management may determine that one process is more acceptable than another; for example, local prejudices or the goal of introducing a new product into existing facilities.

TABLE 3

Polyoxamides Prepared by Gas-Liquid Interfacial Polycondensation[a]

| Diamine | Inherent viscosity (dl/g) | Yield[b] (%) |
|---|---|---|
| 2,2'-Dimethyl-4,4'-diaminobiphenyl | 2.20 | 99 |
| Bis(4-aminophenyl)methane | 1.64 | 97 |
| 1,4-Phenylenediamine | 0.87 | 92 |
| Chloro-1,4-phenylenediamine | 0.97 | 92 |
| *trans*-1,4-Cyclohexylenediamine | 0.76 | 92 |

[a] Polymerization in $CHCl_3$-$H_2O$ (1/1 v/v) with diamine at 0.025M in the total volume, a temperature of 55°C, and pH 6.8-7.2.

[b] Based on diamine.

Toxicity of all materials is more of a problem today on all scales than a few years ago. This is proper but nevertheless presents problems in experimentation and achieving results. Benzene and chloroform may be "out". Hexane is "in" but is often a poor solvent choice. Methylene chloride is generally acceptable but is low boiling. For safety reasons, we now do the Nylon Rope Trick with a high-boiling chlorofluorocarbon solvent.

Some problems with large scale use of interfacial polycondensation in polymer preparation are:

Toxicity of solvents

Incomplete yields in some cases

High energy for stirring

Heat of reaction and stirring

Corrosion of equipment

Disposal of by-products

These problems can be overcome as is shown by the manufacture of polycarbonates in Europe and the U.S., aromatic polyamides in Japan, and a number of polyesters in Russia.

The continued interest in interfacial polycondensation and other interfacial syntheses is demonstrated by the range of topics presented and discussed at this Symposium.

## REFERENCES


[1] H. Schnell, Angew. Chem., 68, 633 (1956); H. Schnell, L. Bottenbruch, and H. Krimm, Belgian Patent 523,543 (1954), assigned to Farbenfabriken Bayer A. G.

[2] A. J. Conix, Ind. Chim. Belge, 22, 1457 (1957); Ind. Eng. Chem., 51, 147 (1959); see also I. Goodman, "Polyesters" in Encycl. Polymer Sci. Tech., ed. by N. M. Bikales, Wiley-Interscience, New York, 11, 96 (1969).

[3] A. J. Conix, Belgian Patent 565,478 (7/1/58), assigned to Gevaert Photo-Producten N.V.; Chem. Abstr., 55, 25356 (1961).

[4] E. L. Wittbecker and P. W. Morgan, J. Polymer Sci., 40, 289 (1959) and following papers; presented in part at the 134th National ACS Meeting, Chicago, Ill., September 1958.

[5] P. W. Morgan, SPE (Soc. Plastics Engrs.) J., 15, 485 (1959).

[6] P. W. Morgan and S. L. Kwolek, J. Chem. Ed., 36, 182, 530 (1959)

[7] P. W. Morgan, "Condensation Polymers by Interfacial and Solution Methods", Interscience, New York, 1965.

[8] H. Schnell, "Chemistry and Physics of Polycarbonates", Interscience, New York, 1964.

[9] W. F. Christopher and D. W. Fox, "Polycarbonates", Reinhold, New York, 1962.

[10] Chem. Eng. News, Nov. 9, 1959.

[11] P. W. Morgan, U.S. Patent 3,068,527 (12/18/62), assigned to the Du Pont Co.

[12] V. Enkelmann and G. Wegner, J. Appl. Polymer Sci., 21 (4), 997 (1977).

[13] (a) W. Sliwka, Angew. Chem. Internat. Ed., 14 (8), 539 (1975); J. E. Vandegaer, U.S. Patent 3,577,515 (1963), assigned to Pennwalt Corp.; (b) Chem. Eng. News, July 29, 1974, p. 15 and June 5, 1972, p. 68.

[14] R. E. Whitfield, L. A. Miller and W. L. Wasley, Textile Res. J., 31 (8), 704 (1961); J. Appl. Polymer Sci., 8, 1607 (1964).

[15] Chem. Eng. News, May 28, 1962, p. 18.

[16] T. M. S. Chang, Chemtech, 5 (2), 80 (1975).

[17] P. W. Morgan, Plast. Rub. Mater. Appl., 4 (1), 1 (1979).

[18] W. P. Weber and G. W. Gokel, "Phase Transfer Catalysis in Organic Synthesis", Springer-Verlag, New York, 1977; C. M. Starks and C. Liotta, "Phase Transfer Catalysis", Academic Press, New York, 1978.

[19] L. Bottenbruch, "Polycarbonates" in Encycl. Polymer Sci. Tech., ed by N. M. Bikales, Wiley-Interscience, New York, 10, 710-764 (1969).

[20] L. Bottenbruch and H. Schnell, German Patent 1,101,386 (1961); K. H. Meyer and H. Schnell, German Patent 1,056,141 (1959), assigned to Farbenfabriken Bayer A. G.; see ref. 8, p. 93.

[21] A. J. Conix, French Patent 1,284,226 (2/9/62) and German Patent 1,162,079 (1/30/64), assigned to Gevaert Photo-Producten N. V.

[22] A. J. Conix and U. L. Laridon, Angew. Chem., 72, 116 (1960); A. J. Conix, French Patent 1,288,306 (3/23/62), assigned to Gevaert Photo-Producten N. V.

[23] B. C. Oxenrider, French Patent 1,299,933 (7/27/62), assigned to Allied Chemical Corp.

[24] Ref. 7, pp. 49, 343, 349-355, 370, 497, 507.

[25] P. W. Morgan and S. L. Kwolek, J. Polym. Sci., 40, 307, 315 (1959).

[26] Ref. 7, p. 49.

[27] Ref. 7, p. 353.

[28] Z. K. Brzozowski, J. Dubczynski, and J. Petrus, J. Macromol. Sci.-Chem., A13 (7), 875 (1979).

[29] P. W. Morgan, in "Macromolecular Syntheses", Vol. 4, ed. by W. J. Bailey, Wiley, New York, 1972, p. 29.

[30] H. W. Hill, S. L. Kwolek, and P. W. Morgan, French Patent 1,199,460 (12/14/59) and U.S. Patent 3,006,899 (10/31/61), assigned to the Du Pont Co.

[31] Ref. 7, pp. 99-101, 191-192, 492-493.

[32] P. W. Morgan, S. L. Kwolek, H. W. Hill, "Synthesis of Condensation Polymers by Interfacial Polymerization in Systems with High Phase Miscibility", presented at the National ACS Meeting, Dallas, TX, April 1973.

[33] M. Wallrabenstein, A. Schöpf and D. Frank, U.S. Patent 3,718,625 (2/27/73) and British Patent 1,286,240 (8/23/72), assigned to Glanzstoff A. G.

[34] L. B. Sokolov, Polymer Sci. (USSR), 7, 662 (1965).

[35] T. V. Kudim and L. B. Sokolov, Vysokomol. Soedin., Ser. A, 20 (8), 1802 (1978); Chem. Abstr., 90, 6721 (1979).

[36] D. W. van Krevelen, Angew. Makromol. Chem., 22, 133 (1972).

[37] L. B. Sokolov, Japan Plastics Age, Nov.-Dec. 1975, pp. 47-49.

[38] T. C. Pletcher and P. W. Morgan, J. Polymer Sci., Chem. Ed., 18, 643 (1980).

[39] L. B. Sokolov, in "Interfacial Synthesis", Vol. 1, ed. by F. Millich and C. E. Carraher, Jr., Dekker, New York, 1977, pp. 141-165.

[40] L. B. Sokolov and A. S. Astakhova, Polymer Sci. (USSR), 5 792 (1963).

# Polymerization of Lipid and Lysolipid Like Diacetylenes in Monolayers and Liposomes

H.H. HUB, B. HUPFER, H. KOCH, and H. RINGSDORF

Institut für Organische Chemie
Johannes Gutenberg-Universität
D-6500 Mainz, West-Germany

ABSTRACT

Since conventional model membranes (Black Lipid Membranes (BLM), liposomes) lack in long time stability, a number of polymerizable lysophospholipid and phospholipid like monomers were synthesized, where the diacetylene moiety was the polymerizable unit. The hydrophilic head groups were simple ammonium, ether, or amine structures as well as phosphorus containing groups also occurring in nature. A first characterization of the monomers was achieved by surface pressure-area diagrams at the gas-water interface. Additionally, these monolayers could be polymerized via UV irradiation yielding stable,totally rigid polymer films. The mechanism of the polyreaction was followed with a new device permitting the recording of monolayer UV spectra. Electron microscopy showed that sonication of aqueous suspensions of phospholipid and lysophospholipid analogs yielded spherical vesicles, which could be polymerized by UV light under complete retention of their structure.

INTRODUCTION

One of the most exciting phenomena in cell and molecular biology is the cell-cell interaction. If mice lymphocytes come in contact with tumor cells, the result can be an interaction of the antigens of the cancer cell with the sensitized lymphocyte, which finally results in a destruction of the tumor cell [1]. This kind of interaction is an interesting problem to mimic. The simplest way to do this is to use phospholipid liposomes as cell models. Unfortunately, these vesicles have a very low stability, and in the collision of a tumor cell and a liposome the latter would be the less stable partner and would be destroyed or endocytosed. Therefore, our attention turns toward more stable cell models. These models can be obtained by connecting via polymerization the molecules forming a spherical liposome [6,7,8].

As initial model compounds we used simple diacetylene carbonic acids with different alkyl chain lengths (I). Since a monolayer at the gas-water interface can be viewed as one half of a biological membrane, we spread these monomers on a Langmuir trough and polymerized the resulting oriented monolayers via UV irradiation yielding very rigid, red colored polymer films (II) of a thickness of approximately 2,5 nm [2]. The same kind of topochemical diacetylene polymerization had formerly been observed in the solid state [3] and in multilayers [4] leading to the same kind of conjugated red colored polymers.

Our monolayer polymer films were also found to form bilayers if deposited onto a porous substrate such as an electron microscope grid. These bilayers can span holes as large as 0,5 mm in diameter, are stable in air and water for months, and are very good, stable cell membrane models.

Simple carbonic acids as (I) are not at all perfect analogs to naturally occurring phospholipids, and thus we synthesized

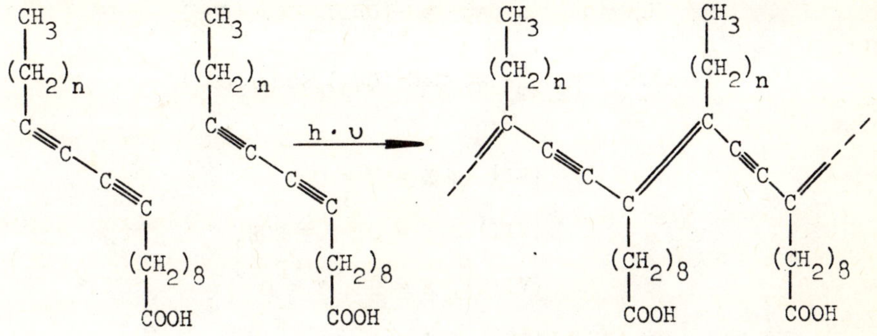

$$n\ \mathrm{CH_3-(CH_2)_n-C{\equiv}C-C{\equiv}C-(CH_2)_8-COOH} \xrightarrow{h\cdot\upsilon} \left[\mathrm{=C(-(CH_2)_n-CH_3)-C{\equiv}C-C(-(CH_2)_8-COOH)=}\right]_n$$

(I) (II)

compounds containing head groups also occurring in natural membranes such as the polymerizable phosphatidic acid analog (III), the cephalin analog (IV), and the lecithin analog (V) [5]. The lactone (XI) was the end product of a reaction, which was expected to lead to betain (XIV) [8]. Since natural phospholipids contain two alkyl chains per molecule, as initial models we prepared polymerizable compounds with two alkyl chains and simple ether (VI, VIII), amine (VII), ammonium (IX) [6,7], and betain (X) [8] head groups. Additionally, the head groups of the lysophospholipid analogs (III - V) were used to prepare the first real phospholipid analogs polymerizable in oriented systems (XII, XIII) [5].

$$\mathrm{H_3C-(CH_2)_{12}-C{\equiv}C-C{\equiv}C-(CH_2)_9-OR}$$

(III): $R = PO_3H_2$

(IV): $R = \mathrm{PO(O^{\ominus})-O-(CH_2)_2-\overset{\oplus}{N}H_3}$

(V): $R = \mathrm{PO(O^{\ominus})-O-(CH_2)_2-\overset{\oplus}{N}(CH_3)_3}$

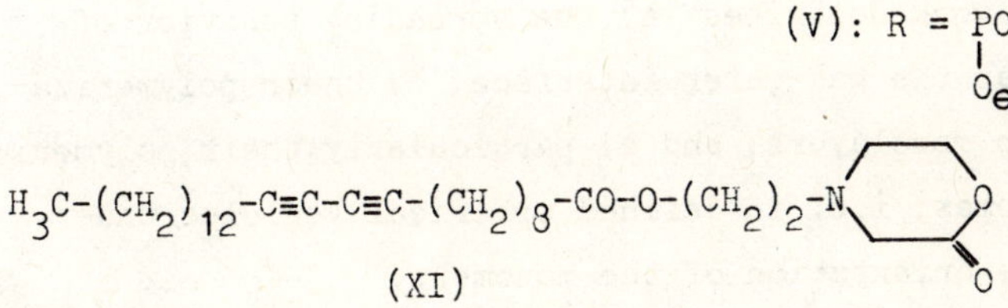

$$\mathrm{H_3C-(CH_2)_{12}-C{\equiv}C-C{\equiv}C-(CH_2)_8-CO-O-(CH_2)_2-N\langle(CH_2)_2-O-CO-CH_2\rangle}$$

(XI)

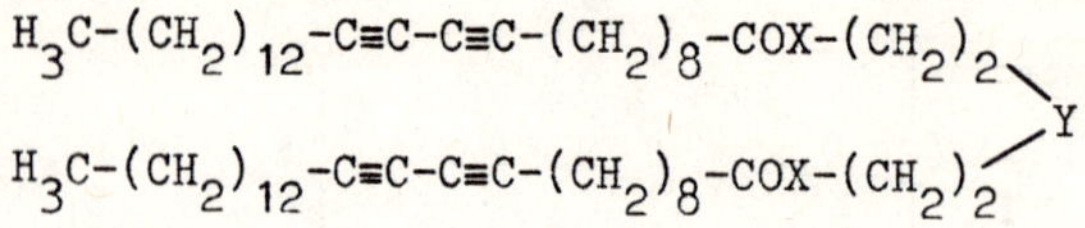

$$\begin{array}{l} H_3C-(CH_2)_{12}-C\equiv C-C\equiv C-(CH_2)_8-COX-(CH_2)_2 \diagdown \\ \qquad\qquad\qquad\qquad\qquad\qquad\qquad\qquad\qquad\qquad\qquad\quad Y \\ H_3C-(CH_2)_{12}-C\equiv C-C\equiv C-(CH_2)_8-COX-(CH_2)_2 \diagup \end{array}$$

(VI) : $X = Y = O$

(VII) : $X = O;\ Y = NCH_3$

(VIII): $X = NH;\ Y = O$

(IX) : $X = O;\ Y = \overset{\oplus}{N}(CH_3)_2\ Br^{\ominus}$

(X) : $X = O;\ Y = N-(CH_2)_2-SO_3H$

(XIV) : $X = O;\ Y = N-CH_2-COOH$

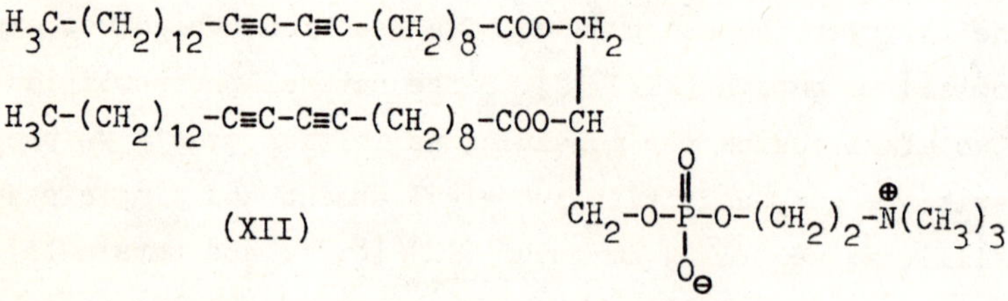

$$\begin{array}{l} H_3C-(CH_2)_{12}-C\equiv C-C\equiv C-(CH_2)_8-COO-CH_2 \\ \qquad\qquad\qquad\qquad\qquad\qquad\qquad\qquad\qquad\quad | \\ H_3C-(CH_2)_{12}-C\equiv C-C\equiv C-(CH_2)_8-COO-CH \\ \qquad\qquad\qquad\qquad\qquad\qquad\qquad\qquad\qquad\quad | \\ \qquad\qquad\qquad\qquad\qquad\qquad\qquad\qquad\qquad\quad CH_2-O-\overset{O}{\overset{\|}{\underset{O^{\ominus}}{\underset{|}{P}}}}-O-(CH_2)_2-\overset{\oplus}{N}(CH_3)_3 \end{array}$$

(XII)

$$\begin{array}{l} \qquad\qquad\qquad\qquad\qquad H_3C-(CH_2)_{16}-COO-CH_2 \\ \qquad\qquad\qquad\qquad\qquad\qquad\qquad\qquad\qquad\quad | \\ H_3C-(CH_2)_{12}-C\equiv C-C\equiv C-(CH_2)_8-COO-CH \\ \qquad\qquad\qquad\qquad\qquad\qquad\qquad\qquad\qquad\quad | \\ \qquad\qquad\qquad\qquad\qquad\qquad\qquad\qquad\qquad\quad CH_2-OPO_3H_2 \end{array}$$

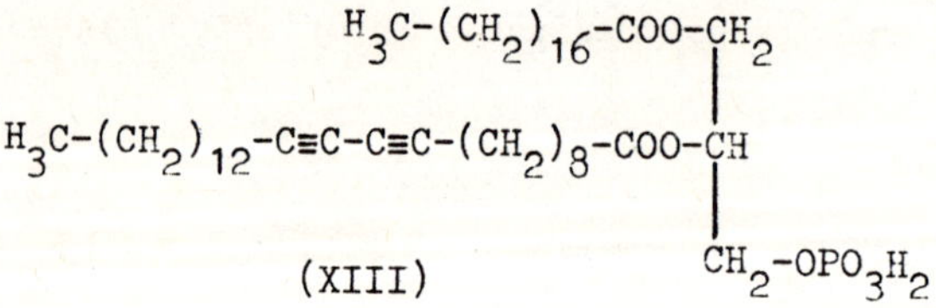

(XIII)

The present work describes a) the spreading behavior of these monomers at the gas-water interface, b) their polymerization behavior in monolayers, and c) particularly their polymerization in liposomes, i.e. in defined spherical vesicles under retention of the orientation of the monomers.

## EXPERIMENTAL

The preparation of the acids and alcohols of the type (I) has already been described [9]. For our experiments we used the acid (F: 68-69°C) and alcohol (F: 59-60°C) with n=12, which can be obtained from 1-Iodopentadecyne-1 and 10-Undecynoic acid and 10-Undecyne-1-ol, respectively as starting compounds. The phosphoric acid ester (III),(F: 79-81°C), was synthesized by the reaction of the corresponding alcohol (see above) with excess $POCl_3$ in dry $CCl_4$ and hydrolysis of the crude reaction product with $H_2O$ [5,10]. In a similar manner, the cephalin (IV) and lecithin (V) analogs were prepared. β-Bromoethyl-phosphoric-acid-dichloride [10] and β-Phthalimidoethyl-phosphoric-acid-dichloride [11] were the phosphorylating reagents. The phthalic acid group was split off with Hydrazine [11] yielding (IV), (F: >230°C); the Bromo compound was quaternized with Trimethylamine in $CHCl_3/CH_3CN$ and bromine ions were removed with $Ag_2CO_3$ in $CHCl_3/CH_3OH/H_2O$ [13] yielding (V), (F: >250°C). Acylation of Diethyleneglycol, N-Methyldiethanolamine, Bis(2-aminoethyl)ether Hydrochloride, and N,N-Bis-(2-hydroxyethyl)-2-aminoethanesulfonic acid (BES), respectively with the chloride of (I), n=12 (from (I) and Oxalylchloride) at 0°C in dry $CHCl_3$ in the presence of pyridine lead to the formation of (VI), (F: 59,5-60,5°C [6]);(VII), (F: 47,5-48,5°C [6]); (VIII), (F: 100-101°C [6]); and (X), (F: 109°C [8]). (IX), (F: 93,5-94,5°C [7] ), was obtained by quaternization of (VII) with $CH_3Br$ in dry acetone at 0°C. Conventional reactions of phospholipid chemistry [11,12] were used for the synthesis of (XII), (F: soft. 58°C, melted 190°C, cp. ref. 12 [5]). (XIII), (F: 37-38°C [5]) was prepared via the iodohydrine method with silver di-t-butyl phosphate as phosphorylating agent [14]. N,N-Bis(2-hydroxyethyl)glycine (BICIN) did not react with the chloride of (I), n=12 under the formation of the expected (XIV), but after one alcoholysis step ring formation occurred and the lactone (XI) (F: 45°C [8]) was obtained.

Monolayers were spread from chloroform solutions all having concentrations of about 1 mg/ml. The films were spread on a LAUDA film balance, where surface pressure and area were automatically recorded. Polymerization was carried out via UV irradiation (254 nm) with an energy of 5 $mW/cm^2$ at the water surface under nitrogen [2]. UV absorbance in the monolayer was measured with a new device described recently [15].

Liposomes were prepared by sonication of aqueous suspensions (1 mg/ml) of the monomers in water under nitrogen at 50°C (BRANSON sonifier, model B 15). Polymerization of liposomes was achieved by irradiation of these solutions with multichromatic light (mercury high pressure lamp) at 18°C with a water filter between beam source and sample. In between, the samples were removed from the lamp and the optical density was measured (BECKMAN model 25 UV spectrometer).

Scanning electron micrographs were obtained with a ISI-40 scanning electron microscope operated at 15 kV at a magnification of 15,000.

## SPREADING BEHAVIOR OF THE MONOMERS

The surface pressure-area diagrams of the lysophospholipid analogs (III, IV, V, XI) exhibit significant differences due to the different head groups. The phosphatidic acid analog (III) at 20°C shows a single condensed phase with a collapse point of 0,25 $nm^2$/molecule, whereas the cephalin analog (IV) exhibits both a condensed and an expanded film. In the diagram of compound (V) in the entire investigated temperature region (2°C to 50°C) only a liquid expanded phase occurs. This is due to the great volume of the trimethylammonium group, which prevents a dense packing of the alkyl chains in the monolayer. The lactone (XI) at 20°C also exhibits only a liquid expanded phase (Fig. 1).

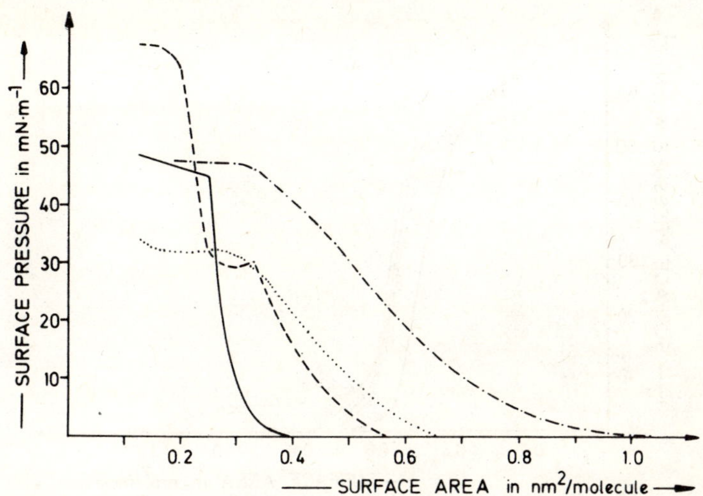


FIG. 1. Surface pressure-area diagrams of lysophospholipid analogs (III, IV, V, XI) at 20°C. (III): (————); (IV): (– – – – –); (V): (–·–·–·–); (XI): (··········).

In contrast to the lysophospholipid analogs the lipid analogs (VI - X) show collapse areas greater than 0,40 nm$^2$/molecule as to be expected from the two long chains per molecule (Fig. 2). Monomers (VI) and (VII) at 20°C exhibit collapse areas of 0,42 and 0,43 nm$^2$/molecule, respectively without real sharp collapse points, whereas (VIII - X) show collapse areas of 0,48, 0,51, and 0,48 nm$^2$/molecule with sharp collapse points similar to monomer (III). The increase in the collapse areas for (VIII) and (IX) may be explained by hydrogen bonding between the molecules of (XIII) and a steric and charge repulsion for (IX) and (X).

## MONOLAYER POLYMERIZATION

The UV initiated polymerization of diacetylenes is a topochemical reaction and does not take place in solution [3]. In the monolayer the polyreaction only occurs, if the temperature of the subphase is below the phase transition temperature i.e. in the

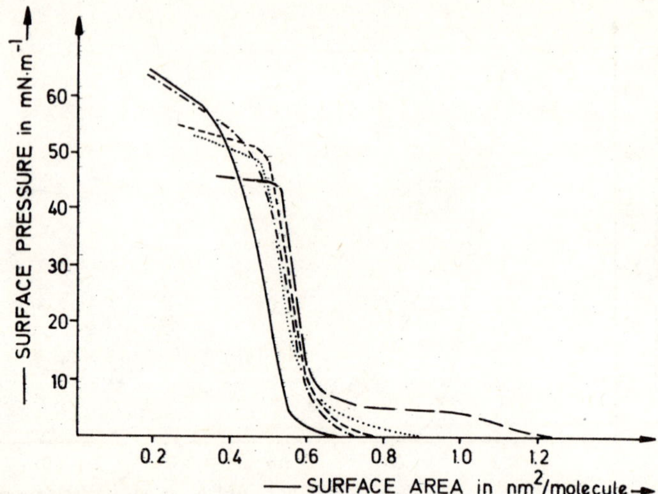


FIG. 2. Surface pressure-area diagrams of lipid analogs (VI - X) at 20°C. (VI): (—·—·—·—); (VII): (————); (VIII): (-------); (IX): (—— ——); (X): (············).

condensed phase, which again points to a topochemical effect. The polymerization under constant surface pressure can qualitatively be followed by the change of surface area [2]. However, an unchanged area after some time does not necessarily mean a complete conversion to the polymer. This can be proved with our new device [15] that permits the recording of monolayer UV spectra during the polymerization process.

As a result of the conjugated nature of the polymer backbone (II) the final polymerized monolayers of all the monomers have a high extinction coefficient in the visible spectrum and can be seen as a reddish tint on the water surface. Because the initial monomers are colorless the degree of absorbance of the monolayer is a criterion for the rate of conversion. With our apparatus the monolayer UV spectra of compounds (VII - IX) were recorded during the polymerization. Fig. 3 summarizes the absorbance curves of (VIII) at various irradiation times.

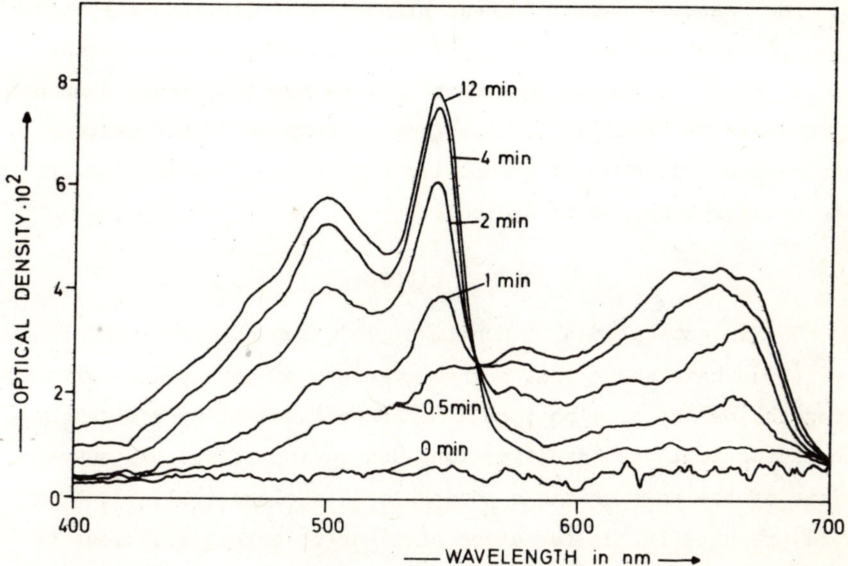


Fig. 3. UV/VIS Multiplot of absorbance of a monolayer of (VIII) versus polymerization time. Constant surface pressure: 10 $mN \cdot m^{-1}$; 20$^oC$, $N_2$ atmosphere.

It becomes obvious that the blue species of the polymer ($\lambda_{max}$: 660 nm) can only be observed in the time interval from 0 up to 2 min. After that time the conversion to the red polymer takes place. The absorption spectrum of the final red polymer monolayer film ($\lambda_{max}$: 500, 540 nm) is identical with the spectra of polymer monolayers [6]. The color change blue→red during polymerization is also observed in the solid state and in multilayers and is probably due to a phase change from a monomer-rich to a polymer-rich solid solution. The blue color most likely results from the polymer dissolved in a monomer matrix. There is probably stress of some sort on the polymer backbone, causing a different electronic structure and therefore a different absorbance. Once enough polymer is formed, a transition from a monomer rich to a polymer rich phase occurs.

The reaction rates of these bifunctional monomers are higher than those of the acids (I), and also the maximum absorbances are slightly different from those of the monofunctional compounds. By further UV irradiation both show a decrease of the maximum absorbance, which could eventually be due to a degradation of the polymer backbone by radicals.

## POLYMER LIPOSOMES

The investigations of Kunitake [16], Deguchi [19], and Fendler [20] have shown that a great variety of amphiphiles with two long chains are able to form a membrane like bilayer assembly. According to these investigations sonication of aqueous suspensions of the polymerizable phospholipid analogs (VII, IX, X, and XIII) results in the formation of clear colorless solutions of monomer liposomes. Electron microscopy proves the formation of bi- and multilayered vesicles of a defined spherical shape and of different diameters (range: 100 nm up to several µm). The diameter and number of bilayers strongly depend on sonication time and intensity. After 30 min of sonication only bilayer liposomes of a diameter of about 100 nm are formed. Further sonication does not alter the size.

We could also confirm the results of other investigators [17,18], who reported the formation of vesicles from a number of single chain amphiphiles. Aqueous suspensions of the acid (I), n=12, the corresponding alcohol, and the lactone (XI) did form liposomes upon sonication with ultrasound. Additionally, (XI) was found to form black lipid membranes [22], if a decane solution of this lactone is deposited on a small aperture of a teflon foil in an apparatus similar to that described by Mueller and Rudin [21]. Investigations concerning the polymerization behavior of these BLM are in progress.

UV irradiation of the monomer liposome solutions results in the formation of polymer liposomes indicated by a color change from colorless via blue to red. The polymerization rate can be followed by the change of optical density. Approximately the same kind of absorption spectra as for the polymer monolayers (Fig. 3) is obtained. Fig. 4 shows the UV/VIS spectrum of a liposome solution of (XI) after 45 min of UV irradiation showing that only traces of the blue polymer ($\lambda_{max}$: 606, 656 nm) are still present.

Electron microscopy shows that the shape of the liposomes remains unchanged during the polymerization. Only the diameters of the spheres are smaller than those of the monomer liposomes. As far as we know, this is the first example for a transformation of a defined monomeric liposome into a defined polymeric liposome

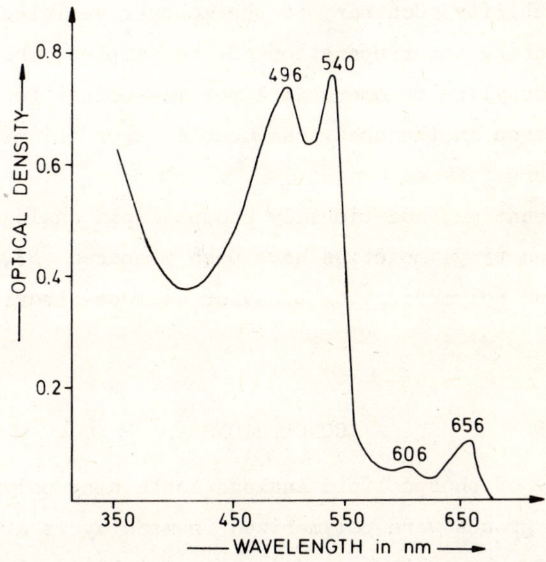


FIG. 4. UV/VIS Spectrum of red polymer liposomes of (XI) after 45 min of UV irradiation. Conc. of (XI): 1 mg/ml.

under complete retention of the orientation and shape of the vesicle.

The polymeric character of our liposomes is also established by their enhanced stability. In contrast to their monomeric analogs aqueous solutions of the polymer liposomes are stable for months. Moreover, they cannot be destroyed by organic solvents. The dilution of an aqueous polymer liposome solution with 50% of ethanol, for instance, does not result in a precipitation. The liposomes retain their structure, what could be shown by electron microscopy. Precipitation of the vesicles can be achieved by the addition of salts, but again it has to be pointed out that in contrast to low molecular weight liposomes the layer structure of our vesicles is not destroyed.

The possibility of obtaining scanning electron micrographs of the polymer vesicles (Fig. 5) is an additional proof of the enhanced stability. Contrary to the monomer vesicles, which are destroyed during the preparation of the samples. The polymerization of diacetylene monomers in liposomes points to a topochemical effect even in the spherical double layer orientation of these monomers.

In the meantime, additionally phospholipid analogs containing butadiene and vinyl moieties have been prepared [23,24]. Their syntheses and polymerization behavior will be described elsewhere.

## CONCLUSIONS

A variety of phospholipid analogs containing polymerizable diacetylene groups were polymerized in monolayers at the gas-water interface as well as in liposomes yielding either extremely rigid planar polymer monolayers or very stable polymer vesicles. Application possibilities of these stable synthetic model

FIG. 5. Scanning electron micrograph of red polymer liposomes of (IX). Magnification: 15,000.

membranes are, for instance, their use as drug carriers or as models for the cell-cell interaction. Investigations on leakage and ruptures in the shell of the polymer vesicles are in progress as are attempts to cosonicate natural and polymerizable lipids with the presence of proteins to obtain totally stable cell models after polymerization.

ACKNOWLEDGMENTS

We thank the Deutsche Forschungsgemeinschaft for financial support (SFB 41).

REFERENCES

[1] L. J. Old, Scientific American, 236 (5), 62 (1977).

[2] D. Day, and H. Ringsdorf, J. Polym. Sci., Polym. Lett. Ed., 16, 205 (1978).

[3] G. Wegner, Makromol. Chem., 154, 35 (1972).

[4] B. Tieke, G. Lieser, and G. Wegner, J. Polym. Sci., Polym. Chem. Ed., 17, 1631 (1979).

[5] B. Hupfer, Ph. D. Thesis, University of Mainz, 1981.

[6] D. Day, H. H. Hub, and H. Ringsdorf, Israel J. Chem., 18, 325 (1979).

[7] H. H. Hub, Ph. D. Thesis, University of Mainz, 1980.

[8] H. Koch, Thesis, University of Mainz, 1980.

[9] B. Tieke, H.-J. Graf, G. Wegner, D. Naegele, H. Ringsdorf, A. Banerjie, D. Day, and J. B. Lando, Colloid Polym. Sci., 255, 521 (1977).

[10] R. Hirt, and R. Bechtold, Pharm. Acta Helv., 33, 349 (1958).

[11] R. Hirt, and R. Bechtold, Helv. Chim. Acta, 40, 1928 (1957).

[12] F. R. Pfeiffer, C. K. Miao, and J. A. Weisbach, J. Org. Chem., 35, 221 (1970).

[13] H. Eibl, D. Arnold, H. U. Weltzien, and O. Westphal, Liebigs Ann. Chem., 709, 226 (1967).

[14] A. F. Rosenthal, in Methods in Enzymology,(J. M. Lowenstein, Ed.), Vol. 35, Academic, New York, 1975, pp. 429.

[15] D. R. Day, and H. Ringsdorf, Makromol. Chem., 180, 1059 (1979).

[16] T. Kunitake, J. Macromol. Sci.-Chem., A13, 587 (1979).

[17] J. M. Gebicki, and M. Hicks, Nature, 243, 232 (1973); J. M. Gebicki, and M. Hicks, Chem. Phys. Lipids, 16, 142 (1976); M. Hicks, and J. M. Gebicki, ibid. 20, 243 (1977).

[18] W. R. Hargreaves, and D. W. Daemer, Biochemistry, 17, 3759 (1978).

[19] K. Deguchi, and J. Mino, J. Colloid Interface Sci., 65, 155 (1978).

[20] J. H. Fendler, Acc. Chem. Res., 13, 7 (1980).

[21] H. T. Tien, Ed., Bilayer Lipid Membranes, Marcel Dekker, New York, 1974, Chapt. 2.

[22] W. Prass, Thesis, University of Mainz, 1980.

[23] K. Dorn, Ph. D. Thesis, University of Mainz, 1981.

[24] H. Schupp, Ph. D. Thesis, University of Mainz, 1981.

[illegible]

[illegible]

[illegible] Rosenthal, in Advances in Psychology [illegible] Academic [illegible]

[illegible]

[illegible]

[illegible]

[illegible]

[illegible]

[illegible]

[illegible]

[illegible]

[illegible]

[illegible]

[illegible]

# Interactions of and Regulatory Requirements of Microcapsules for New Animal Drugs

Russell G. Arnold
Bureau of Veterinary Medicine
Food & Drug Administration
Rockville, Maryland 20857

ABSTRACT

Microencapsulation techniques have been developed to produce microcapsules of nanometer size when necessary. The control of the ratio of wall thickness to capsule diameter and the control of microcapsule diameter size have been improved. Animal digestive systems differ; e.g. Ruminants have a more complex digestive system than other animals. A microencapsulated drug orally administered will remain unaltered as it passes through the rumen and abomasum to the feces. For example, control of fecal breeding flies by retention of a microencapsulated pesticide drug into the manure has been achieved. Unsaturates may also be protected from hydrogenation and be absorbed unaltered in the abomasum by this process. Considerable attention has been given to control the rate, extent, and site of action for pharmaceuticals. These mechanisms are discussed and some examples are given. A description of the Food and Drug Administration requirements for approval of these drug delivery systems are presented.

Microencapsulation techniques have been developed and refined for many years [1-23,25,26]. With the passing of time it has become increasingly clear that the unique advantages that can be obtained by utilizing such specialized processes cannot continue to be cast aside merely because of cost pressures from competing products, technological state-of-the art, or reluctance to abandon the normally accepted drug delivery systems.

Table I depicts the number of all new animal drug applications approved from 1968 to the present time. You will notice the in-

| Table I | New Animal Drug Applications Approved |
|---|---|
| 1968 | 171 |
| 1969 | 304 |
| 1970 | 373 |
| 1971 | 296 |
| 1972 | 419 |
| 1973 | 469 |
| 1974 | 515 |
| 1975 | 528 (45 Original Applications) |
| 1976 | 440 (51 Original Applications) |
| 1977 | 524 (28 Original Applications) |
| 1978 | 508 (24 Original Applications) |
| 1979 (to Nov. 15) | 554 (9 Original Applications) |

crease in the number of applications approved over the years. Hoever, if you will look at the year 1979, you will notice that of the total number approved, only 9 were original drug applications, as defined by a.) use of new drug substance. b.) a novel or new route of administration, c.) use in animal species not previously approved, and/or a new medical claim for the drug, of these categories, the number of approved new drug substances is indeed a small part of these numbers, and shows promise of being even smaller in the future. The decrease in number is because of increasing inflation, and increasing amounts of information required by FDA to be submitted in support of a new animal drug application.

In some cases, because the animal digestive systems differ in their activity, selective approaches must be considered for each animal system. A drug substance may need to be administered in a formulation that will remain unaltered through the rumen into the abomasum for subsequent dissolution and/or excretion. USDA workers found that a microencapsulated pesticide was able to pass into the feces sufficient active ingredient to obtain efficacy against two types of fly larvae. The products "Rabon" and "Dimilin" have also been studied experimentally using this type of delivery system.

Dosing large groups of animals once with a sustained-release, implant, or "depot" drug give some unique advantages[25-26]. Feedlot dosing is made easier by single administrations rather than repeated dosing of the animals. FDA has recently approved the use of the microencapsulation process as a feed additive for cattle. Tallow is encapsulated with a coating of formaldehyde-treated protein which produces a gel of droplets of 1-5 micron diameter. The capsules are then mixed with the grain, which improves flowability of the ration[19]. By delaying digestion until the drug reaches the abomasum deadly soap formation in the rumen is avoided and the cattle can achieve weight gains appreciably below the current average cost to the feedlot producer[22].

Similarly the enrichment of unsaturates in the milk fat content of dairy animals has been reported. The USDA and other scientists[11-12] have found that safflower oil prepared by encapsulation in casein or other proteins treated with formaldehyde and subsequently placed in the grain ration increased the linoleic acid content in cows milk to an average 13.5% of total fat as compared with the preexisting 2.7% of unsaturates. The casein wall protected the oil from being hydrogenated by microorganisms in the rumen of the cattle, thereby creating the higher level of unsaturated fat in the milk.

Newborn fish may be fed microencapsulated food of particle sizes of less than 100 microns[14]. In this manner the fish may be nurtured economically in fish farms or tanks without eutrophication of surrounding water by the breakdown of soluble components, including organic and nitrogen containing compounds from currently used fish chow. Microencapsulated vitamins and drugs similarly ingested would be released directly in the gut, controlling the quantities used.

Another area of application would include stabilization of the active ingredient to achieve a desired shelf life under expected environmental conditions prior to use. Examples of this would be to protect the drug formulation from heat, light, oxidation, moisture, mechanical stress, or a combination of these factors. Such techniques are particularly applicable to the preparation of the medicated feed premixes.

Additionally, temperature-sensing nonreversible systems could be used for the labeling of containers to indicate whether high or low temperatures have been exceeded in cases where temperature restrictions are important[16].

Still another area for the application of microencapsulation is taste-masking or odor avoidance. Several animal species including dogs, cats and horses can be sensitive to unpleasant tastes or odors associated with drug ingredients and/or related impurities, degradation products, formulation excipients. Combination drug products may need protection from interactions between

Recent work by a USDA chemist[20] indicates that starch xanthate polymer may become a popular encapsulation material for pesticides and drugs. The inexpensively derived polymer, as an insoluble but water-permeable starch particle, slowly releases the active ingredient.

I would now like to direct your attention to the procedures to be used for those desiring approval for marketing of a microencapsulated new animal drug. Table 2 provides the normal sequence of events that lead to the approval of a drug for animal use.

In the table, the protocol for studies should be submitted to demonstrate the claim that a microencapsulated product has a special use differing from some other dosage form. Its therapeutic or other medical claims should be delineated. Next, the submission of any Investigational New Animal Drug application (INAD)

Table 2

Sequence for Approval of a
New Animal Drug Application

1. Protocol for studies
2. Filing of INAD
3. Submission of NADA

   a. Safety )
   b. Efficacy ) 21 CFR §514.1
   c. Manufacturing controls )

4. Environmental Impact Analyses Report where applicable.
5. Review by the Bureau and by the Administration.
6. Approval of Application.
7. FEDERAL REGISTER Announcement.

is a request for studies to be conducted and the number of animals to be used. After the investigational trials are completed, a new Animal Drug Application (NADA) is to be generated which provides studies that determine an adequate margin of human and animal safety. The submission should also demonstrate efficacy of the intended microencapsulated drug as administered to the animal. The manufacturing controls should also be included in this submission describing the steps and procedures used for control at each stage thru to the finished dosage form.

Safety and efficacy are of principal importance in the development of a microencapsulated new animal drug and should be considered prior to the submission of an NADA. The following BVM suggestions will be limited to the area of manufacturing controls for the proposed drug:

## Bureau of Veterinary Medicine Suggestions for New Animal Drug Applications Using a Microencapsulation Process

1. Limits of solvents remaining in the formulation, including reaction intermediates of the polymeric or other wall-coating materials, must be described.

2. Particle size, range, and distribution should be specified for the capsules, including ratios of core diameter wall thickness with the percent W/W of each. If a multiwall, colony, or aggregate coating system is to be used, then these should be characterized.

3. Permeability of wall(s) material with its expected environment during administration (e.g. the blood stream for intravenous or the tissue with intramuscular injection, gastric or other digestive fluids following oral administration, etc.) should be demonstrated and permeation rates measured.

4. If controlled-release, taste-masking, or other special property or effect is claimed, adequate test methods and specifications following evaluation of appropriate statistical data demonstrating these properties or special effects should be submitted.

5. Good Manufacturing Practices § CFR 21, Part 200.

    a. In addition to the normal processes necessary for production of premixes and dosage form, microencapsulation requires many precautions, quality control, and specialized techniques. What precautions are taken to prevent electrostatic attraction of dust to the particles after formulation?

    b. What precautions are used to prevent cross contamination of the drugs in the manufacturing facility?

c. What steps are taken to prevent sticking together of the capsules, or their rupture or collapse due to improper conditions of storage and handling prior to formulation?

d. A portion of the Environmental Impact Analyses Report must indicate the nature and levels of materials leaving the manufacturing facilities as air or water effluents and how these effluents are controlled to meet current local, state, and federal pollution control requirements.

6. Stability testing to determine an expiration date for the product stored under expected environmental conditions prior to use should include assays for active ingredient, particle size of the microencapsules, and measurements to demonstrate that the advantage claimed for the microencapsulated product in the finished dosage form have not been destroyed or diminished during storage.[24]

7. Safety of wall materials and solvents used in the micrencapsulation process should be demonstrated. If the wall material persists,environmental impact considerations must be evaluated. An Environmental Impact Statement may be required in these instances.

8. If specific claims regarding the process to differentiate the formulation from competing products, such as "controlled-release" or "stabilized" or "odor-free" are made, then data must be presented to substantiate such claims.

9. The drug content must be expressed in percent such as W/W or W/V. Limits of acceptance of drug content in the formulation should be included as a laboratory control for the finished dosage form. Where controlled release limits are proposed, ranges should be defined utilizing appropriate statistical methods.

10. When microencapsulation process produced by one company are supplied for further formulation, or packaging and labeling, or simply marketing by another firm which will be the sponsor of a New Animal Drug Application, the microencapsulator may submit manufacturing data, controls, etc., in the form of a Master File to the FDA. This Master File may be referenced to support a New Animal Drug Application by a letter of autorization from the microencapsulator to the sponsor and/or FDA.

## REFERENCES


1. J. A. Bakan and J. L. Anderson in Microencapsulation (L. Lachman, ed), pp. 384-407, Liberman.

2. S. A. Bell, M. Berdick, and W. M. Holliday, J. New Drugs, 6 (1966)

3. T. M. S. Chang, F. C. MacIntosh, and S. G. Mason, Can. J. Physiol. Pharmacol, 44 155.(966).

4. L. A. Luzzi, M. A. Zoglio, and J. V. Maulding, J. Pharm. Sci., 59 (1970).

5. L. Luzzi, J. Pharm. Sci., 59, (1970).

6. R. E. Phares and G.J. Sperandio, J. Pharm. Sci., 53, (1964).

7. R. E. Sparks, Paper presented at Microencapsulation Workshop, Hopatcong, New Jersey, September (1968).

8. T. M. S. Chang, Artificial Cell, Charles C. Thomas, Springfield, III. (1971).

9. U. S. Patent 3, 577, 515 Joan E. Vanegaer to Penwalt Corp.

10. R. W. Miller and C. H. Gordon, Economic Entomology, 65, 455, (1972).

11. Chem. Eng. News, July 31, 1971, p. 11

12. J. Bitman, L. P. Dryden, H. R. Goering, T. R. Wrenn, R. A. Yoneos, K. I. and L. F. Edmonson, J. Am. Oil Chem., 50, 93. (1973).

13. R.E. Sparks, Microencapsulation Course, Sept. 26-28, 1971, Woodbridge, N. J.

14. Chem. Eng. New, August 12, 1974.

15. C.B. DeSavgny and E.E. Ivy, "Microencapsulated Pesticides" In Microencapsulation: Processes & Application (J. E. Vandegaer, Ed.) Plenum, New York (1974).

16. Quick Frozen Foods Magazine, November 1971.

17. P. Speiser, "Microencapsulation by Coacervation, Spray Encapsulation and Nanoencapsulation" in Microencapsulation (J. R. Nixon, ed.) Marcel Dekker New York (1976).

18. Abrams, J. and Hinkes, "T. M. Acceptability and Performance of Encapsulated Warfarin." Pest Control, 42, 14, May 1974.

19 Render Magazine, August, 1976.

20 Chem. Eng. News, June 28, 1976.

21. Santos, A. Summary of Results of Laboratory Screening of Compound 443A Against O. quadrasi Contr. Rel. Mollusci. Newsltr. Univ. of Akron, April 1975. pp. 504.

22. Feedstuffs, March 7, 1977.

23. P. L. Madan R. K. Jani, and A. J. Bartilucci "New Method of Preparing Gelatin Microcapsules of Soluble Pharmaceuticals". J. Pharm Sci. Vol. 67, No. 3 (1978).

24. Bureau of Veterinary Medicine Stability Guidelines February 1, 1978.

25. Controlled Drug Release by Polymer Dissolution II, Enzyme - Medicated Delivery Device J. Holly & P. V. Trescony.

J. Pharm Sci, Vol. 68, No. 7 pp. 919-921.

26. Controlled Release of Tetracyline II: Development of and In-Vivo Flow-Limited Pharmacokinetic Model L. Olandoff and J. M. Anderson.

J. Pharm Sci. Vol. 68, No. 9, pp. 1147-1150

[illegible]OSENSORS FOR BIO[illegible]

18 Abrams, [illegible] and [illegible] and Performance of [illegible] Electrodes," [illegible], 14, [illegible]

19 [illegible] Magazine, April [illegible]

20 Chem. Eng. News, June [illegible]

21 [illegible] and Rodney [illegible] Notes on Chemical [illegible] from [illegible] in Cedar [illegible] Mueller [illegible] of [illegible]

[illegible]

22 R. [illegible] and [illegible] method of [illegible] [illegible] of [illegible] [illegible] No. [illegible]

23 [illegible] Formula [illegible]

24 [illegible] [illegible]

[illegible]

25 [illegible] Research [illegible] [illegible] [illegible]

# Interfacial Synthesis in the Preparation of Reverse Osmosis Membranes

J.E. CADOTTE, R.S. KING, R.J. MAJERLE, and R.J. PETERSEN

FilmTec Corporation
15305 Minnetonka Boulevard
Minnetonka, Minnesota 55343

ABSTRACT

Reverse osmosis membranes are currently finding increasing use in the production of potable water from salt-laden water sources, and in the purification and recycle of industrial process waters. Most reverse osmosis membranes are comprised either of polyamide hollow fibers or cellulose acetate sheet films. A new type of reverse osmosis membrane, made by the interfacial synthesis of ultrathin membranes directly on microporous support media, is being explored. Two types of interfacially formed membranes are specifically described. One is based on an ultrathin polyamide barrier layer made by interfacial reaction of polyethylenimine with isophthaloyl chloride. The barrier material is supported by a heat-polymerized intermediate zone of polyethylenimine, supported in turn by a microporous polysulfone substrate. The second membrane consists of a polyamide formed from piperazine and a mixed acyl halide reagent. This nonpolymeric amine leads to a different surface texture and the absence of an intermediate zone between the barrier film and the microporous polysulfone substrate. This latter membrane exhibits excellent water fluxes (as high as 4.0 $m^3/m^2$-day at 100 atm) under seawater reverse osmosis test conditions, and shows very high rejection (99+%) of inorganic salts containing polyvalent anions. Differences in the morphology of the salt barrier layers for these two cases are described, and the resulting effects on membrane desalination properties are discussed.

## INTRODUCTION

Reverse osmosis is a process whereby water is forced by pressure through a semipermeable membrane counter to the normal direction of osmotic flow. In the case of water desalination, the semipermeable membrane must possess a particularly dense barrier layer to reject dissolved solutes in the water, such as sodium chloride. Rather low diffusive permeabilities for water through such dense structures are the expected norm.

Reverse osmosis became a commercially feasible process upon the development of the asymmetric, or anisotropic, membrane by Loeb and Sourirajan.[1] This membrane was formed from a solution of cellulose acetate having about 39 percent acetyl content. It was cast in such a fashion as to develop a dense surface skin about 0.25 μm thick on top of a porous structure 50 to 150 μm thick. The bulk of the membrane contained about 55 to 60 percent void space, with pores ranging from greater than 1000 Å on the bottom (coarse) side of the membrane to less than 100 Å in the zone immediately underlying the surface layer. Considerable variation in these parameters occurs, however, depending upon the method of membrane fabrication. Kesting[2] has provided an excellent review of the various routes to asymmetric cellulose acetate membranes and their resulting physical structures.

Although asymmetric cellulose acetate membranes find broad commercial application today, these reverse osmosis membranes suffer severe drawbacks. These include limited pH stability, susceptibility to biological attack, compaction at high pressures (which reduces water flux), wet-dry irreversibility (in most but not all types), and an upper temperature operating limit at about 35°C (95°F).

One approach toward improved membrane performance was to fabricate the surface skin and the porous support layer as two separate entities, then combine them to achieve maximum performance. By this method, each layer could be optimized separately, the surface layer for reverse osmosis properties and the porous underlayer for durability and compaction resistance. This approach was successfully developed for cellulose acetate barrier layers deposited on microporous polysulfone (poly[oxy-1,4-phenylenesulfonyl-1,4-phenylene]) substrates.[3,4] Qualities of increased water flux and decreased sensitivity to compaction were achieved by this approach, but the other drawbacks of cellulose acetate membranes remained.

Interfacial synthesis appears to be a promising method for deposition of ultrathin barrier films _in situ_ on microporous polysulfone substrates. This paper describes several reverse osmosis membrane compositions that have been successfully prepared by interfacial methods. Also described are results of scanning electron microscopy studies on thin-film polyamide composite membranes, with particular attention to piperazine-based polyamide compositions.

Various thin-film composite membranes can be prepared by interfacial deposition of ultrathin polymeric barrier films on micrporous polysulfone substrates. To be functional in water desalination, however, the interfacially formed barrier layers must have a balance of hydrophilic and hydrophobic groups. They must also consist of a fairly rigid chemical structure overall to resist compaction under high pressure.

Table 1 illustrates several types of such membranes.[5,6] Aromatic polyamines and polyphenolics can be prepared by a gas-liquid interfacial reaction, using formaldehyde vapor. An acid catalyst in the resorcinol solution is needed. These membranes were simply prepared by exposing the impregnated polysulfone supports to formaldehyde vapor in closed Petri dishes. Similarly, aromatic polyester and polyamide membranes can be prepared by exposure to heated vapors of acid chlorides, as illustrated in the case of sorbitol.

Where aromatic polyester and polyamide membranes are involved, however, the best results are generally realized through use of hexane

TABLE I. Examples of Experimental Semipermeable Membrane Compositions Prepared by Interfacial Methods

| Membrane Type | Aqueous Reactant | Crosslinking Agent | Reverse Osmosis Properties* | |
|---|---|---|---|---|
| | | | Salt Rej. (%) | Water Flux ($m^3/m^2$-day) |
| polyamine | 1,3-diaminobenzene | formaldehyde vapor | 97 | 1.8 |
| polyester | resorcinol | isophthaloyl chloride in hexane | 40 | 1.8 |
| polyphenol | resorcinol | formaldehyde vapor | 75 | 1.7 |
| polyamide | piperazine | isophthaloyl chloride in hexane | 98 | 1.1 |
| polyester | sorbitol | terephthaloyl chloride in hexane | 97 | 0.9 |
| polyester | sorbitol | trimesoyl chloride vapor | 60 | 4.0 |

* Data from 16 hour tests with 3.5% synthetic seawater at 100 atm pressure and 25°C.

solutions of the corresponding aromatic acyl halides. Examples of these are also illustrated in Table 1.

From related work in reverse osmosis, it was apparent that polyamides possessed particularly promising semipermeability characteristics necessary for water production and salt rejection.(7) Interfacial formation of polyamides upon porous substrates to produce reverse osmosis membranes has also been described, although salt rejections greater than 85 percent were not achieved.(8) To be commercially useful, reverse osmosis membranes should exhibit salt rejections of at least 95 percent for brackish water treatment and 98.5 percent for seawater desalination, while providing water throughput rates (fluxes) of at least 0.40 $m^3/m^2$-day (about 10 gallons per square foot of membrane area per day) at normal use pressures.

In our research effort, the interfacial formation of ultrathin polyamide membranes was investigated, using microporous polysulfone as a preferred microporous support layer. Table 2 contains data obtained in initial trials of this approach. These data (which actually pre-date the data in Table 1) reveal the difficulty involved in preparing salt-retentive membranes by interfacial polyamide membrane synthesis. In all cases involving monomeric diamines, salt rejections were far below useful levels. Only in the case of a polymeric amine, polyethylenimine, was a membrane obtained showing promise for reverse osmosis desalination.

The findings in Table 2 for membranes made from monomeric diamines were not particularly surprising, however. Morgan, in his review of condensation polymers,(9) stated that washed but undried

TABLE II. Reverse Osmosis Performance of Membranes Prepared by Interfacial Reaction of Terephthaloyl Chloride with Various Amines

| Amine | Reverse Osmosis Performance* | |
|---|---|---|
| | Salt Rejection (%) | Water Flux ($m^3/m^2$-day) |
| hydrazine | ** | ** |
| 1,2-ethanediamine | ** | ** |
| 1,3-propanediamine | 30 | 9.5 |
| 1,6-hexanediamine | 8 | 4.2 |
| 1,3-diaminobenzene | 12 | 1.1 |
| 1,4-diaminobenzene | 38 | 2.2 |
| piperazine | 74 | 2.4 |
| polyethylenimine | 96 | 1.2 |

* Test conditions 100 atm, 3.5% synthetic seawater, 25°C.

** Too brittle to test.

films of 6/10 polyamide prepared by unstirred interfacial synthesis were readily permeable to inorganic salts and small dye molecules. At first glance, therefore, membranes with poor salt rejection capabilities are to be expected via interfacial polyamide synthesis.

The last two cases in Table 2, piperazine and polyethylenimine, have been examined in detail, and conditions have now been developed that provide high salt rejection membranes in both cases. Studies on the polyethylenimine case have been reported by us in recent years.[10-12] Piperazine polyamide reverse osmosis membranes have been studied of late by us, and are the main subject of this paper. Differences in the manner of their behavior (piperazine vs. polyethylenimine) in regard to interfacial formation, membrane morphology, and subsequent properties are discussed in this presentation.

## EXPERIMENTAL METHODS

To prepare the microporous polysulfone substrate, a 15-percent solution of polysulfone (Union Carbide Corporation, Grade P-3500) in dimethylformamide was drawn out on a flat glass plate using a 175 μm knife gap. The plate was immediately immersed in a water bath at room temperature in one smooth motion. Within 30 seconds, a white microporous sheet formed and spontaneously lifted away from the glass surface. After five minutes, the polysulfone support was removed, rinsed with fresh water, then stored in deionized water until used. The top surface (air side) was used as the support surface for membrane formation.

Interfacial membrane synthesis was accomplished using unstirred, nondispersion methods. Starting with polyethylenimine, interfacial polyamide membranes were made using isophthaloyl chloride. This reactant gave higher water fluxes than terephthaloyl chloride (Table 2) while providing equivalent salt rejections. The polyethylenimine used in this study was Tydex 12, made by Dow Chemical Company. It is no longer available from this source, but an equivalent grade may be obtained from Cordova Chemical Company, Sacramento, California, or purchased from laboratory chemical supply houses. A microporous polysulfone support was immersed in a solution of 0.67% by weight of polyethylenimine in deionized water for two minutes, then drained for an equal time. The support film was then covered with a solution of 0.5% isophthaloyl chloride in hexane. After removal from the hexane solution, the coated substrate was heat-cured in an oven at 110°C for 15 minutes.

Piperazine polyamide membranes were formed by similar methods. In one version, a microporous polysulfone substrate was saturated with an aqueous solution of 1.0% piperazine, 1.0% of an inorganic acid acceptor, and 0.5% dodecyl sodium sulfate. Excess reagent was either drained off or was "squeezed" off by means of a low durometer rubber roller. A suitable squeeze roller was fashioned from a length of glass tubing inserted through latex rubber tubing. The impregnated polysulfone substrate was then covered with 0.1 to 1.0 percent by weight isophthaloyl chloride in hexane for 20 seconds. The resulting thin-film composite membrane was drained and air-dried at room temperature. Variations of this membrane were also made wherein part or all of the isophthaloyl chloride was replaced by trimesoyl chloride (benzene-1,3,5-tricarboxylic acid chloride). Potential acid acceptors include sodium hydroxide, sodium carbonate, and sodium phosphate. In one case, N,N'-dimethylpiperazine was used as an acid acceptor. In the absence of any such additive, piperazine was used as an acid acceptor itself.

In a second version, the polysulfone support was coated with a 1:1 water: isopropanol solution containing 2.0 percent sulfonated polysulfone, 1.0 percent piperazine and 2.0 percent N,N'-dimethyl-piperazine. The sulfonated polysulfone was prepared by treatment of ten parts Udel P-3500 resin in dichloromethane with five parts chlorosulfonic acid, decanting, washing the precipitated polymer with additional dichloromethane, then neutralizing with aqueous triethylamine. The coated support film was interfacially reacted with isophthaloyl chloride as before to produce a piperazine polyamide membrane, followed by a heat-cure at 110°C for 15 minutes. This version contained a hydrophilic sulfonated polysulfone sublayer between the support film and the polyamide barrier layer.

These membranes were tested in a high pressure reverse osmosis test loop, decribed elsewhere.[13] Test conditions were 100 atm (1500 psi) feedwater pressure and 25°C temperature, and 3.5 percent by weight synthetic seawater (made with "Sea-Salt", Lake Products Company, St. Louis, MO), except as otherwise stated. Crossflow was maintained at about 1000 fold greater than membrane flux to avoid concentration polarization effects at the membrane surface. Membrane specimens were run under test conditions for 16 to 24 hours before taking flux and salt rejection measurements.

For scanning electron microscopy studies, samples 4 $cm^2$ were obtained from polysulfone supports both without membrane and with interfacial membranes deposited upon them. The samples were stored 24 hours in a vacuum dessicator prior to mounting to insure evaporation of any surface moisture; critical point drying was not necessary. A specimen 8 mm in diameter was cut from each sample and mounted, with the glossy (air or membrane) surface facing up, using conductive silver paint. The mounted specimens were again vacuum dessicated to allow the paint to dry thoroughly. Eight inches of wire with a 1:1 ratio of gold to gold: palladium (60:40) were then evaporated onto the specimens at 40 V in a vacuum of $1.8 \times 10^{-5}$ torr using a Denton Vacuum Evaporator. Specimens were subsequently observed using a Hitachi Model 450 scanning electron microscope.

## REVERSE OSMOSIS TEST RESULTS

Table 3 contains reverse osmosis test data on these various membranes. Polyethylenimine interfacially crosslinked by isoph-

TABLE III. Reverse Osmosis Performance of Membranes from Polyethylenimine and Piperazine Interfacially Reacted with Isophthaloyl Chloride

| Amine Reagent | Acyl Chloride Concentration | Water Flux ($m^3/m^2$-day) | Salt Rej. (%) |
|---|---|---|---|
| 0.67% polyethylenimine | 0.5% | 1.0 | 99.3 |
| 1.0% piperazine, 0.93% NaOH | 1.0% | 0.37 | 96 |
| 1.0% piperazine, 0.93% NaOH | 0.1% | 0.86 | 98 |
| 2.0% sulfonated polysulfone, 2.0% piperazine, 2.0% N,N'-dimethylpiperazine | 0.1% | 0.81 | 98.4 |

thaloyl chloride was found to give salt rejections in excess of 99 percent when tested with synthetic seawater. These salt rejections could be obtained routinely as skill developed in their preparation. Water fluxes for this membrane averaged 1.1 $m^3/m^2$-day under the test conditions stated.

The poly(piperazine isophthalamide) thin-film composite membrane could be made to give seawater salt rejections in excess of 95 percent under carefully controlled conditions. These included using fresh reagents and making sure that the isophthaloyl chloride was pure. Quality of the microporous polysulfone support was also important. The support film had to be fresh, free of pinholes and other defects, and cast from fresh, clear polysulfone solutions. Even so, salt rejections above 97 percent were very difficult to obtain.

Use of selected water-soluble polymers as additives in the membrane synthesis improved membrane salt rejections to as high as 98.5 percent, which represented the lower limit for single-pass seawater desalination. The case in Table 3 involving the use of sulfonated polysulfone is the best example of the phenomenon.

At its best, the piperazine isopthalamide membrane gave fluxes of 1.0 $m^3/m^2$-day. This was considered to be a marginal performance for commercial application. When trimesoyl chloride was substituted for isophthaloyl chloride, however, membrane fluxes were enhanced dramatically. Yet, peculiar salt rejection results were observed. Results are illustrated in Figures 1 and 2. In simulated seawater performance tests, the flux, initially at 1.0 $m^3/m^2$-day for the piperazine isophthalamide membrane, increased to 3.4 $m^3/m^2$-day for the piperazine trimesamide membrane with a peak flow of about 4.0 $m^3/m^2$-day for the 50:50 blend. Salt rejection of the synthetic seawater by the barrier layer dropped immediately at low levels of trimesoyl chloride addition to the isophthaloyl reagent, falling into the range of 60 to 70 percent for most of the membrane combinations.

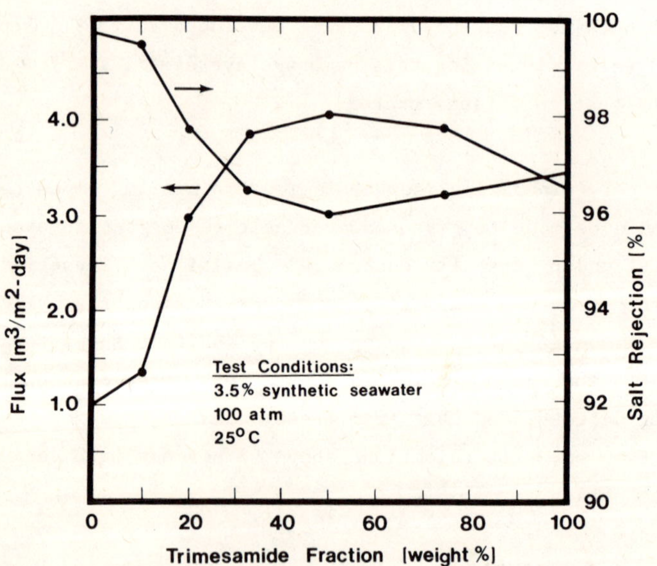


FIGURE I. Seawater Desalination Flux and Salt Rejection as a Function of Isophthaloyl and Trimesoyl Content in Piperazine Polyamide Membranes.

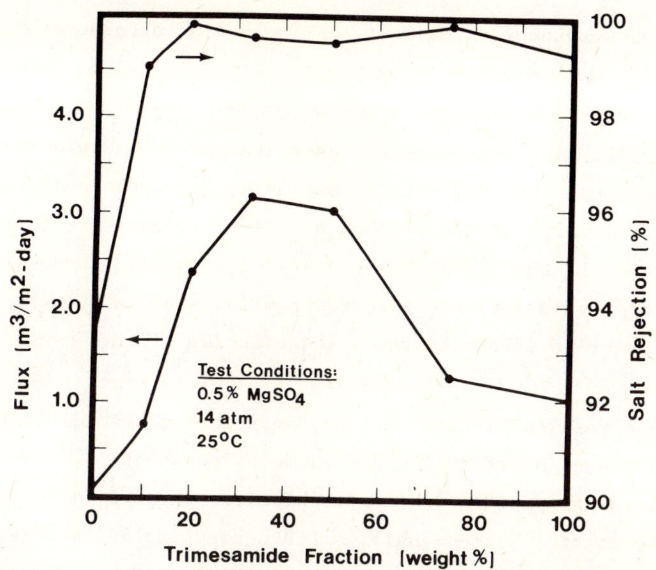


FIGURE II. Brackish Water Desalination Flux and Salt Rejection as a Function of Isophthaloyl and Trimesoyl Content in Piperazine Polyamide Membranes.

Simulated brackish water tests with 0.5 percent magnesium sulfate at 14 atm (200 psi) exhibited the same type of flux curve, showing a maximum of 3.1 $m^3/m^2$-day flux at a weight ratio of 2:1 isophthaloyl chloride: trimesoyl chloride in the hexane solution. Salt rejections, initially good at about 94 percent for the piperazine isophthalamide thin-film composite membrane, actually increased to the 99 percent level upon the substitution of ten percent of isophthaloyl chloride by trimesoyl chloride in the interfacial reactant solution. At greater addition levels of trimesoyl chloride, the magnesium sulfate rejection remained above 99 percent, in direct contrast to the seawater rejection data. Thus, while greatly enhancing flux, trimesoyl chloride also caused an unexpected divergence in salt barrier properties toward simulated seawater versus simulated brackish water.

In a subsequent experiment, a single disc of piperazine trimesamide membrane was sequentially tested toward a series of inorganic salt solutions under reverse osmosis conditions at 14 atm and 25°C. Results of these tests are shown in Table 4. These data show that the salt rejection is highly dependent on the valence charge of the anion. Divalent anions (e.g. sulfate) are rejected by the barrier layer very effectively. Monovalent anions (e.g. chloride) are rejected poorly. The cation charge has little or no effect on the salt rejection of this membrane.

These results implied that the polypiperazineamide barrier layer in these interfacially formed membranes contained pendant carboxylate groups when trimesoyl chloride was used as an interfacial reactant. Subsequently, a 1.0 percent solution of trimesoyl chloride in hexane was stirred rapidly with water for 30 minutes. Hydrolysis of acid chloride groups liberated hydrochloric acid, which was extracted into the water phase. Separation of the water and titration with base showed the presence of 83 percent of

TABLE IV. Rejection of Various Solutes by a Piperazine Trimesamide Interfacial Membrane

| Solute Used in Test Loop | Reverse Osmosis Test Data* | |
|---|---|---|
| | Salt Rejection (%) | Flux ($m^3/m^2$-day) |
| 0.1% $MgSO_4$ | 98.0 | 1.4 |
| 0.1% NaCl | 70 | 1.3 |
| 0.5% NaCl | 50 | 1.7 |
| 0.5% $Na_2SO_4$ | 97.8 | 1.7 |
| 0.5% $MgCl_2$ | 46 | 1.3 |
| 0.5% $MgSO_4$ | 97.9 | 1.3 |

* Test conditions: 14 atm, 25°C.

one equivalent of hydrochloric acid. The diacyl chloride form remained soluble in hexane, maintaining a clear solution, and this solution was active in the preparation of interfacially-formed piperazine polyamide films. It appears, therefore, that facile hydrolysis of one of the acyl halide groups on the trimesoyl chloride ring occurs, but that further hydrolysis is slowed thereafter. Hydrolysis as a competing reaction with polyamide formation in the interfacial contact zone could occur and result in carboxylate ion formation in the piperazine trimesamide membranes.

Table 5 shows the effect of acid acceptors in the aqueous piperazine phase upon reverse osmosis properties of resulting membranes. When trimesoyl chloride was reacted interfacially with piperazine, the tightest membranes with the highest salt rejections occurred as a function of the following acid acceptor order: none $>$ sodium carbonate $>$ sodium hydroxide. For isophthaloyl chloride under identical conditions, this order was reversed. Acid acceptors appeared to promote the hydrolysis side reaction on trimesoyl chloride, leading to poorer salt rejections in the final interfacial membrane as the basicity of the acid acceptor increased.

TABLE V. Effect of Acid Acceptors on Properties of Polypiperazineamide Membranes Formed with Trimesoyl or Isophthaloyl Chloride

| | | Reverse Osmosis Test Data* | |
|---|---|---|---|
| Acyl Halide | Acid Acceptor | Salt Rejection (%) | Flux ($m^3/m^2$-day) |
| trimesoyl Chloride | none | 88 | 1.5 |
| | $Na_2CO_3$ | 54 | 2.3 |
| | NaOH | 28 | 1.2 |
| isophthaloyl chloride | none | 55 | 2.3 |
| | $Na_2CO_3$ | 80 | 0.69 |
| | NaOH | 97 | 0.29 |

* Test conditions: 68 atm (1000 psi), 3.5% synthetic seawater, 25°C.

Addition of sodium dodecyl sulfate surfactant to the aqueous piperazine solution did not significantly change any of the performance data of the membrane formed from trimesoyl chloride and piperazine. For the isophthaloyl chloride-piperazine series, addition of 0.5 percent to 1.0 percent of the surfactant improved salt rejections significantly, but at about a 40 percent loss in membrane flux. Thicker membranes appeared to result.

If carboxylic acid groups are present in polypiperazineamide membranes containing the trimesoyl group, then such membranes should show a pH dependence relating to the ionization state of these pendant groups. Table 6 shows the effect of feedwater pH

TABLE IV. Effect of Feedwater pH on the Reverse Osmosis Properties of the 9:1 Isophthaloyl:Trimesoyl Polypiperazineamide Membrane

| | | Reverse Osmosis Test Data* | | | |
|---|---|---|---|---|---|
| | pH | Salt Rej. (%) | Flux ($m^3/m^2$-day) | Salt Rej. (%) | Flux ($m^3/m^2$-day) |
| | | Example 1 | | Example 2 | |
| Increasing pH | 7.9 | 93 | 0.66 | 96 | 0.46 |
| | 8.8 | 92 | 0.65 | 94 | 0.45 |
| | 9.9 | 92 | 0.63 | 94 | 0.56 |
| | 11.0 | 91 | 0.66 | 94 | 0.44 |
| | 11.9 | 89 | 0.67 | 94 | 0.45 |
| | | Example 3 | | Example 4 | |
| Decreasing pH | 8.0 | 89 | 0.65 | 91 | 0.66 |
| | 6.8 | 90 | 0.31 | 91 | 0.34 |
| | 5.2 | 89 | 0.26 | 91 | 0.33 |
| | 2.6 | 90 | 0.22 | 91 | 0.30 |
| | 2.2 | 91 | 0.21 | 91 | 0.30 |

* Test conditions: 40 atm, 1% sodium chloride, 25°C.

on examples of a 9:1 copolyamide of piperazine with isophthaloyl and trimesoyl chlorides respectively. Two test samples were taken from neutral to pH 11.9; flux and salt rejection were essentially constant through this range. Two other samples were taken from pH 8.0 down to pH 2.2; flux dropped dramatically for both examples between pH 8.0 and 6.8, as the conversion of carboxylate salts to free carboxylic acid groups began to take place, reducing membrane swelling. Again salt rejections remained constant. In another, separate example of polypiperazineamide membrane containing higher carboxylate content (higher trimesoyl substitution), in which salt rejections were initially 60 to 70 percent, the same type of flux decrease occurred upon changing pH from 8.0 to 5.5, but salt rejection increased to 90 percent.

## SCANNING ELECTRON MICROSCOPY

Both the polysulfone support and the polypiperazine membranes were examined in detail and characterized by scanning electron microscopy. Figure 3 shows different aspects of the microporous polysulfone support layer. Figure 3a consists of a cross-section of the upper layers of a thin-film composite reverse osmosis membrane made by interfacial polymerization. In this example the surface barrier layer is about 3000 Å thick. The surface of the polysulfone is indistiguishable from the membrane coating in this photomicrograph, but the porous substructure of the polysulfone can be seen.

Figure 3b is a photograph of the top surface of the polysulfone support material itself, magnified 20,000 X. This surface resembles a layer of spherical particles fused together. Micropores present in this surface layer were measured by both scanning and transmission electron spectroscopy and were found to average about 330 A diameter. Figure 3c illustrates the porous structure underlying this dense surface layer. This cross-section reveals a macroreticular polysul-

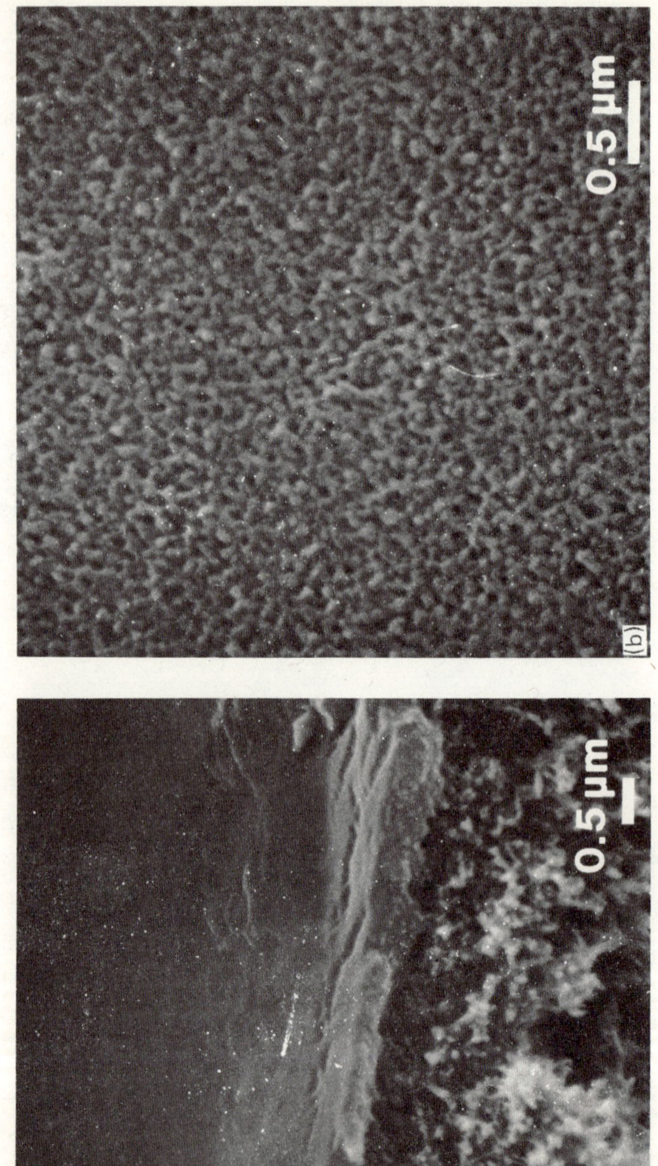
(a)
0.5 μm
(b)
0.5 μm

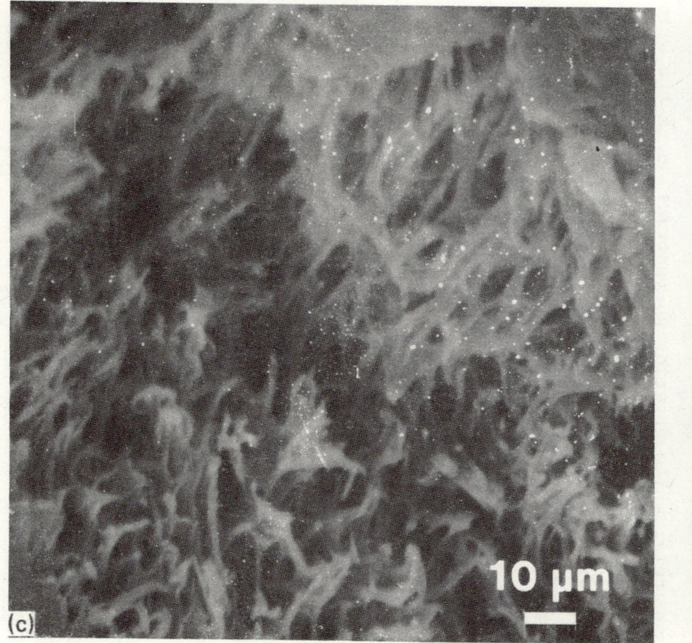


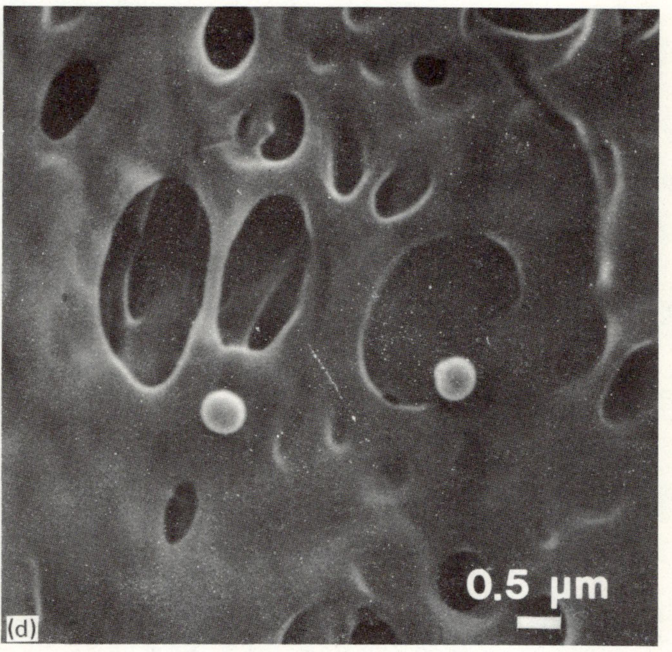


Figure III. Scanning Electron Photomicrographs of Microporous Polysulfone Support Films. a) Cross-section of the upper zone of a thin-film composite reverse osmosis membrane depicting a 300A-thick interfacial membrane deposited upon a microporous polysulfone substrate. b) Top surface of a microporous polysulfone support film showing an array of 330A micropores which support the interfacial membrane. c) Cross-section of the polysulfone support film showing a microporous reticular foam structure. d) Bottom surface of the polysulfone with typical large pores evident.

fone structure resembling an open cell foam. Accurate cross-sections like this were difficult to achieve without smearing or serious deformation. This specimen was prepared by tearing of the polysulfone material while immersed in liquid nitrogen. Figure 3d shows the bottom side of a microporous polysulfone support film. Large pores, diverse in dimensions and typically ranging from 20,000 to 40,000 Å, were present. This side of the film was in contact with a smooth glass plate during quenching of the film in water.

Microporous polysulfone supports for interfacially formed reverse osmosis membranes are observed, therefore, to be highly asymmetric films. The top surface with its 330 Å-diameter microporous proves very effective in supporting ultrathin, interfacially formed membranes. A 2000 Å-thick interfacial membrane can be supported to 100 atmospheres feedwater pressure by this microporous surface layer without breakthrough. The microporous sublayer supports this surface composite zone very adequately, while offering no restriction to the water flow that has come through the membrane. In commercial practice, this polysulfone support would be about 25 to 40 μm thick and backed by a nonwoven polyester web.

Figure 4 contains scanning electron photomicrographs of the surfaces of various piperazine polyamide membranes, along with a polyethylenimine isophthalamide membrane. These surfaces represent the side in contact with the hexane solution of the acyl chloride. The polyethylenimine isophthalamide membrane, Figure 4a, is basically smooth with some occasional longitudinal ridges. White beads of polysulfone detritus were accidentally deposited on the surface of this interfacial membrane during its preparation, and are artifacts in this figure. In Figure 4b, the piperazine isophthalamide membrane, in contrast, is very rugged in surface topography. Although interfacial formation takes place under quiescent conditions, the structure in Figure 4b suggests mixing at the interface, resulting in formation of polymer globules that become cemented together. Surfactant was used in polypiperazineamide membrane preparations. Initial formation at the interface may have involved initially emulsified piperazine isophthalamide, which then became insoluble and fused together. This and the two remaining membranes in Figure 4 were prepared using a 1.0 percent aqueous piperazine solution containing 1.0 percent sodium hydroxide and 0.5 percent sodium dodecyl sulfate. The acyl chloride reagent was 1.0 percent reactant by weight in hexane.

Figure 4c, depicting the surface of the piperazine trimesamide interfacial membrane, shows a completely different surface structure. The membrane surface texture results from a swelling of the membrane during formation. This swelling results in an expansion of the interfacial membrane in the plane of its formation. For the specimen in Figure 4c, development of a bubbled surface pattern resulted. Other patterns such as a series of closely spaced ridges were also seen, each one being indicative of this membrane swelling phenomenon. Between the ridges this membrane presents a smooth surface not at all reminiscent of the piperazine isophthalamide membrane. Swelling is thought to occur via formation of pendant sodium carboxylate groups in the membrane structure, followed by subsequent water absorption. Further reaction apparently takes place within the sowllen membrane, fixing it into this convoluted shape permanently. Thus, even though the membrane was dessicated in the preparation of specimens for scanning electron microscopy, the ridged structure remained extant.

Figure 4d shows an example of a 2:1 copolymer of piperazine isophthalamide-co-trimesamide respectively. This example resembled neither membrane of the respective homopolymers of Figure 4d and 4c, but was more like the polyethylenimine isophthalamide membrane. This copolymer membrane, which represents the case of peak water permeability under reverse osmosis conditions, apparently avoids the excess of carboxylate groups that cause the swelling in the piperazine trimesamide membrane.

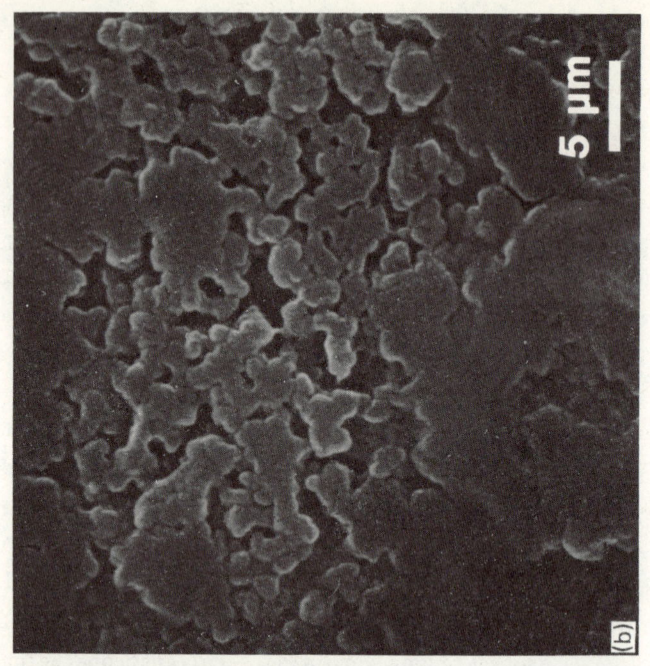
5 μm
(b)

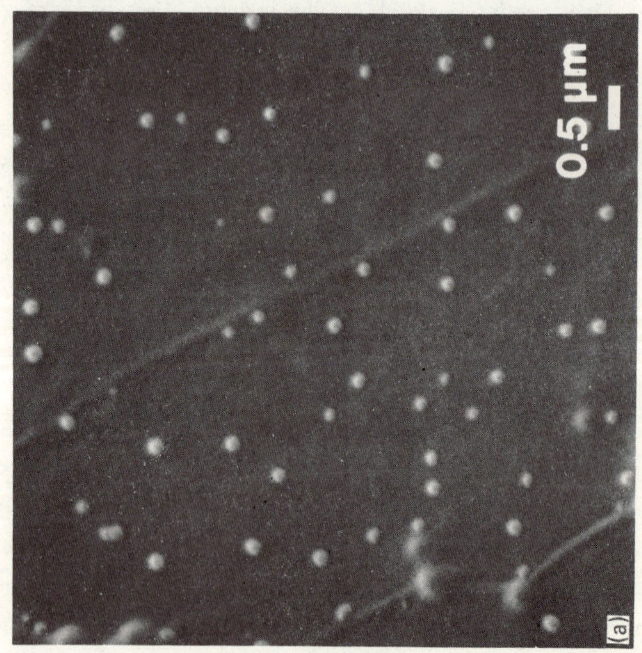
0.5 μm
(a)

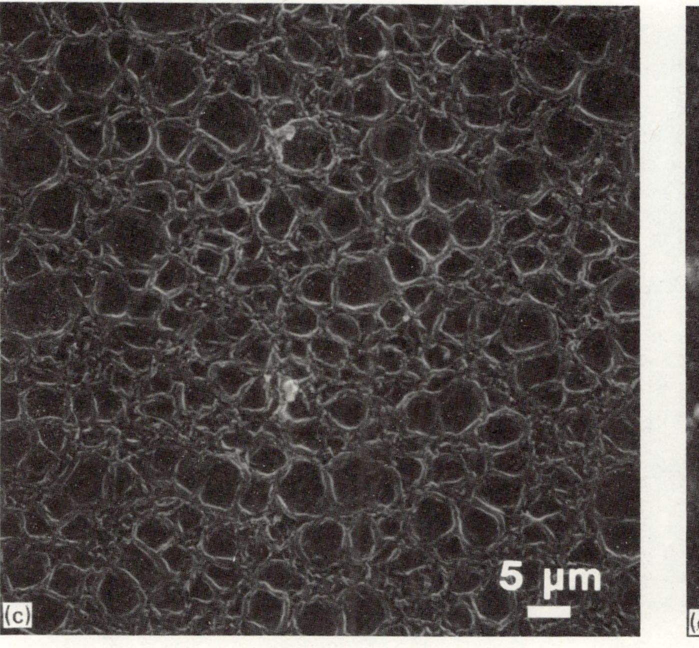


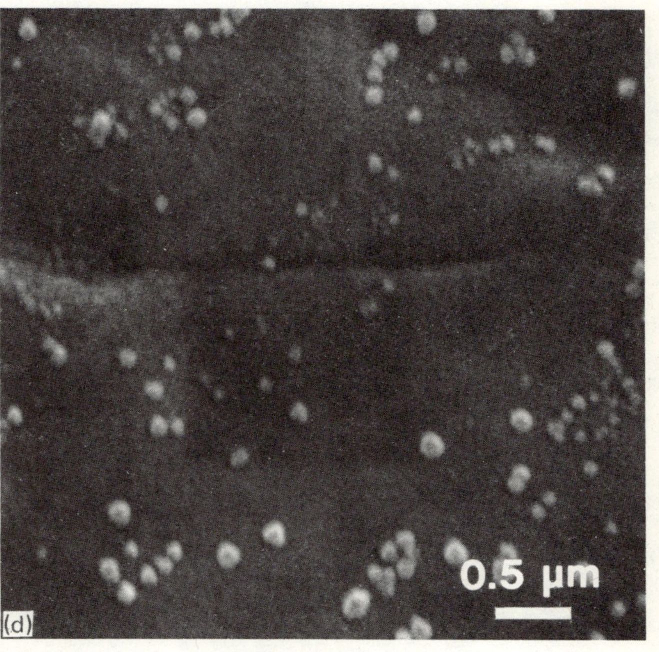


Figure IV. Scanning Electron Photomicrographs of the Surface of Interfacially Formed Polyamide Membranes.

a) Polyethylenimine isophthalamide (white beads are artifacts -- colloidal polysulfone particles). b) Piperazine isophthalamide. c) Piperazine trimesamide. d) 2:1 piperazine isophthalamide-co-trimesamide (white beads are colloidal polysulfone particles).

The backside of the polypiperazineamide membranes, the side in contact with the aqueous piperazine solution impregnated into the surface of the polysulfone support film, was also examined. The polysulfone was dissolved away by playing dichloromethane solvent onto the composite membrane placed face down on a microscopy mounting grid. Figure 5a shows the back surface of the piperazine isophthalamide interfacial membrane. Surface projections corresponding to the size of the polysulfone surface micropores were noted. In Figure 5b, the back surface of the piperazine trimesamide membrane does not show this pattern; rather, a rippled structure was present. This structure was evidence of membrane swelling, and resembled one of the alternate types of ridged surface structures observed in addition to that of Figure 4c. The piperazine isophthalamide membrane as it formed remained in intimate contact with the polysulfone support surface, while the piperazine trimesamide membrane lifted away from the polysulfone surface as it swelled during the interfacial formation step.

## DISCUSSION

A strong body of evidence has been developed in the field of interfacial polyamide synthesis to indicate that polymer formation takes place in the organic phase. Thus, the amine diffuses across the water-solvent interface and reacts with the acyl halide in the organic phase. A smooth polymer film initially forms at the interface. Further polymer formation takes place on the organic solvent side of the deposited film. This latter surface becomes increasingly rough as polymer formation proceeds. The polymer film is normally porous, allowing for the transport of amine reactant and the acid byproducts across the interfacial region. This accounts in part for the difficulty in obtaining high salt-rejecting membranes by interfacial synthesis.

In the case of polyethylenimine, however, the reaction of this polyamine with isophthaloyl chloride at the interface should result

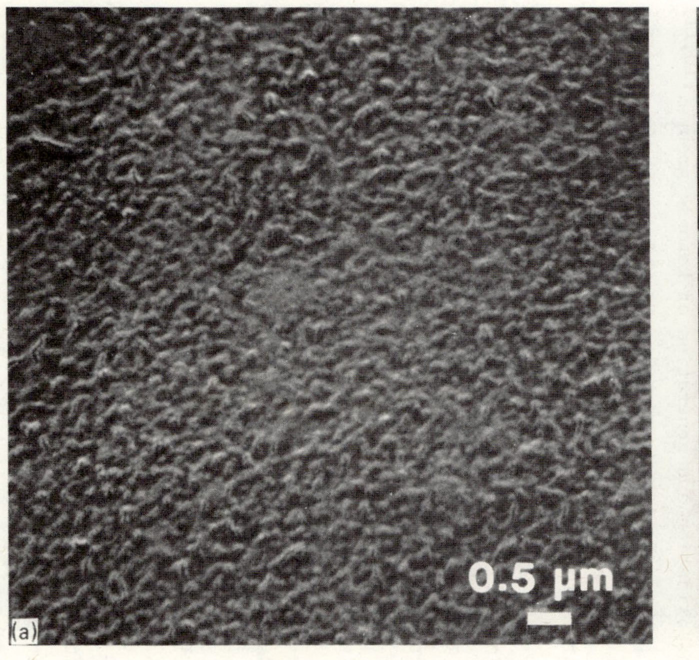


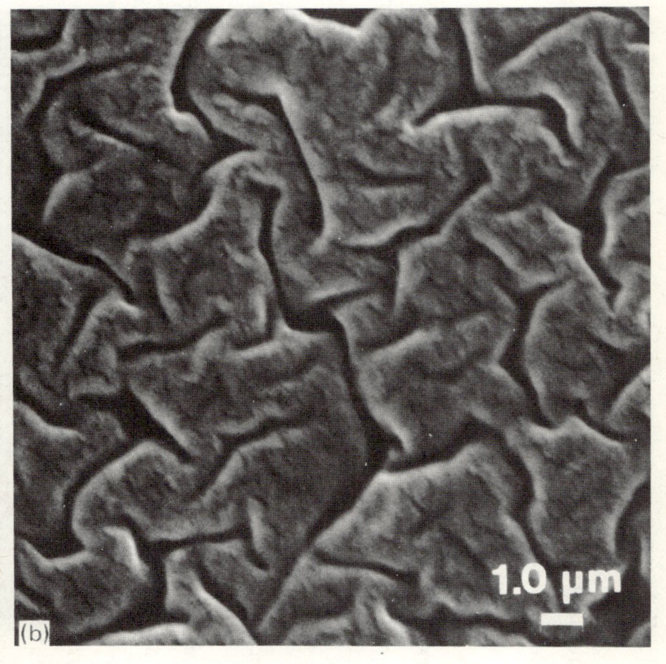


Figure V. Scanning Electron Photomicrographs of Interfacial Membrane Surfaces in Contact with Microporous Polysulfone.

a) Piperazine isophthalamide membrane backside showing template of polysulfone surface micropores.
b) Piperazine trimesamide membrane backside showing effect of swelling and absence of any patterning by the polysulfone surface originally in contact with it.

in the instantaneous formation of a barrier layer through which further polyamine, because of its high molecular weight, cannot penetrate. Thus, after the initial formation of a continuous film, further reaction must take place not in the organic phase but in the aqueous phase. Because of the low solubility of isophthaloyl chloride in the aqueous phase, a thin skin develops which only slowly increases in cross-section thereafter. A graded structure is also developed, ranging from complete acylation on the organic solvent side to the barest insolubilization on the aqueous face.

Actually a layer of unreacted polyethylenimine becomes entrapped between the interfacial polyamide deposit and the microporous support surface. Upon heating in air to greater than 100°C, this unreacted polyethylenimine is both dried and insolubilized. The insolubilization takes place mainly by the splitting out of ammonia, but also involves oxidation and generation of ketonic groups.[10,11]

The final result is a three-layer structure that consists of a microporous substrate, an intermediate layer of insolubilized polyethylenimine, and a surface barrier layer of dense polyamide material. The intermediate layer serves as a transition zone between the polysulfone substrate and the polyamide barrier layer, providing additional support to this extremely thin barrier surface. The final composite thus withstands 100 atmospheres pressure without difficulty.

In contrast to this three layer structure, the piperazine isophthalamide membrane does not form an intermediate zone between the barrier layer and the polysulfone substrate. It is in direct contact with the polysulfone surface, as indicated in Figure 5a. Furthermore, the desalting zone will tend to be "buried" in the polyamide layer, being close to the polysulfone surface. Any significant discontinuities or defects in the polysulfone surface structure would therefore disrupt the salt berrier layer. This affects in a major way the success or failure in achieving high salt rejections for piperazine isophthalamide interfacial membranes. In this respect, we have experienced considerable difficulty in fabricating polypiperazineamide membranes by machine processes, and noted on several occasions that substandard polysulfone support material greatly lowered resulting membrane salt rejections.

An intermediate zone can be built into the polypiperazineamide membrane construction, as illustrated by the use of sulfonated polysulfone. When heated to 130°C, this sulfonated polysulfone polymer undergoes insolubilization by intramolecular sulfone formation. Thus, a three layer composite can be formed, consisting of microporous polysulfone as the base layer, insolubilized sulfonated polysulfne as the intermediate layer, and a piperazine isophthalamide interfacial membrane as the salt barrier overlay. This composite membrane exhibits salt rejections approaching the 99 percent level, as shown by the example in Table 3.

The large change in properties of the polypiperazineamide membranes as a function of trimesoyl content deserves comment. The scanning electron photomicrographs displayed a striking change in morphology as isophthaloyl chloride was replaced by trimesoyl choride in the membrane preparation. Other conditions, which included use of sodium dodecyl sulfate as surfactant and sodium hydroxide as acid acceptor, remained constant. We believe that the results show a two-fold mechanism of action by trimesoyl chloride beyond that of isophthaloyl chloride -- its trifunctionality and the generation of free carboxylate groups in the final membrane.

The trifunctionality of trimesoyl chloride results in the instantaneous formation of a three-dimensional crosslinked polyamide structure at the interface upon exposure of the hexane solution of the acyl chloride to the aqueous piperazine solution. Thus, a smooth ultrathin membrane forms at the interface. Swelling and thickening by further polyamide formation takes place on this interfacial membrane with further time exposure, leading to the convoluted structure shown in Figure 4b. This additional behavior, particularly the swelling, is promoted by the hydrolysis and salt formation of residual carboxylic acid chloride groups on the trimesoyl linkages in the polyamide, and by premature hydrolysis of unreacted trimesoyl chloride.

The formation of piperazine isophthalamide is not like this. Because isophthaloyl chloride is only difunctional, linear oligomers are formed in the first instant of exposure. These oligomers are partially soluble in the aqueous phase, especially those that are amine-terminated,[6] and will also be emulsified to some extent by the surfactant. When molecular weights of the oligomers exceed solubility limits and surfactant emulsification potential, precipitation and fusion of the polyamide particles takes place. The interfacial membrane at this point would be relatively rough and quite porous. Further diffusion of reactants through this structure would eventually produce a thick, dense membrane having satisfactory salt rejection characteristics. The resulting membrane would thus have the appearance of Figure 4a.

When trimesoyl and isophthaloyl chlorides are combined, the trifunctionality of the trimesoyl chloride contributes to instantaneous formation of a smooth, ultrathin interfacial film. The rugged texture of the piperazine isophthalamide membrane is essentially eliminated in the 2:1 copolymer of piperazine isophthalamide-co-trimesamide.

The fact that water flux reaches a peak in brackish water tests with the 2:1 copolymer (see Figure 2) is of interest. Two explanations seem plausible. First, the piperazine trimesamide homopolymer membrane may have too much carboxylate ion functionality, such that it begins to behave as a soft, compressible gel structure when wet. Compaction of this gel structure under the hydrostatic pressure involved would limit water flux.

The second explanation is more complex. Hydrolysis and membrane swelling appears to take place concurrently with interfacial membrane formation in the piperazine trimesamide membrane. It is likely that reactants diffuse into this swollen membrane and react in situ in the membrane structure. Thus, although the piperazine trimesamide membrane was originally a swollen, permeable polymer network, it subsequently became dense by deposition of new polyamide within the swollen network. This explanation is supported by the fact that the convoluted surface texture of the membrane in Figure 4b remains fixed even after dessication. A flat film merely swollen with water would shrink back to a tight, stretched film upon drying. For the high flux copolymer, however, carboxylate anion formation presumeably occurs at a slower rate, so that the membrane fabrication process is finished before this phenomenon begans to assert itself. Then, under reverse osmosis test conditions on the finished membrane, carboxylate salt formation and membrane swelling reaches a state of completion, generating a high flux membrane. We believe that this latter explanation fits the data better.

In summary, thin-film composite reverse osmosis membranes can be readily prepared by interfacial polymerization techniques. This approach behaves well in the preparation of polyethylenimine isophthalamide membranes. In fact, it has reached a commercial stage in the form of a polyetheramine isophthalamide membrane version, tradenamed "PA-300".[14] The polypiperazineamide membrane system is more difficult insofar as fabrication of high salt rejection membranes is the goal. Large variations in both physical structure and reverse osmosis properties occur in this system as the content of the trifunctional trimesoyl chloride is increased. Piperazine trimesamide and its copolymers with the isophthalamide behave like hybrid membranes, possessing both cation exchange character and reverse osmosis desalination properties. They may find usefulness in the treatment of waters containing polyvalent anions, or in the separation of larger constituents such as in the concentration of sugar solutions.

## ACKNOWLEDGEMENT

Portions of this work were supported by the Office of Water Research and Technology, U.S. Department of the Interior, under Contract Numbers 14-34-0001-2883, 14-34-0001-6512 and 14-34-0001-8512. The research support by this agency is gratefully acknowledged.

## REFERENCES

1. Loeb, S., and Sourirajan, S., Advan. Chem. Ser, 38, 117 (1962).

2. Kesting, R.E., "Asymmetric Cellulose Acetate Membranes", in Reverse Osmosis and Synthetic Membranes, S. Sourirajan, Ed., National Research Council of Canada, Ottawa, 1977, Chapter 5.

3. Rozelle, L.T., Cadotte, J.E., and McClure, D.J., Appl. Polymer Symposia, 13, 61 (1970).

4. Riley, R.L., Lonsdale, H.K., Lyons, C.R., and Merten, U., J. Appl. Polymer Sci., 11, 2143 (1967).

5. Cadotte, J.E., Cobian, K.E., Forester, R.H., and Petersen, R.J., "Continued Evaluation of In Situ-Formed Condensation Polymers for Reverse Osmosis Membranes", available from National Technical Information Center, Springfield, VA, Report No. PB-253193, April 1976.

6. Cadotte, J.E., Steuck, M.H., and Petersen, R.J., "Research on In Situ-Formed Condensation Polymers from Reverse Osmosis Membranes", ibid., Report No. PB-288387, March 1978.

7. Blais, P., "Polyamide Membranes", in Reverse Osmosis and Synthetic Membranes, S. Sourirajan, Ed., National Research Council of Canada, Ottawa, 1977, Chapter 9.

8. Scala, L.C., Ciliberti, D.F., and Berg, D., U.S. Pat, 3,744,642, July 10, 1973.

9. Morgan, P.W., "Condensation Polymers by Interfacial and Solution Methods", Interscience Publ, New York, 1965, page 51.

10. Cadotte, J.E., Kopp, C.V., Cobian, K.E., and Rozelle, L.T., "Progress Report on In Situ-Formed Condensation Polymers for Reverse Osmosis: Second Phase", available from National Technical Information Service, Springfield, VA, Report No. PB-234198, June 1974.

11. Rozelle, L.T., Cadotte, J.E., Cobian, K.E., and Kopp, C.V., "Nonpolysaccharide Membranes for Reverse Osmosis: NS-100 Membranes", in Reverse Osmosis and Synthetic Membranes, S. Sourirajan, Ed., National Research Council of Canada, Ottawa, 1977, Chapter 12.

12. Petersen, R.J. and Cobian, K.E., Plating and Surface Finishing, 63, 51 (1976).

13. Rozelle, L.T., Cadotte, J.E., Nelson, B.R., and Kopp, C.V., Appl. Polymer Symposia, 22, 223 (1973).

14. Riley, R.L., Fox, R.L., Lyons, C.R., Milstead, C.E., Seroy, M.W., and Tagami, M., Desalination, 19, 113 (1976).

# Synthesis and Structural Characterization of Titanium Polyoximes

Charles E. Carraher, Jr.
Larry P. Torre
H. Michael Molloy

Wright State University
Department of Chemistry
Dayton, Ohio 45435

ABSTRACT

The synthesis and structural characterization of titanium polyoximes derived from bis(cyclopentadienyl)titanium dichloride is described. Dioximes were selected for study based on geometrical dissymmetry, potential biological activity and the potential for extended conjugation through the resulting polyoxime chain. Product yield and chain length appears dependent on the electronic character of the oxime. Product chain length also appears dependent on polymer solubility. Structural characterization using elemental analyses, infrared spectroscopy, light scattering photometry, control reactions and coupled thermogravimetric-mass spectroscopy is described.

INTRODUCTION

The synthesis and characterization of organometallic polymers is an ongoing portion of our research program. Recently we reported the synthesis of titanium and zirconium polyamidoximes of form I [1,2 ].

$$Cp_2MCl_2 + \underset{H_2N}{\overset{HO-N}{}}\!\!\!\!\!\!\!\!\!\!\!\!\!\!\!\!\!\!\!\! \quad C\!-\!R\!-\!C \underset{NH_2}{\overset{N-OH}{}} \longrightarrow \;\leftarrow\!\underset{Cp}{\overset{Cp}{M}}\!-\!O\!-\!N\!=\!\underset{H_2N}{C}\!-\!R\!-\!\underset{NH_2}{C}\!=\!N\!-\!O\!\rightarrow$$

I

While reaction is known to almost exclusively occur only at the oxime sites, reaction at the amine sites does occur when excess $Cp_2MX_2$ is employed leading to branched and crosslinked products. Group IVB containing polymers have generally shown poor solubilities in both extent and range. Even small amounts of crosslinking could permit this generally poor solubility to become complete insolubility.

Here we report the synthesis of titanium polyoximes of form II. The use of dioximes allows the use of the oxime functional group yet eliminates the formation of branched and crosslinked products.

$$Cp_2TiCl_2 + HO\!-\!N\!=\!\underset{R}{C}\!-\!R'\!-\!\underset{R}{C}\!=\!N\!-\!OH \longrightarrow \;\leftarrow\!\underset{Cp}{\overset{Cp}{Ti}}\!-\!O\!-\!N\!=\!\underset{R}{C}\!-\!R'\!-\!\underset{R}{C}\!=\!N\!-\!O\!\rightarrow$$

II

## EXPERIMENTAL

The following chemicals were used as received without further purification: bis(cyclopentadienyl)titanium dichloride (Strem Chemicals, Inc., Davers, Mass.), menadione (Vitamin $K_3$; United States Biochemical Corp., Cleveland, Ohio), Progesterone ($\Delta^1$-pregnan-3,20-dione; U.S.B.C.), Androstadiendione ($\Delta^{1,4}$-androstadien-3,17-dione, U.S.B.C.), 3,20-Pregnanedione (Eastern Chemicals), Phenindione (2-phenyl-1,3-indanone; Pfaltz-Bauer, Inc., Stamford, Conn.), 1,4-cyclohexanedione (Aldrich Chemical Company, Milwaukee, Wis.), p-benzoquinone dioxime (Aldrich), and hydroxylamine hydrochloride (Fisher Scientific Company, Fairlawn, N.J.).

The oximes were synthesized as described in reference 3. The resulting solids were washed with 1:1 ethanol-water after the product was recovered by suction filtration. The product was then recrystallized twice from 1:1 ethanol-water.

2-Methyl-1, 4-naphthoquinone dioxime (menadioxime, Vitamin $K_3$ dioxime),: the yellow solid [4] was obtained in 49% yield; m.p. = 164-167°C, (lit. 166-168°C; IR (KBr) 3260 $cm^{-1}$ (O-H), 1620 $cm^{-1}$ (C=N), 980 $cm^{1}$ (N-O).

$\Delta^1$-Pregnan-3,20-dioxime (progesterone dioxime),: the white solid [5] was obtained in 86% yield; m.p. = 248-252°C, (lit. 247-256°C; IR (KBr) 3280 $cm^{-1}$ (O-H), 1640 $cm^{-1}$ (C=N, 915-980 $cm^{-1}$ (N-O).

$\Delta^{1,4}$-Androstadien-3,17-dioxime (androstadiendioxime): the white solid [6] was obtained in 90% yield; m.p. = 232-235°C, (lit. 229-230°C); IR (KBr) 3260 $cm^{-1}$ (O-H), 1660 $cm^{-1}$ (C=N), 870-980 $cm^{-1}$ (N-O).

3,20-Pregnanedioxime (pregnanedioxime): the white solid [7] was obtained in 72% yield; m.p. = 174-176°C (lit. 260°C); IR (KBr) 3300 $cm^{-1}$ (O-H, 1695 $cm^{-1}$(C=N), 900-1010 $cm^{-1}$ (N-O),

Anal. Calcd. for $C_{21}H_{34}O_2N_2$: C, 73.0; H, 9.9; O, 9.3%
Found: C, 76.2; H,10.1; O, 9.2%

1,4-Cyclohexanedione dioxime: the white crystalline solid [8] was obtained in 70% yield; m.p. = 189-192°C, (lit. 188°C), IR (KBr) 3200 $cm^{-1}$ (O-H), 1660 $cm^{-1}$ (C=N), 925-1000 $cm^{-1}$ (N-O.

2-Phenyl-1,3-indanone dioxime (phenindioxime): the white solid [9] was obtained in 68% yield; m.p. = 198-200°C, (lit. 193-196°C); IR (KBr) 3200 $cm^{-1}$ (O-H), 1650 $cm^{-1}$ (C=N), 957, 1045 $cm^{-1}$ (N-O).

Reaction sequences were similar to those described elsewhere [10,11]. Briefly the $Cp_2TiCl_2$ contained in an organic (for interfacial polycondensations) or aqueous (for aqueous solution polycondensations) liquid is added to rapidly stirred aqueous solutions containing neutralized dioxime. The polyoxime precipitates from the reaction solution and is recovered using suction filtration. The product is washed repeatedly with water and chloroform. The reaction assembly is described elsewhere [12].

Infrared spectra were obtained using KBr pellets employing a Perkin-Elmer 457 Grating Spectrophotometer. Light scattering photometry was conducted employing a Brice-Phoenix BP-3000 Universal Light Scattering Photometer. Refractive index increments were determined using a Bausch and Lomb Abbe Refractometer Model #3-L.

Mass spectral data was obtained using a Model 21-491 DuPont Mass Spectrometer which was coupled through a jet separator to a DuPont 951 Thermogravimetric Analyzer controlled through a DuPont 990 Thermal Analyzer console. The TG-MS assembly was interfaced with a Hewlett-Packard 2116C computer. The assembly is described in detail elsewhere [13].

Elemental analyses were performed by Galbraith Laboratories, Knoxville, Tenn. Elemental analyses for the metals were performed by thermal degradation of the samples to $TiO_2$.

## RESULTS AND DISCUSSION

A number of dioximes could have been utilized since they are readily synthesized from corresponding diones. The utilized dioximes were chosen based on geometrical dissymmetry (to discourage crystalline formation enhancing product solubility), potential biological activity (a number of these compounds are currently undergoing biological study) and the potential for extended conjugation through the polymer chain (electrical characterization is currently occurring). Dioximes can be considered as weak acids with Lewis acid-base reaction characteristics similar to carboxylic acids, such as forming salts on addition of sodium hydroxide. The diones and dioximes utilized in this study are pictured in Figure 1.

Bis(cyclopentadienyl)titanium dichloride contains two readily identifiable chromophores: the red Ti-Cl chromophore and the

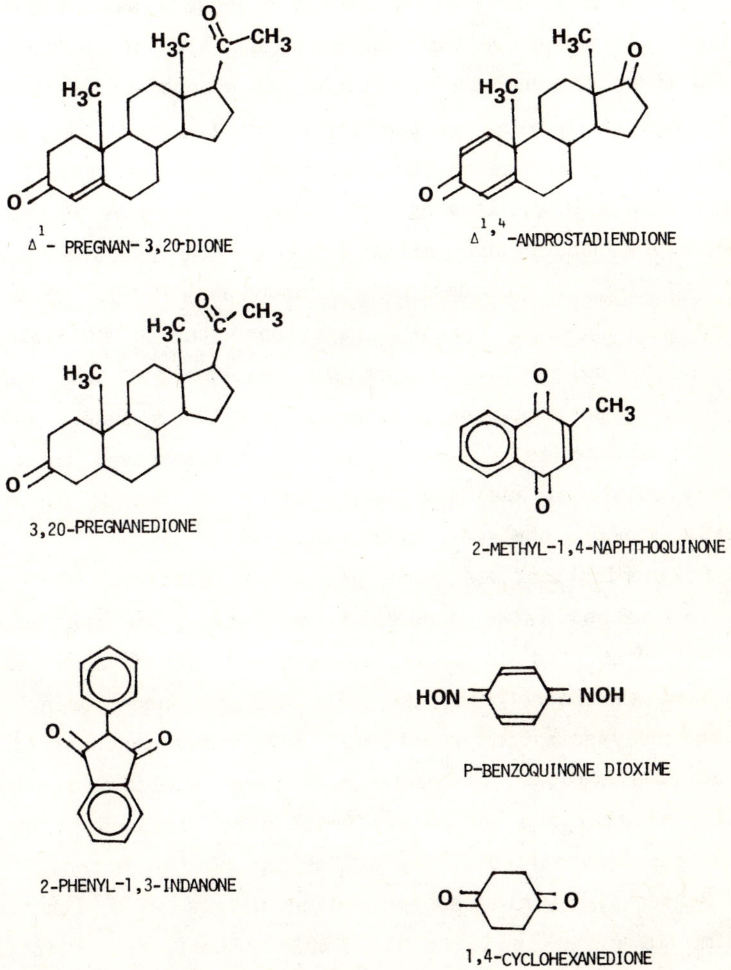

$\Delta^{1}$- PREGNAN-3,20-DIONE

$\Delta^{1,4}$-ANDROSTADIENDIONE

3,20-PREGNANEDIONE

2-METHYL-1,4-NAPHTHOQUINONE

2-PHENYL-1,3-INDANONE

P-BENZOQUINONE DIOXIME

1,4-CYCLOHEXANEDIONE

FIGURE 1. Diones and dioximes utilized to synthesize titanium polyoximes.

yellow Ti-Cp chromophore [14]. Carraher and co-workers [10] have noted that in all polymer families synthesized involving the titanium metallocene studied (except the titanium polyester-ethers where the Lewis base coloration mask any color change) that the initial color of $Cp_2TiCl_2$ in organic solvents is deep red, which gives way to a yellow color upon polycondensation.

The interfacial polymerization procedure was as described earlier. It was noted that the deep red color of $Cp_2TiCl_2$ (dissolved in chloroform) faded after 10-15 seconds of reaction (stirring) to leave a yellow color for the remaining time of reaction. This indicates the loss of the chlorine atoms from the metallocene moiety, leaving the Ti-Cp chromophore as the only color site, hence, the yellow color of the reaction mixture. (The red Ti-Cl is the dominant chromophore [14]). This color change was observed for all reactions studied utilizing the interfacial polymerization process, except for the reactions run using p-benzoquinone dioxime as the Lewis base. p-Benzoquinone dioxime is a deep brown color (dissolved in a basic aqueous solution) and thus masks any color change due to the $Cp_2TiCl_2$ monomer that may occur during reaction.

Polymerization was also performed utilizing the aqueous solution method (also described earlier). As the color of $Cp_2TiCl_2$ dissolved in water is yellow (rather than red when dissolved in chloroform), no color changes were observed during the polymerization reaction. Polyoximes synthesized using the aqueous solution technique gave lower yields and otherwise similar thermal and solubility results as compared to the same polyoximes synthesized employing the interfacial method.

Percentage yield and molecular weight as a function of dioxime is given in Table 1. Table 2 contains a listing of trends as a function of various parameters. There appears to

TABLE 1.

Results as a function of dioxime.

| Dioxime (Product Identification Number) | Yield(%)[a] | %Ti Found | %Ti Theory | Solubility[b] (% by weight) | dn/dc[c] | $\bar{M}_w$[c] | $\overline{DP}$ |
|---|---|---|---|---|---|---|---|
| $\Delta^1$-Pregnane-3,20-dione(1) | 40 | 8.9 | 9.2 | 0.60 | -0.217 | $3.0 \times 10^5$ | 590 |
| 3,20-Pregnanedione(2) | 12 | | | 0.067 | -2.9 | $9.7 \times 10^3$ | 19 |
| $\Delta^{1,4}$-Androstadiendione(3) | 35 | 10.0 | 9.8 | 0.13 | -1.9 | $2.0 \times 10^4$ | 41 |
| 2-Methyl-1,4-Naphthoquinone(4) | 35 | 12.3 | 12.7 | 0.27 | -1.1 | $9.1 \times 10^4$ | 240 |
| 2-Phenyl-1,3-Indanone(5) | 62 | 10.9 | 11.2 | 0.21 | -0.66 | $1.6 \times 10^5$ | 370 |
| p-Benzoquinone(6) | 64 | 14.6 | 15.2 | 0.11 | -1.9 | $2.1 \times 10^4$ | 67 |
| 1,4-Cyclohexanedione(7) | 28 | 14.7 | 15.0 | 0.20 | -1.2 | $7.8 \times 10^3$ | 25 |

a. Reaction conditions: 1.0 mmole $Cp_2TiCl_2$ in 25.0 ml chloroform added to a rapidly stirred (23,000 rpm no load) solution containing 1.0 mmole dioxime and 2.0 mmole NaOH in 25.0 ml water.

b. Solubility of product in 1-methylimidazole at 25°C.

c. Determined in 1-methylimidazole at 25°C.

TABLE 2

Trends as a Function of Various Parameters (Using the Numerical Designations Given in Table 1)

| | |
|---|---|
| Yield (decreasing) | 6 > 5 > 1 > 3 ≃ 3 > 7 > 2 |
| Chain Length (decreasing) | 1 > 5 > 4 > 6 > 3 > 7 > 2 |
| Steric (increasing) | 6 ≃ 7 > 4 ≃ 3 > 1 ≃ 2 > 5 |
| Nucleophilicity of Oxime (decreasing) | 6 ≃ 4 > 3 > 1 > 5 > 7 ≃ 2 |
| Product Solubility (decreasing; in 1-methylimidazole) | 1 > 4 > 5 ≃ 7 > 3 ≃ 6 > 2 |
| Size (decreasing) | 1 ≃ 2 > 3 > 5 ≃ 4 > 7 ≃ 6 |

be a general correlation between product average chain length and product yield and the factor of nucleophilicity of the oxime emphasizing a reasonable correlation where both yield and chain length growth are favored by a more nucleophilic oxime. There also exists an even better fit of chain length with polyoxime solubility. This also is reasonable if product insolubility is in fact an important factor in determining polyoxime chain length. (It must be noted that the solubility trend is for 1-methylimidazole and not the polymerization system.)

Elemental analyses for nitrogen and hydrogen were generally consistent with expected (calculated) values. Values found for carbon were typically 30-40% lower than expected. (Elemental analyses of products of $Cp_2TiCl_2$ and $\Delta^{1,4}$-androstadiendioxime, %C=71.0 calc., 39.8 found; %H-7.0 calc., 5.6 found; %N-5.7 calc., 4.5 found; 2-methyl-1,4-naphthoquinone dioxime, %C-66.7 calc., 44.8 found; %H-4.8 calc., 3.8 found; %N-7.4 calc., 4.8 found; 2-phenyl-1,3-indanone-70.1% calc., 57% found; %H-4.7 calc., 4.2 found; %N-6.5 calc., 8.3 found; b-benzoquinone diocime-%C=61.2 calc., 42.2 found; %H-4.5 calc., 4.0 found; %N-8.9 calc., 9.9 found). Other workers have also found poor elemental analysis results to be typical of Group IVB metallocene containing polymers. It is thought that this behavior is due, at least in part, to the moderate to good high ( >1000°C) temperature stability of the thermal degradation residue of the polymers. Elemental analyses for titanium were in agreement with expected values based on form II (Table 1).

Infrared spectral bands characteristic of the cyclopentadienyl, Cp, moiety from the metallocene typically occur at 1440, 1010, and 830 $cm^{-1}$ [15]. The presence of the Cp moiety in the IR spectra of most of the polyoximes cannot be unambiguously identified as some of the dioxime monomers have bands at about 1440, 1010, and/or 830 $cm^{-1}$, which mask the major contributions from the Cp moiety.

All of the polyoxime products exhibited bands characteristic of the dioxime monomer used. For instance, IR spectra of the condensation product of 2-methyl-1,4-naphthoquinone dioxime and $Cp_2TiCl_2$ shows bands at 1430 and 1010 $cm^{-1}$ characteristic of the Cp moiety. The region at about 830 $cm^{-1}$ which is also characteristic of the Cp moiety is masked by a medium to strong band at 820 $cm^{-1}$ derived from the dioxime itself. A strong broad band is observed in both dioxime monomer and polyoxime product at about 3600 to 3100 $cm^{-1}$. This was believed to be due to the presence of O-H stretching [15] in the dioxime. This assignment appears to be incorrect as the strong band from 3600 to 3100 $cm^{1}$ appears in the IR spectra of the polyoxime. A salt of the dioxime monomer was made [5]. Inspection of the resulting IR spectra showed the strong broad peak to be present. The following are bands present in the polyoxime product; Bands at about 2980 to 2900 $cm^{-1}$ which are attributed to the presence of aliphatic C-H (stretching) groups; Bands at about 1630 to 1600 $cm^{-1}$ are characteristic of C=N stretching; Bands at about 1020 to 950 $cm^{-1}$ are characteristic of N-O stretching; Bands at 820, 770, and 700 $cm^{-1}$

are characteristic of the presence of disubstituted benzene moieties; Other bands present in both monomer and polymer appear at 1385, 1350, 1315, 1270, 1240, and 895 $cm^{-1}$. A new non-assigned band appears at 495 $cm^{-1}$ in the polyoxime product. Thus, the presence of components from each reactant is indicated by the infrared spectrum of the product.

Two polyoxime products were studied using coupled thermogravimetric-mass spectroscopy, TG-MS. Briefly, as the TG temperature increases, volatile components resulting from sample degradation are carried, employing helium gas, into the MS, resulting in a mass spectrum characteristic of the evolved fragments. Mass spectrum were recorded for each 8C$^{o}$ increment of temperature from ambient to about 900$^{o}$C.

The product from 2-methyl-1,4-naphthoquinone dioxime exhibits a weight loss beginning at 40$^{o}$C with an almost constant rate of weight loss to 230$^{o}$C resulting in a total weight loss of about 10% of the sample weight. This is followed by a more rapid weight loss (about 6%) from 230$^{o}$C to 285$^{o}$C. From 285$^{o}$C to 900$^{o}$C the weight loss (about 27%) was almost constant. The degradation occurring at about 230$^{o}$C to 285$^{o}$C gave volatile products which yielded 23 major mass-to-charge, m/e, peaks all present in the scan at 250$^{o}$C (Table 3; Normalized intensity > 1; normalized to mass 28 (nitrogen) = 100.0%) The masses at 66, 65, 39, 40, 38, 62, 31, and 37 amu and the respective intensities of these mass peaks is consistent with the standard [16] mass spectra of the Cp moiety.

Mass peaks characteristic of the Cp moiety reach a maximum at 250$^{o}$C, then decrease ending by 500$^{o}$C. Thus, masses characteristic of the Cp moiety are the major components of the initial thermal degradation occurring from about 100$^{o}$C to 500$^{o}$C. These mass peaks then decrease as temperature continues to increase, and larger mass fragments indicative of the dioxime moiety appear.

TABLE 3

TG-MS Data from the Polycondensation Product of 2-Methyl-1,4-naphthoquinone dioxime and $Cp_2TiCl_2$.

| Mass Peak (m/e) | Possible Assignments | Normalized Intensity at (°C) 127.3 | 181.0 | 250.0 | 472.3 |
|---|---|---|---|---|---|
| 20 | | 1.501 | | | |
| 26 | HC = CH | | | 1.355 | |
| 29 | | 1.280 | | | |
| 31 | (Cp) | | | 2.576 | |
| 38 | (Cp) | | | 1.914 | |
| 39 | (Cp) | | 1.908 | 8.289 | |
| 40 | (Cp) | | 4.297 | 10.156 | |
| 41 | $CH_3$-C-$CH_2$ | | | | 2.952 |
| 43 | O-N=CH | 1.721 | | | 1.611 |
| 50 | HC-C=C-CH | | | 1.269 | |
| 51 | HC=C-CH=CH | | | 1.838 | |
| 52 | HC=CH-CH=CH | | | 1.374 | |
| 55 | ON=C-CH | | | | 4.232 |
| 56 | O-N=C-$CH_2$ | | | | 1.524 |
| 57 | O-N=CH-$CH_2$ | 1,228 | | | |
| 61 | HC=C-C=CH-CH | | | 1.374 | |
| 63 | HC=CH-CH-C-C | | 1.339 | 3.685 | |
| 65 | (Cp) | 1.186 | | 10.894 | 1.028 |
| 66 | (Cp) | 2.760 | 4.769 | 22.689 | |
| 67 | | | | 1.137 | 1.994 |
| 68 | O-N=C-HC-CH | | | | 2.186 |
| 69 | O-N=C-C-$CH_3$ | | | | 1.341 |
| 77 | ONC-C-CH-CH-C | | | | 1.071 |
| 79 | | | | | 2.552 |
| 80 | [structure], ON=[structure] | | | 3.335 | |
| 81 | [structure] | | | | 2.386 |
| 91 | | | | | 2.404 |
| 94 | ON=[structure] | | | | 2.186 |
| 95 | | | | | 1.620 |
| 105 | | | | | 2.334 |
| 107 | | | | | 1.193 |
| 117 | | | | | 1.742 |
| 121 | NC-CH=CH-C-C(NO) -C-$CH_3$ | | | | 1.959 |
| 131 | ON=[structure]=NO, $CH_3$ | | | | 1.054 |

The absence of significant (normalized intensities > 1) masses at 47 (Ti), 79 ($TiO_2$), 112 (CpTi), and 170-180 ($Cp_2Ti$) indicates titanium containing, volatile species are not present, which is consistent with the metal remaining within the residue.

The polyoxime derived from $\Delta^1$-pregnone-3,20-dioxime exhibits a constant weight loss of about 10% over the temperature range of ambient to 265°C. A comparatively rapid degradation occurs from 265°C to 310°C resulting in a weight loss of 11%. Weight loss is again almost constant to 470°C (≃10%) and is followed by another more rapid degradation of 8% to 525°C. Weight loss from 525°C to 900°C is again almost constant reulting in a total weight loss of 56%. The relative ion intensity reached a maximum at about 260°C within the temperature range of ambient to 265°C. The MS scan at 160°C showed only three fragments with normalized intensities greater than one. All these are indicative of the presence of the Cp moiety (Table 4; e/m = 66, 65, 40) indicating that as in the case of the previously described polyoxime the Cp moiety is the initial species released during thermal degradation.

The degradation from 265°C to 310°C yielded 19 major mass peaks all present in the scan at 278°C (Table 4). Again, masses at 66, 65, 39, 40, 38, 62, 31, and 37 were present, indicating a major source of degradation products in this range was the Cp moiety. Masses at 80, 63, 44, 41, 32, and 29 are also present, and can be tentatively assigned to fragments of the steroid moiety (Table 4).

During the third range of degradation (470°C-585°C), masses indicative of the Cp moiety have all but disappeared. Larger fragments (81, 91, 105, 117, 121, 131) have appeared and can be tentatively assigned to larger fragments of the progesterone dioxime moiety (Table 4).

TABLE 4
TG-MS Data from the Polycondensation Product of a $\Delta$-pregnane-3,20-dione dioxime with $Cp_2TiCl_2$

| Mass Peak (m/e) | Possible Assignment | Normalized Intensity at (°C) 158.3 | 278.0 | 342.3 | 472.7 |
|---|---|---|---|---|---|
| 27 | $H_2C-CH$ | | 1.847 | | 2.160 |
| 29 | Cp | | 1.339 | | |
| 37 | | | 1.129 | | |
| 38 | Cp | | 1.549 | | |
| 39 | | | 5.758 | | |
| 40 | Cp | 1.808 | 9.250 | 2.456 | |
| 41 | $HC{=}NO, H_2C-CH_2-CH$ | 4.122 | 3.202 | 2.952 | |
| 43 | | | | | 1.611 |
| 55 | $H_2C-CH_2-C-CH_3$ | | | | 4.232 |
| 56 |  [structure], $ONC-CH_2$ | | | 1.524 | |
| 63 | | | 1.129 | | |
| 65 | Cp | 1.648 | 6.852 | | 1.028 |
| 66 | Cp | 4.276 | 14.221 | 1.577 | |
| 67 | $H_2C-CH_2-C-CH-CH_2$ | | | 1.994 | |
| 68 | $H_2C-CH_2-\overset{CH_3}{C}-CH$ | | | | 2.186 |
| 69 | $ON-C-CH_2-CH$ | | | | 1.341 |
| 77 | $ON-C-CH_2-CH$ | | | | 1.071 |
| 79 | $HC-C-CH_2-CH_2-CH-CH$ | | | 1.193 | 2.552 |
| 80 | | | 1.146 | | |
| 81 | 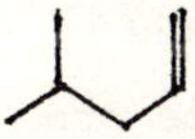 [structure], $ON{=}C-CH{=}C-CH_2$ | | | | 2.386 |
| 91 | | | | | 2.404 |
| 94 | O-N= 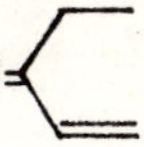 [structure] | | | | 2.186 |
| 95 | | | | | 1.620 |
| 105 | | | | | 2.334 |
| 107 | | | | | 1.193 |
| 117 | | | | | 1.742 |
| 121 | ON= 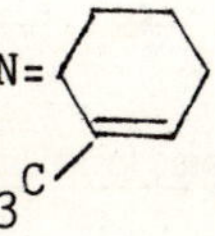 [structure], $H_3C$ | | | | 1.959 |
| 131 | | | | | 1.054 |

Again, no mass peaks indicative of titanium containing fragments (47, 79, 170-180) are present, consistent with the metal remaining within the residue.

An interesting mass peak of 343 amu was observed at a temperature of about 550°C. The molecular weight of the $\Delta^1$-pregnan-3,20-dioxime moiety is 344.5 amu. Thus, the peak observed is consistent with the dioxime moiety stripped of its two acidic hydrogens, as it would exist in the polyoxime backbone.

Reactions were attempted excluding either monomer, dioxime or $Cp_2TiCl_2$ with all other reaction conditions held constant. The reactions resulted in clear solutions as compared to the gelatinous or foamy material formed when both monomers are present.

In summation, the evidence presented within this section is consistent with the formation of polyoximes of the form II. Control reactions indicate that the product is the result of both monomers being present. TG-MS data show the presence of the Cp moiety from the metallocene in the product and is consistent with the presence of the dioxime moiety. Elemental analyses for the presence of the dioxime moiety. Elemental analyses for hydrogen, nitrogen, and titanium are consistent with expected values for products of the form II. IR data show the presence of the dioxime moiety in the respective polyoxime product and for many compounds, also the presence of the $Cp_2M$ moiety.

## REFERENCES

[1] C. Carraher, and R. Frary, J. Polymer Sci., Polymer Chemistry Ed., 12, 799 (1974).

[2] C. Carraher, and R. Frary, Die Makromolekular Chemie, 175, 2307 (1974).

[3] W. Sorenson, and T. Campbell, Preparative Methods of Polymer Chemistry, 2nd Ed., Interscience, New York, (1968), 118.

[4] Handbook of Tables for Organic Compound Identification, (3rd Ed.), compiled by ZVI Rappoport, pub. CRC, Cleveland, Ohio, 181 (1967).

[5] Handbook of Tables for Organic Compound Identification, (3rd Ed.), compiled by ZVI Rappoport, pub. CRC, Cleveland, Ohio, 2779 (1967).

[6] E. Hershberg, J. Org. Chem.. 13, 542 (1948).

[7] M. Jarrot, M.F. Laine, K. Qui, and R. Goutarel, Bull. Soc. Chim. France, 111 (1962).

[8] Dictionary of Organic Compounds, (4th Ed.), Vol. IX, Oxford University Press, New York, 785 (1965).

[9] Dictionary of Organic Compounds, (4th Ed.), Vol. IX, Oxford University Press, New York, 2699 (1965).

[10] C. Carraher, Interfacial Synthesis, Vol. 11 (Ed. by F. Millich and C. Carraher), Chap. 21, Marcel Decker, New York, 1977.

[11] C. Carraher, and S. Bajah, Polymer (Br.), 14, 42 (1973) and 15, 9 (1974).

[12] C. Carraher, J. Chemical Ed., 46, 314 (1969).

[13] C. Carraher, M. Molloy, M. Taylor and T. Tiernan, Organometallic Polymers, (Edited by C. Carraher, J. Sheats and C. Pittman), Marcel Dekker, N.Y., 1980.

[14] J. Chien, J. Phys. Chem., 67, 2477 (1963).

[15] C.N.R. Rao, Chemical Applications of Infrared Spectroscopy, Academic Press, N.Y., 1963.

[16] E. Stenhagen, S. Abrahamsson, F.W. McLafferty (Eds.), Atlas of Mass Spectral Data, Wiley-Interscience, N.Y., 1969.

# Organotitanium Polydyes Derived from Phenylsulfonphthalein Dyes, and Congo Red, Eriochrome Black T, Nigrosine and Indigo Carmine-Synthesis and Doping Characteristics

Charles E. Carraher, Jr.
Richard A. Schwarz
Jack A. Schroeder
Macy Schwarz

Department of Chemistry
Wright State University
Dayton, Ohio 45435

ABSTRACT

The synthesis of titanium polydyes emphasizing the interfacial technique is described. Emphasis is placed on applications of the polydyes taking advantage of the potential permanent, nonleaching nature of such materials compared with the monomeric dyes themselves. The polydyes are fluorescent and can be impregnated into paper, cloth and paint and "doped" into plastics giving fluorescent materials.

## INTRODUCTION

Carraher and co-workers have reported the synthesis of titanium polyesters (I), polyethers (II) and polyamines (III) (for instance 1-6). The reactivity of the $RSO_3^-$ moiety appears analogous to that of $RCO_2^-$ towards metal halides. Here we report the synthesis of titanium polydyes derived from several dyes including phenylsulfonphthalein dyes (IV), Nigrosine (III), Indigo Carmine (V), Eriochrome Black T, Congo Red and 3-(2-pyridyl)-5,6-diphenyl-1,2,4-triazine-p, p'-disulfonic acid (PDTD, V).

$$Cp_2TiCl_2 + {}^-O_2CRCO_2{}^- \rightarrow \left( \underset{Cp}{\overset{Cp}{Ti}}-O-\overset{O}{\overset{\|}{C}}-R-\overset{O}{\overset{\|}{C}}-O \right) \quad I$$

$$Cp_2TiCl_2 + HO-R-OH \rightarrow \left( \underset{Cp}{\overset{Cp}{Ti}}-O-R-O \right) \quad II$$

$$Cp_2TiCl_2 + H_2N-R-NH_2 \rightarrow \left( \underset{Cp}{\overset{Cp}{Ti}}-NHR-NH \right) \quad III$$

$$Cp_2TiCl_2 + HO-R-SO_3{}^- \rightarrow \left( \underset{Cp}{\overset{Cp}{Ti}}-O-R-\underset{O}{\overset{O}{S}}-O \right) \quad IV$$

$$Cp_2TiCl_2 + {}^-O_3S-R-SO_3{}^- \rightarrow \left( \underset{Cp}{\overset{Cp}{Ti}}-O-SO_2-R-SO_2-O \right) \quad V$$

## EXPERIMENTAL

The following chemicals were used as received: $Cp_2TiCl_2$ (Strem Chemicals, Inc.; Danvers, Mass.), Nigrosin (certified; Matheson, Coleman and Bell; Norwood, Ohio), Bromophenol Blue (Fisher Scientific Co.; Fairlawn, N.J.), Eriochrome Black T (Eastman Chem. Corp., Rochester, N.Y.); Bromothymol Blue (Fisher Sci. Co.; Fairlawn, N.J.), Indigo Carmine (MCB,

Norwood, Ohio), Congo Red (MCB) and Ethidium Bromide (Chemalog).

Polycondensations were carried out as described in ref. 1. Briefly aqueous solutions containing dye and any added base were added to rapidly stirred aqueous (for aqueous solution systems) or organic (for interfacial systems) solutions containing $Cp_2TiCl_2$. The polymerization equipment, a one pint Kimex Emulsifying Jar fitted onto a Waring Blendor (700 model 31BL46, is fully described in reference 6.

Infrared spectra were obtained using KBr pellets and Perkin-Elmer 621 and 457 Grating Infrared Spectrophotometers. Ultraviolet-visible spectra were obtained utilizing a Cary Model 14 Spectrophotometer.

Solubility studies were performed by adding about 1mg of the polydye product to 2ml of solvent contained in a test tube. The tubes were stoppered, periodically shaken, and viewed over a week's period. Elemental analyses for titanium were obtained utilizing a 951 DuPont Thermal Gravimetric Analyzer fitted onto a DuPont 990 Thermal Analyzer console. Softening ranges were obtained using a Fisher-Johns Melting Point Apparatus.

Light scattering was conducted utilizing serial dilutions employing a Brice-Phoenix 2000 Universal Light Scattering Photometer. Refractive index increments were determined using a Bausch and Lomb Abbe Refractometer Model #3-L. Viscometry measurements were obtained employing serial dilutions of polymer solutions using a Cannon-Ubbelohde Semimicro Dilution Viscometer. Timing was accomplished utilizing a Wescan Instruments Inc. (Santa Clara, Calif.) Model 1221 Viscosity timer.

## RESULTS AND DISCUSSION

The present study is part of a larger study involving inclusion of dyes into polymers. Selection of dyes was based on two primary requirements. First, the dye must possess two (acceptable) Lewis base sites (such as alcohol, acid, amine). Second, the dye (in the correct form to allow the desired condensation to occur) was to be water soluble. Phenylsulfonphthalein (PSP) dyes exhibit two nucleophilic sites in water solution over a wide pH range, generally from below a pH of 5 to about 14, existing as noted in form VI.

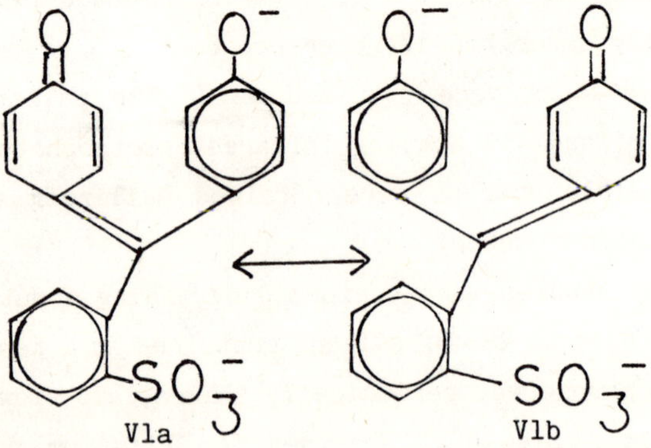

VIa VIb

Polymers derived from condensation with $Cp_2TiCl_2$ and the various dyes will be referred to simply as polydyes.

Polydye structure for condensation with dyes exhibiting only two "active" Lewis base sites will be linear with the connective bonds being the amine and/or imine (for Nigrosine; analygous to form III), sulfone-ether (for PSP dyes, analogous to form IV and depicted by form VII) and sulfone (PDTD, form V).

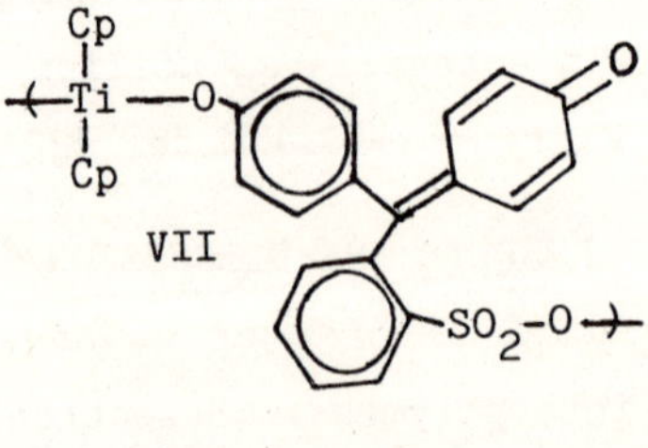

VII

Polydyes derived from dyes exhibiting more than two reactive sites may contain a complex branched and crosslinked structure. For instance Eriochrome Black T contains three potentially active reactive sites under the employed reaction conditions. The resulting polydye is soluble in such dipolar aproatic solvents as dimethylformamide, triethylphosphate, and dimethylsulfoxide, eliminating the presence of excessive crosslinking. Elemental analysis is consistent with a highly branched structure approaching form VIII.

```
      Ti        Ti        Ti
      |         |         |
-Ti - D - Ti  - D - Ti  - D -
```

where D = Eriochrome Black T moiety

VIII

T = $Cp_2Ti$ moiety

Indigo Carmine is tetrafunctional and gives products which are insoluble in all tested solvents consistent with the presence of some crosslinking. Thus the structure of products derived from such multifunctional dyes is complex and a mixture of several forms, the proportion and distribution surely dependent on a number of factors including reaction conditions.

The utilized dyes are pictured in Figure 1 with the representative structures depicting the purchased form of the dye.

The infrared spectra of the polydyes are consistent with forms such as V, VII as representing the repeating unit. For instance all of the polydye products exhibit bands at 1440, 1010 and 830 $cm^{-1}$ characteristic of the Ti-Cp moiety. The PSP dyes react at one alcohol site and the sulfate site. There is a shift in the assymetric stretching band for S=O from a doublet at 1170 $cm^{-1}$ for the dye to a single band at 1160 $cm^{-1}$ for the polydye and a shift in the S-O symmetric stretching band

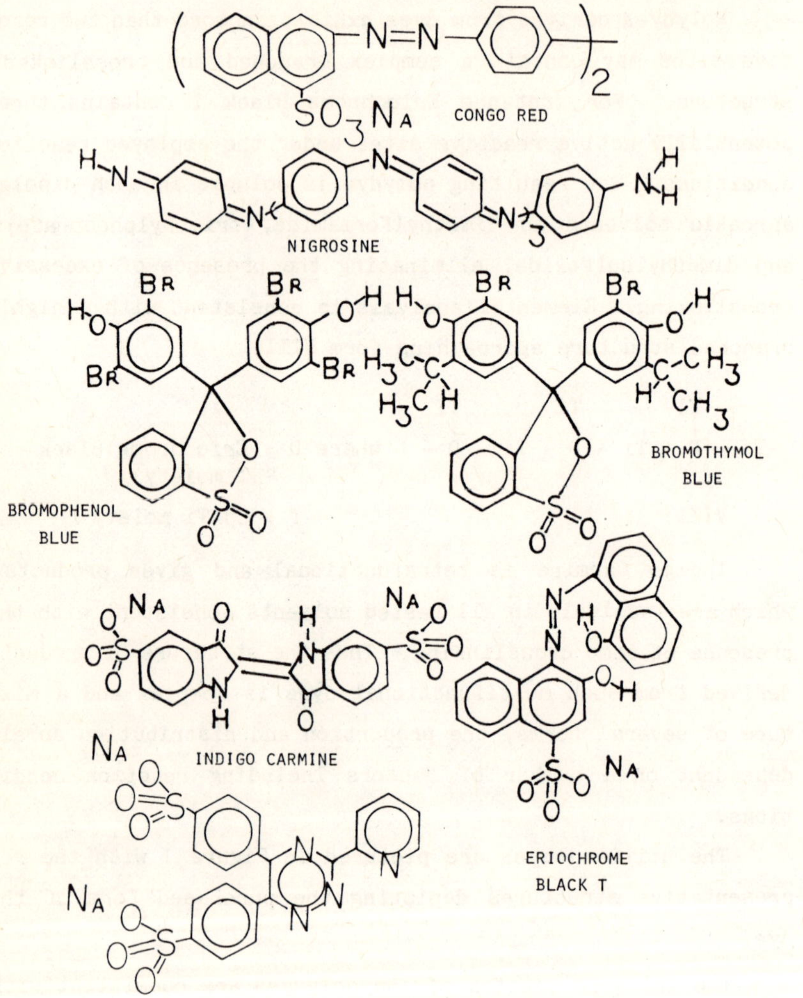

FIGURE 1. STRUCTURES OF UTILIZED DYES

from 1060 $cm^{-1}$ for the dye to 1080 $cm^{-1}$ for the polydye [8]. Spectra of the polydyes exhibit no bands about 3100 to 3500 $cm^{-1}$ consistent with an absence of the alcohol moiety. The polydye products do exhibit a band about 1600 $cm^{-1}$ characteristic of a ketone carbonyl conjugated with alkenes [8]. A shoulder was present at 1580 $cm^{-1}$ characteristic of alkene stretching vibrations in conjugated systems [8,9]. Elemental analyses were also consistent with a structure of form VII. For instance the percentage titanium found for the product derived from Bromophenol Blue was 5.9, calculated was 5.7.

Further control reactions are consistent with the precipitated material being derived from both the dye and $Cp_2TiCl_2$. Reactions were attempted excluding either monomer (dye or $Cp_2TiCl_2$). No precipitation occurred.

Thus spectral, elemental analyses and control reaction results are consistent with a structure of form VII as the repeating unit for the polydyes derived from PSP, etc.

The polymeric nature of the polydye products is indicated by both viscometry and light scattering photometry results. For instance the Bromophenol Blue product which gave an LVN of 7 ml/g exhibited a dn/dc of 0.24 and a weight average molecular weight of $5.3 \times 10^5$.

Table 1 contains a list of polydyes synthesized and reaction conditions. Generally polydyes are formed utilizing only the interfacial technique though analogous titanium polyamines, polyethers and polyesters are formed using both the interfacial and aqueous solution techniques. None of the aqueous solution systems were extensively studied so polydye formation might occur under other sets of reaction conditions employing the aqueous solution systems.

Product yield and molecular weight are sensitive to both stirring time and amount of added base. Results given in Table 2 illustrate such dependencies. Thus product yield decreases with stirring time. The decrease in product with increase in stirring times beyond 10 seconds is typical for systems employing more than an exact equivalence of base and for many systems such a decrease is typical even when only an equiva-

TABLE 1
Results as a function of dye.

| Dye | Reactant Amount(mmole) Dye:Ti:NaOH | Reaction Time (Secs) | Yield (%) | Reaction System | Visual Inception of Degradation ($^{o}C$) | Final Color ($250^{o}$) |
|---|---|---|---|---|---|---|
| Indigo Carmine | 0.6:0.42:0.12 | 60 | 7 | IF | 80-90 | dk brown |
| Nigrosine | 0.4:0.42:0.6 | 60 | 29 | IF | 130-150 | dk brown |
| | 0.3:0.3:0.6 | 30 | 29 | IF | | |
| | 0.5:0.5:1.0 | 60 | 0 | S | | |
| | 1.0:1.0:1.5 | 60 | 54 | IF | | |
| | 0.5:0.7:1.0 | 300 | 0 | S | | |
| | 0.4:0.7 | 40 | 66 | IF | | |
| Eriochrome Black T | 0.5:0.7:1.0 | 60 | 0 | S | | |
| | 0.5:0.7:1.0 | 60 | 48 | 1F | 170-180 | dk brown |
| | 1.0:1.0:2.0 | 60 | 65 | IF | | |
| Bromothymol Blue | 0.5:0.7:1.0 | | | | | |
| | 0.5:0.7:1.0 | 60 | 0 | S | | |
| | 0.5:0.7:0.0 | 60 | 0 | IF | | |
| Bromophenol Blue | 0.5:0.5:1.5 | 13 | 51 | IF | 100-120 | brown-black |
| | 0.5:0.7:1.0 | 90 | 0 | S | | |
| | 0.5:0.7:1.0 | 40 | 19 | IF | | |
| Congo Red | 0.5:0.7 | 90 | 8 | IF | | |
| PDTD | 0.4:0.8 | 15 | 5 | 1F | | |

Reaction Times: $Cp_2TiCl_2$ in 30 ml of chloroform added to rapidly stirred (18,600 rpm no load) solution of dye in 30 ml of water containing any added base (sodium hydroxide) at $25^{o}C$ for the interfacial systems (IF). The aqueous solution (S) systems were identical to those employed for the IF except the $Cp_2TiCL_2$ is in 30 ml of water. PDTD is 3-(2-pyridyl)-5,6-diphenyl-1,2,4-triazine-p,p'-disulfonic acid, disodium salt.

TABLE 2

Results as a function of reaction variables for the synthesis of polydye products utilizing Bromophenol Blue.

| Reaction Amount(mmole) | Reaction Time (secs) | Yield (%) | LVN (ml/g) |
|---|---|---|---|
| Dye: Ti: NaOH | | | |
| 0.5:0.5:0.0 | 60 | 0 | -- |
| 0.5:0.5:0.50 | 40 | 32 | |
| 0.5:0.5:1.0 | 7 | 43 | 14 |
| 0.5:0.5:1.5 | 1.3 | 51 | 17 |
| 0.5:0.5:2.0 | 50 | 12 | 44 |
| 0.5:0.5:1.0 | 60 | 19 | |
| 1.0:1.0:2.0 | 16 | 80 | |
| 1.0:1.0:2.0 | 20 | 65 | |
| 1.0:0.7:2.0 | 40 | 19 | 9 |

Reaction Conditions: $Cp_2TiCl_2$ in 30 ml of chloroform added to rapidly stirred (18,600 rpm no load) aqueous solution of Bromophenol Blue with any added sodium hydroxide at 25°C.

lence of base is present. Thus 0.085 grams of a polydye product (a titanium polyetherester) was added along with 0.5 mmoles of sodium hydroxide to 30 ml of water and stirred (23,000 rpm no load) for 30 seconds. Only 14% of the polydye product was recovered. This was repeated except omitting the base. A recovery of 65% of the polydye product was made.

Use of excess base also typically reduces product yield, probably through hydrolysis of formed polydye chains.

The use of electrophillic metal containing monomers such as $Cp_2TiCl_2$ is important in reactions utilizing $R\text{-}SO_3^-$ functional groups (such as the PSP dyes) since organic acid chlorides do not condense with the $R\text{-}SO_3^-$ moiety under the conditions employed in this study.

The polydyes exhibit thermograms similar to other organometallic polymers where stability plateaus (TG) and exothermic (in air, DSC) degradation occur [3]. Most products exhibited poor weight retention with initial degradation beginning below 200°C. For instance the polydye derived from Eriochrome Black T begins losing weight about 100°C, losing about 10% to 600°C whereupon an additional 60% is lost by 650°C in air. The product from Bromophenol Blue begins losing weight about 100°C with 5% weight loss to 300°C, followed by loss of an additional 60% to 550°C in air. The final degradation product at 800 to 1000°C in air is titanium dioxide.

Visual observations of the physical changes of the polydye products and the dyes were made from 25°C to 300°C utilizing the Fisher-Johns Melting Point Apparatus. The polydyes generally did not melt in the normal sense for organic compounds. Color changes accompanied the transitions. TGA thermograms of the polydyes showed that weight losses accompanied these physical changes. For compounds such as the $Cp_2TiCl_2$ condensation products with Bromophenol Blue, the product appeared to melt between 100 and 200°C. There was an accompanying change in color from a dark red to a brownish color which became dark brown to black by 300°C. A majority of the compounds did not melt below 300°C but did gradually change color. Onset of the color change often was accompanied by an apparent change in the solid structure (i.e. breakup of the larger layered crystals to smaller "crystals" and changes in the texture of the solid) which appeared to correspond within about 50°C of the noted initial degradation. The temperature ranges and texture changes were recorded in Table 1 as visual inception of degradation and final color at 250°C.

## APPLICATIONS (POTENTIAL AND ACTUAL)

The advantages of including dyes as part of the backbone of polymer chains is the (potential) permanent, nonleaching nature of such polydyes compared with the monomeric dyes themselves where movement even within apparently solid lattices (such as within plastics) occurs over extended time.

The following miniexperiments were conducted utilizing polydyes derived from Nigrosine and Bromophenol Blue contained to 0.01 to 1.0% weight per volume solutions of DMSO, DMF and HMPA. The polydye solutions readily impregnated paper such as Whatman #1 filter paper and #235 No-Rivl (Fort Howard Paper Co., Green Bay, Wis.) commercial paper towels to give highly colored, fluorescent paper. Thus such polydyes may be of use in the construction of specialty fluorescent posters and other areas requiring permanent, nonleaching coloring agents.

Cotton, Denim and 50%-50% polyester-cotton fabric were impregnated with the dyes. Typically a 3 inch by 3 inch piece of white cloth became highly colored by addition of 1.0 to 0.1 mg. of polydye product. The coloration was not permanently fixed, washing out after two to three machine washings employing typical bleach and fabric soap. A fixing agent or procedure will have to be found to more permanently fix the dye to the fabric before they can be used in the fabric and textile industries as coloring agents.

The following plastics were doped by gently heating about 3 grams of plastic above its $T_m$ and then adding dropwise 3 to 10 drops of 0.1 to 1.0% by weight polydye solution: polyethylene (with 12% vinyl acetate; Union Carbide; Bakelite), polycarbonate (DuPont, Lexan Grade 131, 112 Natural), polypropylene (Hercules, ProFax), high density polyethylene (U.S.I., Petrothene NA-144), nylon (DuPont, Zytel), polyvinyl chloride (B.F. Goodrich, Geon type 102E PF 5), polystyrene (Monsanto,

Lustrax HH-101-2020 Crystal 300) and SAN-styrene copolymer (Monsanto, Lustran A 21-2020 Crystal Run M1527). The plastics accepted the "dope" retaining the polydye product throughout the plastic. The greatest problem was experienced with polyvinyl chloride where the heated polyvinyl chloride often partially decomposed giving HCl which appeared to often change the color of the polydye.

Polydyes that were fluorescent before they were utilized as "doping" agents in the plastics yielded fluorescent plastic plugs. Good fluorescence was noted at the 10 ppm concentration level for most of the doped plugs. (Some of the tested plastic samples were fluorescent so added caution had to be excercised in evaluation of these doped plugs.) The plugs remained fluorescent over the six month test period. Again the polydye products could be employed in plastics to give permanent coloration and/or fluorescence to the plastic.

Impregnation of DuPont white latex paint was carried out by simple addition of about 0.1 to 1.0 ml of 0.01 to 1.0% polydye product contained in DMSO, HMPA, and DMF solutions to 10 ml of latex paint. These doped paints were applied to wood and teflon coated surfaces. The wood applied paints dried to give a typical latex coated surface, while the teflon coated surface yielded flexible, coherent films. The colors were bright and appeared to be commercially acceptable. Of particular interest was the pleasing blue coloration derived from the Bromophenol Blue polydye product because of the difficulty often experienced in obtaining good blue colored coatings. Good coloration was achieved using about 10 to 500 ppm of the polydye product. Again the coatings were fluorescent under both short and long ultraviolet radiation.

The polydyes may not only offer desirable coloring, but desirable ultraviolet "sink" activity due to the presence of the $Cp_2Ti$ moiety. This may lead to better weatherability for exterior coatings through retardation of ultraviolet-associated coatings failures.

Other potential uses include as lazer dyes and as energy exchangers permitting greater depth of penetration of coatings curing.

## REFERENCES

[1] C. Carraher, J. Polymer Sci., A-1, 9, 366 (1971).

[2] C. Carraher, CHEMTECH, 744 (1972).

[3] C. Carracher, Interfacial Synthesis, Vol II (Edited by F. Millich and C. Carraher), Marcel Dekker, N.Y., Chpt. 21, 1977.

[4] C. Carraher and S. Bajah, Polymer (Br.), 15, 9 (1974) and 14, 42 (1973).

[5] C. Carraher and P. Lessek, Eup. Polymer J., 8, 1339 (1972).

[6] C. Carraher, J. Chem. Ed., 46, 314 (1969).

[7] J. Brzywacz and H. Smagowski, Z. fuer Naturforschung, 20 (10), 1358 (1965).

[8] L. Bellemy, The Infrared Spectra of Complex Molecules, John Wiley, N.Y. (1958).

[9] C. Rao, Chemical Applications of Infrared Spectroscopy, Academic Press, N.Y., 1963.

[illegible]

[illegible] failures.

[illegible] include [illegible] and [illegible]

## REFERENCES


1. [illegible]
2. [illegible]
3. [illegible]
4. [illegible]
5. [illegible]
6. [illegible]
7. [illegible]
8. [illegible]
9. [illegible]

# Colloid and Interface Effects of Additives in Interfacial Polycondensation

Ethel Zaiser Casassa

Chemical Engineering Department
Carnegie-Mellon University
Pittsburgh, PA. 15113

ABSTRACT

Interfacial polycondensation reactions are routinely conducted with empirically developed methodology which might be improved if some well-known principles of colloid and surface chemistry were invoked. An example in the interfacial synthesis of polymers is the use of additives such as catalysts and emulsifiers to enhance the degree of polymerization. In this paper possible modes of action of surfactants and quaternary ammonium catalysts in stirred interfacial polycondensation are examined. The agents are considered variously as ions, molecules, and micellar aggregates, acting upon each component of the system, i.e. monomers, oligomers, polymers, and solvents. The functions delineated include emulsification of the solvent phases, solubilization of monomers in regular or inverse micelles, solubilization of oligomers and polymers, phase transfer catalysis, and micellar catalysis.

## INTRODUCTION

In colloidal dispersions, since the ratio of surface to volume of the dispersed particles is very high, the contribution of surface energy to the total free energy is large. Most so-called colloid phenomena are merely the consequence of ordinary processes of diffusion, adsorption, and electrostatic and van der Waals interactions, which take on special aspects because of the extra free energy and the small particle size (0.01 to 10 μ) of colloids. Heterogeneous reaction systems, which are of obvious interest to surface scientists, become all the more fascinating when the different phases are highly dispersed into units of colloidal size. Nowhere is this interest more evident than in free-radical emulsion polymerization where both the formation of a latex and the colloidal properties of latex particles have been widely studied [1]. One aspect of emulsion polymerization which has been clarified by thorough investigation is the role of the surfactant or emulsifier which is nearly always employed. A surfactant, as an amphiphile, tends to be distributed at a polar/nonpolar interface in a concentration in excess of the bulk value in solution, and has in addition a tendency to aggregate at a characteristic concentration, the critical micelle concentration (CMC). These properties, which are of course related, are put to excellent use in emulsion polymerization.

Current understanding of the functioning of surfactant in emulsion polymerization is in part the outcome of intensive research into the colloid and surface science of such systems. Interfacial polymerization, which has received far too little attention from colloid chemists, differs from free-radical emulsion polymerization in important ways. This paper offers an overview of the range of phenomena that principles of colloid and surface chemistry suggest are likely to be found in interfacial reaction systems, with special emphasis on the behavior of surfactants and other additives.

## COLLOID AND SURFACE ASPECTS OF INTERFACIAL POLYCONDENSATION

Interfacial synthetic methods are particularly suited to polycondensation, as in the production of polyamides and polyesters where they are valued for speed, low energy requirements, and the high degree of polymerization they afford. The reactions, which are generally nucleophilic substitutions involving pairs of reactants of dissimilar chemical nature and solubility, require careful selection of system variables to ensure the growth of long chains and to minimize side reactions. Interfacial reaction systems have the following elements in common:

(1) Two immiscible solvent phases, usually water and a nonpolar liquid.

(2) Two monomers, each insoluble in the phase that dissolves the other.

(3) Monomers that are bifunctional, reacting to form regular chains of alternating monomer residues.

In a polyester synthesis, for example, one monomer is the enolate anion of a bisphenol in strongly alkaline water solution, the other a diacid chloride in a chlorinated hydrocarbon. The condensation reaction occurs either at the phase boundary, or within either liquid phase near that surface, hence stirring facilitates reaction by creating a dispersion of high interfacial area. Virtually complete reaction is so much the rule that efforts to improve interfacial preparative methods have dwelt upon achievement of a high degree of polymerization rather than upon improvement of the yield. Early work in the field offers many good examples of organic industrial chemical research directed toward the empirical achievement of a desired product. System variables that have been explored in depth are the concentrations and relative amounts of reactants, the selection of the nonpolar solvent, the chemical structure of the monomers, rate of stirring, sequence of addition of reactants, and the use of additives such as catalysts or emulsifiers [2,3,4]. More recently, reviews have dealt with the transport of reactants across phase boundaries [5], interface effects on reaction rates [6], and the fluid mechanics and transfer aspects of stirring [7]. Mechanistic studies have not been concerned with the functioning of the additives which are generally used. Although the heterogeneity and complexity of interfacial polycondensation systems is daunting, insofar as they are colloidal dispersions, it may be profitable to consider the colloid phenomena which can occur when additives are used to enhance the molecular weight of the product. These effects can be classified as emulsification, solubilization, and catalysis. Wherever specific examples of systems are mentioned, the reference will be to polyesterification as a typical polycondensation reaction. With minor modifications this approach may of course also be applied to other polycondensation reactions.

## EMULSIFICATION IN INTERFACIAL POLYCONDENSATION

An emulsion, i.e. a liquid-liquid colloidal dispersion, is the most intimate mixture possible for the two mutually insoluble solutions which comprise an interfacial reaction system. Because of surface-tension lowering by organic solutes, even in the absence of additives, surface tensions of both solvents in interfacial polymerization systems can be expected to be less than those of pure solvents by reason of the dissolved monomer in each phase. This reduction, which varies with different solute-solvent pairs, and is larger for the aqueous phase than for the organic one, undoubtedly aids in the dispersion of one phase in the other when the system is stirred. As the surface-tension lowering is the result of a surface excess of solute, droplets of either phase dispersed in the other will have a slight surface excess of whichever monomer is dissolved in the liquid of the droplet. This surface excess should favorably affect the availability of that monomer for reaction with the second monomer in the other liquid phase. Emulsions formed in the absence of surfactant additives, however, are unlikely to persist if stirring is discontinued as most monomers are not markedly amphiphilic. Also for lack of amphiphilic character symmetrical quaternary ammonium salts (tetramethyl- or tetraethylammonium bromide, for example) which are often added as catalysts should not materially enhance emulsification.

Surfactant "emulsifiers," often added to polycondensation systems are effective stabilizers of emulsion droplets regardless of the charge type of the surfactant. If we restrict discussion here to surfactant concentrations below the CMC, we need not be concerned with the action of micellar aggregates of surfactants but only with stabilization conferred by adsorption of individual amphiphilic molecules or ions at the polar/nonpolar liquid interfaces of emulsion droplets. Since the polar portions of amphiphiles thus adsorbed are attracted to the aqueous phase and the nonpolar portions to the organic liquid, the posture of an adsorbed amphiphile will depend on which liquid makes up the droplets and which the continuous phase of the emulsion. In any case, cationic, anionic, and nonionic surfactants will confer not only stability but positive, negative or zero charges, respectively, upon droplets on which they are adsorbed. If such charges attract or repel charged reactive monomer species (e.g. enolate ions in polyesterification), they will aid or hinder the mutual approach of reactive species residing in the different phases. Just such a charge effect may account for the superiority we have found for cationic over anionic and nonionic surfactants at low concentrations in a linear aromatic polycondensation [8]. Thus while emulsification in itself may be an aid in interfacial polycondensation, the charge type of the dispersal agent can be an important consideration. We cannot neglect to mention that the emulsifying action of surfactants at concentrations in excess of the CMC is, if anything, stronger than at lower concentrations, but it is difficult to separate this from other phenomena which fall in the category of solubilization.

## SOLUBILIZATION BY SURFACTANTS IN INTERFACIAL SYSTEMS

Solubilization, the enhancement of the solubility of a substance in a liquid by action of a third substance, may figure in two ways in interfacial polymerization. The most often cited role of this phenomenon in stirred interfacial polymerization is the conferral of solubility, or partial solubility, on oligomers as well as growing polymer chains by the action of surfactant. The prevention of premature precipitation of polymer must of course aid in attainment of high molecular weight, although there is evidence that polymerization continues after precipitation of the polymer [9]. The most likely cause of such solubilization is adsorption by the polymer of surfactant ions or molecules, or perhaps even micellar aggregates. Another form of partial solubilization is the swelling of the polymer by the nonpolar solvent of the reaction system. It is well known that such solvent interaction aids in attaining a high degree of polymerization, possibly because swollen polymer remains permeable to monomers.

The form of solubilization most familiar to the colloid chemist, however, is micellar solubilization, which becomes manifest only at concentrations in excess of the CMC in surfactants [10,11]. Micellar aggregates which appear at and above the CMC contain typically 50 to 100 surfactant ions or molecules, and exhibit a solubilizing effect on species ordinarily sparingly soluble in the medium. Note that micelles formed in water have polar groups on the outside, and hydrocarbon-type groups on the interior, while <u>inverse</u> micelles, which may form in nonpolar media, have polar groups and a small amount of water on the inside. The interior environment of a micelle, which thus is very different from the medium in which it is formed, is the most likely location for solubilized material that is not ordinarily dispersible in the liquid medium. Micelles of surfactant in the aqueous phase of interfacial polycondensation systems are capable of solubilizing any nonpolar component of the system, e.g. monomer from the organic phase, the organic solvent itself, or even oligomers. Note that such solubilizates are molecularly dispersed, a condition not to be confused with an emulsion type of dispersion where the droplets are made up of solvent containing dissolved monomer. In polyesterification, diacid chloride monomer solubilized in micelles in the aqueous phase might be available for reaction with the enolate monomer of the aqueous phase; but as a given micelle can accommodate very few solubilizate molecules, it is not certain that such dispersal would be helpful beyond the early stages of polymerization. Furthermore, diacid chloride monomer in the aqueous phase is vulnerable to base-catalyzed hydrolysis, a side reaction which may be important if the rate is comparable to that of the condensation reaction. Inverse micelles, if formed in the organic phase, would carry with them a little solubilized water and water-soluble monomer species, but this is a less likely occurrence than formation of micelles in the aqueous phase.

As with surfactant-stabilized emulsion droplets, the charge type of the surfactant is important to its effect as a micellar solubilizing agent. Micelles of cationic surfactants which may carry solubilized diacid chloride into the aqueous phase are positively charged and thus attract oppositely charged enolate ions. Micelles of anionic and nonionic surfactants would not be helpful in this respect, and whatever advantages they may furnish in polyesterification must arise solely from their action as dispersal agents in increasing the interfacial surface area of the emulsion and in solubilizing oligomer and polymer chains.

It is generally found that there is an optimal concentration of surfactant, different for each polymerizing system, above which the degree of polymerization of the product decreases [2,3]. This may be the consequence of extreme dispersal of diacid chloride monomer either in ultra-fine emulsion droplets of the organic solution phase, or, more likely, in surfactant micelles. With extreme dispersal, polymerization may begin at so many sites at once that the consequent severe depletion of monomers slows chain growth and permits side reactions to become important.

## CATALYSIS IN INTERFACIAL POLYCONDENSATION

Catalytic effects of micelles of surfactants on organic reactions have been intensively studied [12]. The mechanism of a reaction and the charge type of the reacting species in the rate-limiting step are important determining factors in the effect of micelles upon reaction rates. There are many examples of nucleophilic substitution in which the two reacting species are respectively an uncharged molecule and a negatively charged ion, and in which catalysis by micelles has been demonstrated. Micelles of cationic surfactants are good catalysts in such reactions, while those of anionic and nonionic ones are not. Similar effects might be predicted for polyesterification reactions where enolate anions react with uncharged diacid chloride molecules. Other types of reactions may be found to be accelerated by anionic or nonionic micelles. Micellar catalysis takes place only at concentrations above the CMC in solutions containing amphiphiles which exhibit hydrophobic bonding to the point of aggregation. Many of the additives or "accelerators" used in polycondensation are quaternary ammonium compounds which are not micelle-formers, however, and a different explanation must be advanced for catalytic effects which they produce.

A second form of catalysis which may play a role in interfacial polycondensation is phase-transfer catalysis [13]. The usual phase-transfer catalysts are quaternary ammonium cations which form ion pairs with negatively-charged reacting species in nucleophilic substitutions. In a two-phase system these ion pairs are better able than the anion alone to penetrate the nonpolar environment where the uncharged reacting species is found. Thus the reaction is accelerated, while the catalyst emerges unchanged. This kind of catalytic effect in polyesterification could facilitate passage of enolate ion into the organic phase. Note that it is the quaternary ammonium <u>ion</u> which performs this function, not micelles of quaternary ammonium surfactant. Indeed micellization, by limiting the concentration of free cation to the value of the CMC, may be a detriment, even though the amphiphilic character of cationic surfactants might favor ion-pair formation. Phase-transfer catalysts generally used in organic synthesis, such as tetramethyl- or tetraethyl- or benzyltrimethyl-ammonium bases or salts, do not form micelles.

## CONCLUSIONS

The action of additives such as surfactants and catalysts may give rise to numerous phenomena of emulsification, solubilization and catalysis in the complex heterogeneous reaction systems that are common to stirred interfacial polycondensations. The agents that must be considered responsible are the ions, molecules, and, where applicable, the aggregates of the additives. All components of the systems must be considered as acted upon, not merely the monomers but oligomers, polymers, and the solvent components as well.

## ACKNOWLEDGMENT

The support of the Eppley Foundation for Research is gratefully acknowledged.

## REFERENCES

[1] J.L. Gardon, Interfacial, Colloidal, and Kinetic Aspects of Emulsion Polymerization, Chapter 9 in reference 4.

[2] P.W. Morgan, Condensation Polymers by Interfacial and Solution Methods, Interscience, New York, 1965.

[3] V.V. Korshak and S.V. Vinogradova, Polyesters, English translation, Pergamon, Oxford, 1965.

[4] F. Millich and C.C. Carraher, Jr., Interfacial Synthesis, Volume I: Fundamentals, Volume II: Polymer Applications and Technology, Dekker, New York, 1977.

[5] J.H. Bradbury and P.J. Crawford, Chapter 5 in reference 4.

[6] F. MacRitchie, Chapter 6 in reference 4.

[7] J.H. Rushton, Chapters 2 and 3 in reference 4.

[8] Unpublished data of E.Z. Casassa et al.

[9] P.W. Morgan, J. Polym. Sci., Part C, 4, 1075 (1964).

[10] K. Shinoda, T. Nakagawa, B. Tamamushi, and T. Isemura, Colloidal Surfactants, Academic Press, New York, 1963.

[11] K. Shinoda, ed., Solvent Properties of Surfactant Solutions, Dekker, New York, 1967.

[12] J.H. Fendler and E.J. Fendler, Catalysis in Micellar and Macromolecular Systems, Academic Press, New York, 1975.

[13] C.M. Starks, J. Amer. Chem. Soc., 93, 195 (1971); J. Dockx, Synthesis, 441 (1973); M. Makosza, Pure and Applied Chem., 43, 439 (1975).

# Cationic Surfactants in Interfacial Synthesis of Linear Aromatic Polyester

Ethel Z. Casassa, Dean-Yang Chao and Mark Henson

Chemical Engineering Department
Carnegie-Mellon University
Pittsburgh, PA. 15213

ABSTRACT

The polycondensation of terephthaloyl chloride and bisphenol A was used as a model reaction for the production of linear aromatic polyesters in stirred interfacial polymerization. The evaluation of several catalytic and surfactant additives to this system was based upon yields and intrinsic viscosities of the products obtained with both low and high concentrations of cationic surfactants of the quaternary ammonium type, of an anionic surfactant, and of a non-micelle forming quaternary ammonium salt. Although yields were similar in most cases, viscosity differences were marked. The highest intrinsic viscosities, hence highest degrees of polymerization, were found for preparations where cationic surfactant in excess of the critical micelle concentration was employed. Modes of action for such surfactants are suggested. The possibilities include, but are not limited to, solubilization of the product, solubilization of either or both monomers, emulsification of the liquid phases, catalytic phase transfer of one monomer, and micellar catalysis.

## INTRODUCTION

Stirred interfacial reaction as a rapid, low-energy synthetic method has been extremely useful in applications-oriented explorations for novel polymers. Both polyamidization and polyesterification have been subjected to extensive attempts to optimize the yield and, more importantly, the degree of polymerization [1,2,3]. As a result the literature is full of empirical data on effects upon the polycondensation product of such variables as the chemical structure of the monomers, the concentrations and relative amounts of the reactants, the identity of the nonpolar solvent, the rate of stirring, and the presence of additives. The heterogeneity of the systems, the complexity of the reaction mixtures, and the extreme rapidity of the reactions have not encouraged conventional thermodynamic and kinetic analyses. Nevertheless by isolating elements of the problem a closer understanding may be possible, as at least one study of polyamidization has shown [4].

One factor which has been insufficiently explored is the effect of additives upon interfacial polymerization. A surface-active agent, sodium dodecylsulfate (SDS), is known to enhance the degree of polymerization in the interfacial production of polymers and indeed the use of SDS or other emulsifiers has been the subject of optimization studies [2,5]. Sometimes interfacial polycondensations, as types of nucleophilic displacement, are carried out with phase-transfer catalysts [6,7,8], generally small symmetric quaternary ammonium cations [9,10]. It has been suggested [11] that both surfactant and catalytic functions could be combined through the use of cationic surfactants which, as quaternary ammonium compounds, could also aid in transport of monomer between phases. Indeed the use of cationic surfactants has been the subject of patents for polycarbonate and other syntheses [12], but few details have been published. The choice of an aromatic polyesterification reaction for the present study was influenced by current interest in polyarylesters for the desirable thermal and mechanical properties which they display at high degrees of polymerization. The selection of a model polymerization system for the preparation of a linear aromatic polyester, and the differences in the reaction products obtained with the use of additives are described in this paper. The performances of different cationic surfactants, of the anionic SDS, and of a quaternary ammonium nonsurfactant compound are evaluated in terms of yield and intrinsic viscosity of the product, and the results are discussed in the light of some aspects of the colloid chemistry of the reaction systems.

## MODEL POLYMERIZATION SYSTEM

The model system adopted for this study of stirred interfacial polycondensation is the two-phase one of bisphenol A (BPA) dissolved in aqueous sodium hydroxide, and terephthaloyl chloride (TPC) dissolved in chloroform. These monomers were selected for chemical stability and ease of purification, and, with a view to future phase-partitioning studies, for their suitability for spectroscopic analysis. The polar aqueous phase must contain enough NaOH to dissolve BPA completely as the reactive phenolate ion, leaving as little as possible undissociated BPA which could be extracted by the organic phase. The nonpolar chloroform phase is an excellent solvent for terephthaloyl chloride and oligomers formed in the reaction, although not for polyesters of higher molecular weight. The reactive monomer in each phase is quite insoluble in the other phase. The choice of chloroform was dictated by the need to avoid solvents having double bonds which would interfere with projected spectroscopic studies. Because the use of the <u>para</u> isomer of phthalic acid dichloride renders unlikely the formation of cyclic oligomers the polymer is completely regular in structure:

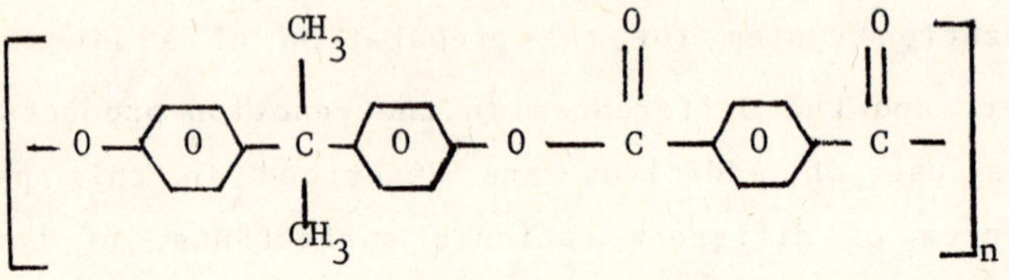

$$\left[ -O-\langle O \rangle - \underset{CH_3}{\overset{CH_3}{\underset{|}{\overset{|}{C}}}} - \langle O \rangle - O - \overset{O}{\overset{\|}{C}} - \langle O \rangle - \overset{O}{\overset{\|}{C}} - \right]_n$$

Two possible side reactions are the alkaline hydrolysis of TPC and dichlorocarbene formation by the action of aqueous NaOH on chloroform. The low solubility of TPC in water serves to protect it from hydrolysis by the alkali. Moreover, the very slow rates of the side reactions, as compared to the polycondensation, insure that neither the course of polymerization nor the yield are adversely affected.

All conditions of the system and procedure, including the amount of each reactant, the phase volumes, stirring rate and time, polymer isolation and purification procedures, were maintained as nearly the same as possible for different preparations, so that the only variable was the kind and amount of additive used. The conditions used were similar to those employed in other polymerizations of this type [5,13,14,15). Exploratory work with a similar system [16] indicated a higher molecular weight of the product when cationic surfactants were used, and on the basis of that work decyl-, dodecyl-, and hexadecyl-trimethylammonium bromides were tested at concentrations both above and below the critical micelle concentration (CMC). For comparison the anionic SDS was also tested, above and below the CMC, and tetramethylammonium bromide as typical of the non-aggregating quaternary ammonium salts often employed as catalysts. Non-ionic surfactants were not used because of the poor results noted with them in the past [1,2,5,16].

## EXPERIMENTAL

### Materials

Bisphenol A (Pfaltz and Bauer) was recrystallized from toluene, and terephthaloyl chloride (Eastman) from *n*-hexane. The aqueous phase was composed of boiled distilled water and sodium hydroxide (Baker). Chloroform (Baker Analyzed) was washed repeatedly with water to remove the alcohol stabilizer and dried over calcium sulfate. Hexadecyltrimethylammonium bromide (B.D.H.) as received was free of surface-active impurity (by the surface tension versus concentration criterion), and it, as well as sodium dodecyl sulfate (B.D.H.) and tetramethylammonium bromide (Eastman), were used without further purification. Decyltrimethylammonium bromide and dodecyltrimethylammonium bromide (both Eastman) as received showed depressions in surface-tension versus concentration plots, and were recrystallized from acetone/ether to remove the surface-active impurities. The solvent for intrinsic viscosity measurements was *o*-chlorophenol, 98+% (Aldrich). Other solvents used for recrystallization of reagents and washing of polymer were reagent grade (Baker).

### Polymerization Procedure

The reaction was carried out in an electric blender with a rheostat to control the speed. An initial charge of 0.025 mole of NaOH dissolved in a total of 72.5 ml of water was stirred at a moderate speed with 0.0125 mole of BPA until solution was complete, when an appropriate amount of additive (surfactant or other) dissolved in 10 ml of water was added. The organic phase was prepared by dissolving 0.0125 mole terephthaloyl chloride in 35 ml of chloroform. The stirring speed of the blender was raised to the maximum value, and the organic phase was introduced rapidly. Mixing was continued thereafter for 5 minutes in cases where the reaction system remained emulsified. In other instances the reaction mass turned to a semisolid paste which adhered to the blender jar walls and stirring was discontinued as ineffective. Nevertheless, a total of 5 minutes was allowed to elapse so that the reaction presumably could continue for the full time, as is known to occur in precipitates in similar systems [17]. At the end of the 5-minute period, stirring (if still in progress) was interrupted and 500 ml of acetone was added to the blender jar. After one minute of gentle stirring the polymer was filtered off and washed with acetone to remove unreacted monomers and chloroform. The moist product was returned to the blender jar, stirred with distilled water to wash out excess alkali and salts, then filtered. This water wash was repeated twice more. The polymerized product was dried to constant weight in a vacuum oven at 40°C.

Examination of Products

The polymeric products produced in 85 to 95% yield were white, granular, and high-melting (> 300°C). Although insoluble in common organic solvents they were completely soluble at room temperature in m-cresol/symtetrachloroethane mixtures (50:50 by volume) and in o-chlorophenol as well as in more highly acidic solvents. Flow times were measured at 30°C in Ubbelohde viscometers for o-chlorophenol solutions containing from 0.1 to 0.5% polyester. Extrapolation to zero concentration to obtain the intrinsic viscosity was carried out from combined Huggins and Kraemer plots of the reduced and inherent viscosities. In the concentration range studied these plots show no signs of polymer aggregation in o-chlorophenol, nor in the m-cresol/symtetrachloroethane medium. Although the intrinsic viscosity is the best viscosity function for evaluating the degree of polymerization, at least on a comparative basis, many investigators have used other viscosity functions measured for the single polymer concentration of 0.5%. The best of these alternatives is the inherent viscosity, much better than the relative viscosity which is regrettably sometimes still used.

## RESULTS AND DISCUSSION

The concentration-dependent aggregation of surfactants was taken into consideration in planning polymerization experiments to test the effects of additives. The substances tested were mainly employed at two different concentrations. The first was the constant low molality of $5 \times 10^{-4}$ in the aqueous phase, a figure chosen so as to permit comparison of effects of ions only, as this concentration is below the critical micelle concentration of all the surfactants used in this study. A second series of preparations employed much higher concentrations where most of the surfactant is in micelles (except of course in the one non-micelle-forming quaternary ammonium salt). These higher concentrations were chosen so that there would be in each case an excess molality of 0.024 over the critical micelle concentration, i.e. 0.024 M surfactant in the micellar form. The total concentrations were different, of course, as were the concentrations of surfactant ions, because of the differences in the CMC's of the various additives. The concentrations of additives, both absolute and relative to the CMC, the polymer yields and intrinsic viscosities are given in Table I.

It is immediately apparent, as others have noted [1,2] that the yield varies very little with the additive chosen. As a

TABLE I

Additive effects in interfacial polymerization of bisphenol A and terephthaloyl chloride

| Additive in alkaline aqueous phase | | | | | Polymer product | |
|---|---|---|---|---|---|---|
| Identity | CMC $10^4$ M | Total $10^4$ M | Multiple of CMC | Micellar $10^4$ M | Ionized $10^4$ M | Percent Yield | [η] |
| None | n.a. | 0 | n.a. | 0 | 0 | 93.4 ± 2.0 | 0.50 ± 0.03 |
| TMAB | n.a. | 5 | n.a. | 0 | 5 | 90.5 | 0.48 |
| SDS | 82 | 5 | < 1 | 0 | 5 | 89.3 | 0.48 |
| DTMAB | 630 | 5 | < 1 | 0 | 5 | 90.5 | 1.24 |
| DDTMAB | 142 | 5 | < 1 | 0 | 5 | 90.1 | 0.58 |
| HDTMAB | 9.2 | 5 | < 1 | 0 | 5 | 89.6 | 0.48 |
| DTMAB | 630 | 100 | < 1 | 0 | 100 | 93.7 | 1.50 |
| DDTMAB | 142 | 100 | < 1 | 0 | 100 | 92.6 | 1.05 |
| TMAB | n.a. | 240 | n.a. | 0 | 240 | 90.6 | 0.62 |
| SDS | 82 | 320 | 4 | 240 | 82 | 85.3 | 1.00 |
| DTMAB | 630 | 870 | 1.2 | 240 | 630 | 88.6 | 1.60 |
| DDTMAB | 142 | 382 | 2.7 | 240 | 142 | 89.6 | 1.27 |
| HDTMAB | 9.2 | 248 | 27. | 240 | 9.2 | 94.4 | 0.79 |

Key:

CMC — critical micelle concentration
M — moles/liter
[η] — intrinsic viscosity, dl/g
TMAB — tetramethylammonium bromide
SDS — sodium dodecylsulfate
DTMAB — decyltrimethylammonium bromide
DDTMAB — dodecyltrimethylammonium bromide
HDTMAB — hexadecyltrimethylammonium bromide

5% difference in the yield represents only 220 mg of polymer, variations of less than 5% are not significant in light of the difficulty of quantitative recovery of polymer in the several steps involved in isolating and washing it. The intrinsic viscosity, however, as a criterion of polymer molecular weight, shows important differences under the influence of additives. These differences can be believed if the polymerization procedure outlined above is rigorously followed. For example, the preparation without additive was repeated by two different operators, months apart, with different batches of purified reagents, to give the same value of the intrinsic viscosity $[\eta]$ with an uncertainty of $\pm$ 0.03. Thus we can say that the extremely low level of $5 \times 10^{-4}$ M in tetramethylammonium bromide (TMAB), sodium dodecyl sulfate (SDS), or hexadecyltrimethylammonium bromide (HDTMAB) has, within experimental error, no demonstrable effect on the intrinsic viscosity. Both decyltrimethylammonium bromide (DTMAB) and dodecyltrimethylammonium bromide (DDTMAB), however, give increased intrinsic viscosities, the decyl compound affording a more than two-fold increase. This must be ascribed to specific effects of the cations of these surfactants, as the additives at $5 \times 10^{-4}$ M have no effect as emulsifiers when the reaction systems are stirred -- in fact, such reaction mixtures turn into semisolid polymer pastes after only 20 seconds of mixing. It was in just such preparations that stirring was discontinued after 20 seconds as ineffectual, and the 5-minute time of contact of the reagents was permitted to elapse without further attempt at mixing.

The two most interesting additives, DTMAB and DDTMAB, were tested at a still higher concentration of $100 \times 10^{-4}$ M in the aqueous phase. The homologous HDTMAB was not tested at this concentration because we wished first to isolate effects of the cations, and HDTMAB (with a CMC of only $9.2 \times 10^{-4}$ M) would be present largely in the micellar form at $100 \times 10^{-4}$ M. The twenty-fold increase in $DTMA^+$ and $DDTMA^+$ from $5 \times 10^{-4}$ to $100 \times 10^{-4}$ M gave further increases in intrinsic viscosity of 3-fold and 2-fold times the respective values in the absence of additive. Although $[\eta]$ was still in the order DTMAB > DDTMAB at the higher of the two concentrations, the proportional increase in $[\eta]$ in going from $5 \times 10^{-4}$ to $100 \times 10^{-4}$ M additive was greater for DDTMAB. It is moreover of some interest that $100 \times 10^{-4}$ M of either of these surfactants, though well below the CMC's, proves an effective emulsifier, the reaction mixtures remaining stirrable throughout the 5-minute reaction time. This emulsification may influence the course of polymerization in ways to be discussed later.

The series of polyesters prepared with additive concentrations of CMC + 0.024 M show enhanced intrinsic viscosity in every case, compared with the results for lower additive concentrations. This is true even of the non-aggregating quaternary salt TMAB, although the effect even at 0.024 M is still not as large as that given by $5 \times 10^{-4}$ M DTMAB and DDTMAB. Thus quaternary cations in sufficient dosage do result in higher molecular weight of the product, but the specific effects of the long decyl and dodecyl substituents are more important, and apparently not wholly dependent on emulsifying properties. In the preparations made with CMC + 0.024 M additives, only TMAB gave a pasty, unstirrable reaction mixture -- all the others were stirrable emulsions. It is noteworthy that the emulsifying effect of the anionic SDS did enhance the degree of polymerization of the product, even though the dodecylsulfate anion at $5 \times 10^{-4}$ M had no effect whatever. Finally, the intrinsic viscosities of products with all three surfactant quaternary salts at

CMC + 0.024 M exceeded those found with lower concentrations of the same additives. The effects here are undoubtedly complex, as these higher concentrations represent not only equal molalities of micellar material but unequal (and in varying degrees larger) concentrations of surfactant cations than in the experiments at $5 \times 10^{-4}$ and $100 \times 10^{-4}$ M surfactant. Considering first the HDTMAB case, little or no effect would be expected for $9.2 \times 10^{-4}$ M cation, since $5 \times 10^{-4}$ M cation has no effect. Thus the entire effect of HDTMAB at CMC + 0.024 M must be ascribed to the micelles. The increased intrinsic viscosity of the polymer when DDTMAB is increased from $100 \times 10^{-4}$ M to CMC + 0.024 M may be due in part to the concomitant increase of the cation concentration to $142 \times 10^{-4}$ M and in part to a micellar effect. It is surprising, however, that the highly specific DTMAB shows relatively small improvement of the product when the concentration is increased from $100 \times 10^{-4}$ M to CMC + 0.024 M, as this represents a 6-fold increase in cation concentration as well as providing 0.024 M in the micellar form. It appears that opposing effects are in operation, and that perhaps a maximum intrinsic viscosity would be achieved with some intermediate DTMAB molality.

## CONCLUSIONS

In the study of a model interfacial polyesterification reaction, consideration of the contributions of additives in the aqueous phase both as ions and aggregates offers at least a partial understanding of the complex effects upon the degree of polymerization as measured by the criterion of intrinsic viscosity. At very low additive concentrations, highly specific enhancement of the viscosity by asymmetric quaternary ammonium cations is noted, with maximal effect for decyl and dodecyl homologs of $R(CH_3)_3N^+$. The anion of the surfactant SDS, on the other hand, has no such effect even though it bears a dodecyl chain. Because of their amphilic character, surfactant ions are present at polar/nonpolar interfaces at higher concentration than in the bulk phase. Electrostatic forces from ions adsorbed at such interfaces in stirred polymerization systems have of course no effect on the uncharged monomer TPC in the organic phase. Adsorbed cations, however, will attract (and anions repel) the negatively charged monomer $BPA^-$ in the aqueous phase. Whether the diffusion of $BPA^-$ into the nonpolar phase to react with TPC is aided merely by this attraction and the consequently increased concentration of $BPA^-$ at the phase boundary, or whether the surfactant cations also act as phase-transfer catalysts which actually penetrate the nonpolar phase, has yet to be demonstrated.

When surfactants are present at higher concentrations where they act as "emulsifers" (but still not so high as to form micelles) they may be expected to exert two effects. One is merely to adsorb upon and solubilize the growing polymer chains, keeping them in suspension and accessible to further addition of monomers from the liquid phases. The other is to aid dispersal of one liquid phase in the other as emulsion droplets, thus increasing the surface area available for diffusion of monomer from one phase into the other phase where reaction can occur. (Note that we deliberately refrain from specifying which is the continuous and which the dispersed phase.) Both effects can be ascribed to all surfactants, whether cationic, anionic, or nonionic. However, cationic surfactants are likely to have the same enabling effects at the surfaces of emulsion droplets as they display at the lesser expanses of surface generated merely by stirring, and may thus manage still to outstrip mere emulsion-formers.

The fact that emulsification can go too far is well known, as there is generally an emulsifier concentration beyond which the degree of polymerization of the product falls off [1,2]. This is possibly due to extreme dispersion of the nonpolar phase in drops so small that the diacid chloride is insufficiently protected from hydrolysis, or that too many different polymer chains are started and the reactants are depleted before a satisfactory degree of polymerization is reached. The latter possibility may account for the success enjoyed by some workers with stepwise addition of monomers to build up high molecular weights [16,17,18].

Another category of surfactant effects is the dispersal of monomers as solubilized molecules within micelles of surfactant. The latter might be ordinary hydrophobic micelles in which TPC could be solubilized in the aqueous phase, or even inverted micelles in which bisphenolate could be solubilized in the chloroform phase. In both cases, one monomer would be placed within easy reach of the other, but perhaps not in sufficient concentration to encourage long-chain growth. Such solubilization might give rise to micellar catalysis, particularly if micelles of cationic surfactant were to catalyze the nucleophilic displacement that is central to the polymerization mechanism. The results presented here, however, give no indication that a micellar catalytic effect is present, and indeed if it did occur it might well have a negative effect on the degree of polymerization analogous to the case of "over-dispersal" by emulsification.

## ACKNOWLEDGMENT

The support of the Eppley Foundation for Research is gratefully acknowledged.

## REFERENCES

[1] P.W. Morgan, Condensation Polymers by Interfacial and Solution Methods, Interscience, New York, 1965.

[2] V.V. Korshak and S.V. Vinogradova, Polyesters, trans. from Russian by B.J. Hazzard, Pergamon Press, Oxford (1965).

[3] F. Millich and C.E. Carraher, Jr., Interfacial Synthesis, Volume I: Fundamentals, Volume II: Polymer Applications and Technology, Dekker, New York, 1977.

[4] J.H. Bradbury, P.J. Crawford, and A.N. Hambly, Trans. Faraday Soc., 64, 1337 (1968).

[5] W.M. Eareckson, J. Polymer Sci., 40, 399 (1959).

[6] C.M. Starks, J. Amer. Chem. Soc., 93, 195 (1971).

[7] J. Dockx, Synthesis, 441 (1973).

[8] M. Makosza, Pure and Applied Chemistry 43, 439 (1975).

[9] A. Conix, Ind. Eng. Chem., 51, 147 (1959).

[10] H. Schnell, Ind. Eng. Chem., 51, 157 (1959).

[11] P.W. Morgan, Reference 1, p. 352.

[12] B.E. Jennings (to Imperial Chem. Ind.) British Patents 897,640, 902,021, and 907,647 (1962); A.J. Conix (to Gevaert Photo-Prod.) French Patent 1,284,226 (1962); L. Bottenbruch and H. Schnell (to Farbenfabriken Bayer A.-G.) Ger. Patent 1,046,311 (1958).

[13] W.R. Sorenson and T.W. Campbell, Preparative Methods of Polymer Chemistry, Interscience, New York (1968).

[14] P.W. Morgan, J. Polym. Sci., Part A, 2, 437 (1964).

[15] P.W. Morgan, Macromolecules, 3, 536 (1970).

[16] E.Z. Casassa and G. Craun, unpublished data on interfacial reaction of BPA with 1:1 mixture of phthaloyl and isophthaloyl chlorides.

[17] P.W. Morgan, J. Polym. Sci., Part C, 4, 1075 (1964).

[18] K. Chujo, Y. Harada, S. Tokuhara, and H. Morita (to Daicell Co. Ltd.) Jap. Patent 7,637 (1967); see Chem. Ab. 68, 3380p (1968).

R.G. [illegible] [illegible] [illegible] [illegible] [illegible] 1953 [illegible]

16. [illegible] Chem. [illegible] [illegible] [illegible] [illegible] and [illegible] [illegible]
Zaffer [illegible] [illegible] [illegible] [illegible] [illegible] [illegible] [illegible]
[illegible] (1951).

# The Reaction of Vinyl Monomers with "Trapped Free Radicals"

Raymond B. Seymour
Department of Polymer Science
University of Southern Mississippi
Hattiesburg, MS 39401

G. Allan Stahl
B. F. Goodrich R & D Center
9921 Brecksville Road
Brecksville, OH 44141

ABSTRACT

Previous attempts to prepare copolymers by the reaction of vinyl monomers, such as styrene, with "trapped free radicals" such as acrylonitrile macroradicles has been unsuccessful because of the large differences in solubility parameters between the liquid and the solid. However, acrylonitrile will react with styrene macroradicals in the absence of oxygen, and yield block copolymers. Block copolymers may be produced from other vinyl monomers and other macroradicals when the difference in the solubility parameter of the monomer and macroradicals is less than 3.2H.

INTRODUCTION

It has been observed, for many years, that some polymers, such as polyacrylonitrile (PAN), polyvinyl chloride, (PVC), polyvinylidene chloride (PVDC), and polytrifluorochloroethylene are insoluble in their monomers and that these polymers precipitate when the monomers are polymerized by free radical-initiated polymerization, in the absence of solvents.

While the macroradicals continued to grow, as evidenced by an increase in the degree of polymerization ($\overline{DP}$), other vinyl monomers, such as styrene, did not add to these macroradicals. This investigation was undertaken in order to ascertain proper conditions for such addition reactions and thus to produce novel copolymers by the reaction of vinyl monomers and "trapped free radicals".

## EXPERIMENTAL

Macroradicals were obtained by heating a 10 percent solution of freshly distilled monomers in a selected solvent at 50°C in the presence of 20 percent azobisisobutyronitrile (AIBN) (based on monomer) and in the absence of oxygen. Oxygen was eliminated from the system by bubbling zero grade-nitrogen through the solution for 30 min. at room temperature.

The mixture was heated for seven days at 50°C to assure the absence of primary free radicals before adding the new monomer to the slurry in a glove bag type dry box in the absence of oxygen.

The products were obtained by filtering in Gooch crucibles and washing the residual solids with new solvent. The washed solids were dried at 100°C for 24 hours and cooled in a vacuum desiccator for 12 hours.

The solvent-free samples were dissolved in a good solvent and the homopolymers and block copolymers were precipitated separately by the addition of selected nonsolvent, filtered and dried. The solvent-free polymeric products were characterized by pyrolysis gas chromatography, turbidimetric titration, infra red spectroscopy, and differential thermal analysis.

## DISCUSSION

It was observed previously that the rate of polymerization of vinyl chloride[1] and acrylonitrile was accelerated in the presence of polymer which had precipitated because it was insoluble in the monomer[2]. This effect was not noted when dead PAN was present and the acceleration was attributed to the presence of "trapped free radicals"[3].

The latter were characterized by electron spin resonance spectroscopy (ESR) and by coupling with diphenyl picryl hydrazyl (DDPH)[4]. It was also shown that the "trapped free radicals" or macroradicals were terminated by the presence of traces of oxygen and were not produced when acrylonitrile was polymerized in a good solvent, such as dimethylformamide (DMF).

Acceleration of polymerization was also observed when styrene was polymerized in a poor solvent, such as methanol[5] or 1-hexane[6]. Other macroradicals have also been obtained by the polymerization of acrylonitrile[7], methyl acrylate[8] and equal mixtures of styrene and maleic anhydride[9] in poor solvents.

The criterion for a poor solvent was that the difference in solubility parameter between the polymer and solvent be at least 1.8 Hildebrand units (H)[10, 11, 12]. The rate of polymerization was fastest when the difference was 1.8H and decreased as this difference in solubility parameters increased[13].

Acrylonitrile and vinyl chloride will diffuse into macroradicals of these monomers and produce higher molecular weight homopolymers. However, other monomers will not diffuse into these macroradicals unless the difference in the solubility parameters of the monomer and macroradical or polymer is less than 3.2H[14].

This explains why it is not possible to produce a block copolymer by adding styrene monomer (9.2H) to acrylonitrile (13H). Yet, some block copolymer is produced when DMF is added dropwise to a slurry of acrylonitrile macroradicals in benzene and styrene. Of course, only dead polymer is obtained when acrylonitrile is polymerized in DMF.

Since styrene and acrylonitrile form a charge transfer complex which has a solubility parameter of 10H, a block copolymer ie, poly (acrylonitrile-b-styrene-co-acrylonitrile) is produced when an equimolar mixture of styrene and acrylonitrile is added to a slurry of acrylonitrile macroradicals in benzene. Likewise, while neither styrene nor maleic anhydride will form a block copolymer with acrylonitrile macroradicals, a block i.e., poly (acrylonitrile-b-styrene-co-maleic anhydride) is produced when equimolor quantities of styrene and maleic anhydride (11.0H) are added to a slurry of acrylonitrile macroradicals. When an excess of styrene monomer is present, the product is poly (acrylonitrile-b-styrene-co-maleic anhydride-b-styrene)[15].

While styrene (9.2H) will not diffuse into macroradicals of acrylonitrile (13.H), acrylonitrile (10.8H) will diffuse into styrene macroradicals (9.1H) dispersed in 1-propanol. The resulting block copolymer, of course, is identical to that which cannot be readily prepared by the addition of styrene to acrylonitrile macroradicals

Styrene macroradicals may be produced in a poor solvent, such as 1-propanol (11.9H) and in a nonpolar solvents, such as 1-hexane (7.3H)[16]. Block copolymers are readily produced when vinyl monomers are added to these macroradicals.

Block copolymers have also been produced by the addition of vinyl monomers to macroradicals of methyl methacrylate[17], vinyl acetate[18], and styrene-maleic anhydride[19]. The length of the styrene block of the latter is limited to about 30 percent of that of the macroradical when the block is prepared in benzene. However, larger blocks may be prepared in solvents which are both nonsolvents for styrene-maleic anhydride copolymer and polystyrene.

It is difficult to prepare stable macroradicals of monomers, such as methyl acrylate, with low Tg values. This restriction, based on mobility, also applies to blocks of monomers with low Tg values. Likewise, blocks cannot be prepared at temperatures above the theta temperature. Thus, while good yields of poly (methyl methacrylate-b-acrylonitrile) are obtained by heating a slurry of methyl methacrylate macroradicals and acrylonitrile in 1-propanol at $70^{o}$, the yields decrease as the temperature is increased. The theta temperature of this system is $82^{o}$.

Macroradicals are also stable in viscous good solvents and block copolymers may be produced by the addition of vinyl monomers to viscous solutions of macroradicals. The yield of block copolymers in poor solvents is also increased when the polymerization takes place in viscous poor solvents[20].

The yield and molecular weight of styrene macroradicals was increased when these macroradicals were produced in silicone oils of increasing viscosities[21]. Likewise, the degree of dispersity decreased as styrene continued to add to these styrene macroradicals. The lowest value for the degree of dispersity was noted after long time polymerization in the highest viscosity silicone oils.

## CONCLUSION

Block copolymers are readily produced by the reaction of vinyl monomers with stable macroradicals in poor solvents. The efficiency and rate of formation of block copolymers is increased in viscous systems.

## REFERENCES

1. W. I. Benough and RGW Norrish PROC Roy Soc (London), A210 301 (1950).

2. C. H. Bamford and A. D. Jenkins PTOC Roy Soc (London) A216 515 (1953).

3. A. Schindler and V. W. Breitenbach, RIC, Sci Suppl 25 34, (1955).

4. C. H. Bamford and A. D. Jenkins, PTOC Roy Soc (London) A228 220 (1955).

5. A. Chapiro, J. Chim Phys 47 747, 764 (1950).

6. R. B. Seymour, G. A. Stahl, and H. Wood Appl Polym Sci Symp 26 246 (1975).

7. R. B. Seymour, D. R. Owen, G. A. Stahl, H. Wood, and W. N. Tinnerman, Appl Polym Sci Symp 25 69 (1974).

8. R. B. Seymour, P. D. Kincaid, and D. R. Owen Adv Chem Ser 129 230 (1973).

9. R. B. Seymour, G. A. Stahl, D. R. Owen, and H. Wood Adv Chem Ser 142 309 (1975).

10. J. Hildebrand and R. Scott Regular Solutions, Prentice-Hall, Englewood Cliffs, NJ (1962).

11. H. Burrell Encyclopedia of Polymer Science and Technology, vol 12. H. F. Mark, N. G. Gaylord and W. M. Bikales, eds. Wiley-Interscience New York, 1970.

12. R. B. Seymour and J. M. Sosa Nature 248 (5451)759 (1974).

13. R. B. Seymour, P. D. Kincaid, and D. R. Owen, J. Paint Technol 45 (580) 35 (1973).

14. R. B. Seymour, P. D. Kincaid and D. R. Owen, Polymer Preprints 13 (1) 512 (1972).

15. R. B. Seymour and J. P. Kispersky U. S. Pat 2,439,229 Apr 6, 1948.

16. R. B. Seymour and D. R. Owen, and G. A. Stahl, Polymer 14, 324 (1973).

17. R. B. Seymour, G. A. Stahl, D. R. Owen, and H. Wood, Advances in Chemistry Series 142 309 (1975).

18. R. B. Seymour and G. A. Stahl, J. Polym Sci 14 2545 (1976).

19. R. B. Seymour, P. D. Kincaid and D. R. Owen, Polymer Preprints 13 (1) 522 (1972).

20. R. B. Seymour and G. A. Stahl J. Macromol Sci/Chem A1 (11) 53 (1977).

21. R. B. Seymour and G. A. Stahl Polym Sci Technol 10 217 (1977).

4. [illegible] and A. B. [illegible], Phil. Mag. [illegible] 22B [illegible]

5. [illegible] Fluid Phys. [illegible] (1958).

6. [illegible] Seymour, G. A. Stahl, [illegible] Appl. Polym. Sym. 26, [illegible]

7. [illegible] Seymour, D. R. Owen, G. A. [illegible], and [illegible] Polym. Sci. [illegible] 25 (1978).

8. [illegible] Seymour, [illegible] Kincaid, and D. R. Owen, [illegible] 150 (1973).

9. R. B. Seymour, G. A. Stahl, D. R. Owen, and [illegible] Chem. Ser. [illegible] (1979).

10. [illegible] and [illegible] Polymer Solutions, [illegible] Reinhold Pub. [illegible]

11. [illegible] in Polymer Science and Technology, [illegible] and N. M. Bikales, eds., Wiley-Interscience, New York, [illegible]

12. [illegible] Nature 248 [illegible] (1974).

13. [illegible] Seymour, [illegible] Kincaid, and D. R. Owen, [illegible] 14 [illegible] (1973).

14. [illegible] Seymour, [illegible] Kincaid and [illegible] Owen, [illegible] 18 [illegible] (1973).

15. [illegible] Seymour and [illegible]

16. [illegible] Seymour [illegible] and G. A. Stahl, [illegible]

17. [illegible] G. A. Stahl, D. R. Owen, and [illegible] (1978).

18. [illegible] Seymour and [illegible] Polym. Sci. [illegible]

19. R. B. Seymour, [illegible] and D. R. Owen, Polymer [illegible]

20. R. B. Seymour and [illegible] Polymer [illegible]

21. [illegible] Seymour, and [illegible] Polymer [illegible]

# Synthesis of N-Methylated Polyamides and Copolyamides

SAMUEL J. HUANG and JOSEPH KOZAKIEWICZ

Department of Chemistry and
Institute of Materials Science
University of Connecticut
Storrs, Connecticut 06268

ABSTRACT

N-Methylated polyamides and copolyamides derived from sebacic acid, N,N'-dimethylethylenediamine, and ethylenediamine were synthesized. Both interfacial polycondensation and melt polycondensation were examined, with the interfacial generally giving the higher molecular weights. The choice of organic solvent was important in the copolyamide synthesis with hexane giving random copolyamides and the more polar chloroform yielding copolyamides with a more blocklike structure, comparable to that obtained from the melt. The softening point of the N-methylated polyamide was between 65-75°C.

## INTRODUCTION

This paper is concerned with the synthesis of N-methylated polyamides and copolyamides with flexible backbone chains. N-Alkylated polyamides were first synthesized in the 1940's and were reported to be low melting, amorphous polymers with good elastic properties(1-5). This is in contrast to unsubstituted polyamides which are know to be highly crystalline, high melting materials. This difference is in part due to the fact that the hydrogen bonding found between the chains of the unsubstituted polyamides is not present in N-alkylated polyamides(2,4).

N-Alkylated polyamides have received relatively little attention compared to unsubstituted polyamides because their properties preclude their use in the preparation of strong fibers. The solubility and low melting points of these polymers, coupled with their thermal stabilities, make these polymers ideal candidates for other uses such as plasticizers, binders, or as soft blocks in elastomeric block copolymers(1-16). It has been reported that it is difficult to find a nonsolvent for some N-alkylated polyamides(8). The polyamides are soluble in most organic solvents and can be expected to interact strongly with a wide range of materials.

The majority of the work on N-alkylated polyamides has been done in an attempt to elucidate the effect of hydrogen bonding on polyamides. The two major methods of synthesis used have been polymerization of prepolymer nylon salts in the melt and interfacial polycondensation of diacid chlorides and diamines. Of the two methods, interfacial polycondensation has generally given the higher molecular weights, but gives poorer yields and is synthetically more difficult to work up. Both synthetic routes resulted in soluble polyamides with the expected reduced melting points caused by the reduction of hydrogen bonding.

Most of the previous work involved the synthesis of N-alkylated polyamides containing rigid aromatic rings in the backbone thereby increasing the transition temperatures. If N-alkylated polyamides were synthesized with flexible backbone chains, this would cause a reduction in the glass transition temperature and the melting point of the polymers. We report here the synthesis of the considerably more flexible N-methylated polyamides and copolyamides derived from sebacic acid, N,N'-dimethylethylenediamine, and ethylenediamine. Both melt polycondensation and interfacial polycondensation were examined.

## EXPERIMENTAL

The diamines and the acid chloride were distilled under Argon prior to use. Sebacic acid was recrystallized twice from ethanol/water and dried over $P_2O_5$. Chloroform was purified by washing with $H_2SO_4$ and water, drying over $CaCl_2$, and distilling. Hexane was purified by washing with $H_2SO_4$, $HNO_3$, and water, drying over $CaCl_2$, and distilling. Benzene was purified by washing with $H_2SO_4$ until clear and then distilling.

The interfacial polycondensation of the methylated polyamide was carried out in a 2-l resin kettle by adding sebacoyl chloride (11.95g, 0.050 mole) in 500ml organic solvent to the aqueous phase consisting of N,N'-dimethylethylenediamine(4.41g, 0.050 mole) and sodium hydroxide (4.00g, 0.100 mole) and stirring rapidly for 5 minutes. The resulting polymer was filtered, washed throughly with water, and dried under vacuum. The copolyamides were synthesized as above with the exception that the aqueous phase contained both the ethylenediamine and the N,N'-dimethylethylenediamine along with sodium hydroxide.

The melt polycondensation of the methylated polyamide was carried out by first preparing the nylon salt by dissolving an equimolar mixture of sebacic acid and N,N'-dimethylethylenediamine in ethanol and heating at reflux for one hour. The solution was cooled, the salt filtered, washed with cold ethanol, recrystallized twice from ethanol, and dried under vacuum. The salt was transferred to a polymerization tube whose atmosphere was then exchanged with Argon and sealed under vacuum. The polycondensation was then carried out by heating the sealed tube for 3 hours at 210°C and then heating an additional 18 hours under vacuum. The copolyamides were prepared as above by first preparing the prepolymer nylon salts, mixing the salts together, and carrying out the polycondensations at 250°C for the same time periods.

The model compound studies were conducted exactly like the copolymerizations using 1 mole N,N'-dimethylethylenediamine and 1 mole ethylenediamine to 2 moles decanoyl chloride.

The viscosities were measured in TFA at 30°C in the Ubbelohde Viscometer. The NMRs' of all polymers were carried out in TFA in a Varian EM360A, 60MHz spectrometer. The softening points were recorded as the temperature range from the point of initial flow to complete sample flow on a Reichert, Austria melting point apparatus and DuPont 990 Thermal Analyzer.

## RESULTS

All of the homopolymers prepared by interfacial polycondensation had approximately the same softening point. The differences caused by varying the organic solvent can be seen in the viscosities and the yields (Table 1). Hexane and benzene gave polymers with similar viscosities but with different yields due to the increased solubility of the polymer in the benzene as compared to the hexane. Only the precipitated polymer which was filterable with a medium porocity sintered glass funnel was collected with these two solvents. The viscosity and percent yield of the homopolymer from chloroform was quite different from that obtained from the other two solvents. This is due to the complete solubility of the polymer in this solvent at the reaction concentration. This made it necessary to separate the chloroform layer and strip the solvent to collect the polymer.

TABLE 1. Poly(N,N'-dimethylethylene sabacamide)

| Polymerization Technique | Solvent System | Percent Yield | Softening Point(°C) | [η][a] |
|---|---|---|---|---|
| interfacial | $CHCl_3/H_2O$ | 92.5 | 68-73 | .110 |
| interfacial | hexane/$H_2O$ | 47.3 | 68-71 | .216 |
| interfacial | benzene/$H_2O$ | 24.3 | 65-70 | .188 |
| melt | | 89.0 | b | .106 |

[a]Measured in trifluoroacetic acid at 30°C.
[b]Oil, solidifies slowly on standing.

The result is a larger yield and lower viscosity due to the isolation of the low molecular weight fraction along with the high molecular weight fraction which is primarily collected by filtration in the other two solvents. The melt polycondensation yielded a polymer which was initially a viscous liquid at room temperature, but solidified on standing after several days.

Both melt and interfacial polymerization gave copolyamides with compositions identical to that contained in the monomer feed (Tables 2,3,4). The viscosities of the copolyamides generally

TABLE 2. Melt Polymerization of Methylated and Unmethylated Ethylenediamine With Sebacic Acid

| % Methylated Diamine | % Yield | Softening Point (°C) | $[\eta]^a$ |
|---|---|---|---|
| 0 | 92.3 | 275-280 | b |
| 25 | 99.3 | 245-250 | .181 |
| 50 | 79.7 | 195-210 | .153 |
| 75 | 96.0 | 50-80 | .131 |
| 100 | 89.0 | c | .106 |

[a]Measured in trifluoroacetic acid at 30 C.
[b]Insoluble in trifluoroacetic acid.
[c]Oil, solidified losely on standing.

TABLE 3. Interfacial Polymerization of Methylated and Unmethylated Ethylenediamine With Sebacoyl Chloride In Hexane/$H_2O$

| % Methylated Diamine | % Yield | Softening Point (°C) | $[\eta]^a$ |
|---|---|---|---|
| 25 | 58.0 | 212-216 | .143 |
| 50 | 70.6 | 130-145 | .158 |
| 70 | 71.2 | 105-110 | .316 |
| 100 | 47.3 | 68-71 | .216 |

[a]Measured in trifluoroacetic acid at 30°C.

TABLE 4. Interfacial Polymerization of Methylated and Unmethylated Ethylenediamine with Sebacoyl Chloride In $CHCl_3/H_2O$

| % Methylated Diamine | % Yield | Softening Point (°C) | $[\eta]^a$ |
|---|---|---|---|
| 25 | 57.6 | 250-253 | .191 |
| 50 | 45.6 | 193-198 | .219 |
| 75 | 25.0 | 85-90 | .210 |
| 100 | 92.5 | 68-73 | .110 |

[a]Measured in trifluoroacetic acid at 30°C.

decreased as the methylation percent increased in the melt polymerization but increased as the methylation percent increased in the interfacial polymerization. When the percent methylation was 50 percent or less, chloroform gave polymers with larger viscosities than hexane, but when the percent methylation was further increased, hexane gave the polymers with the higher viscosities. Interfacial polycondensation generally resulted in higher molecular weights, but lower yields than the melt polycondensation.

## DISCUSSION

It was not surprising to find that the polyamides prepared interfacially had higher viscosities than those prepared via the melt. Interfacial polymerization does not require the monomer purity or exactness of stoichiometry that melt polymerization requires.

The decrease in viscosity with the increase in percent methylation in the melt is probably due to the much greater condensation rate of primary diamines versus secondary diamines(17). With the interfacial systems, the reverse trend seems to indicate that the methylated diamine was reacting faster. To confirm this reversed trend, a model compound study was undertaken using 2 moles of the monofunctional decanoyl chloride to 1 mole of both N,N'-dimethylethylenediamine and ethylenediamine and running the reaction under conditions identical to that used in the copolymerizations. In both hexane and chloroform, the N,N'-dimethylated diamide was obtained in better than 90% yield, indicating that at least initially, the methylated diamine is reacting faster. This type of reactivity difference of diamines to an acid chloride in interfacial systems has been observed by Ogata, et. al. (18). He credited their differences in partition coefficients into the organic phase, where interfacial polymerizations are known to occur, as the controlling factor. The difference in the partition coefficients of the N,N'-dimethylethylenediamine and the ethylenediamine into the hexane and chloroform is probably quite different and is the likely cause of the methylated diamine reacting faster.

In the plot of melting point versus percent methylation for the copolyamides (Fig. 1), we would expect to find a concave upwards curve if the methylated and nonmethylated units were randomly situated throughout the system. This is what is found in the hexane/water interfacial system suggesting that this system does indeed give a random copolymer, despite the diamine reactivity difference. When chloroform/water is used as the interfacial solvent system, however, a convex upwards curve is found. On going from 0 percent to 50 percent methylation, only an approximately 70°C drop in Tm is realized whereas a nearly 110°C reduction in Tm occurs when the methylation percent is further increased from 50 to 75 percent. The same results are found with the copolyamides synthesized from the melt, indicating that the structures of the respective copolyamides are fairly similar. Saotome and Komoto observed the same convex upwards curves for a different set of alkylated copolyamides and suggested that this could only be explained by a block copolymer structure (8). It appears that both the melt and interfacial polycondensation using chloroform gave block copolyamides.

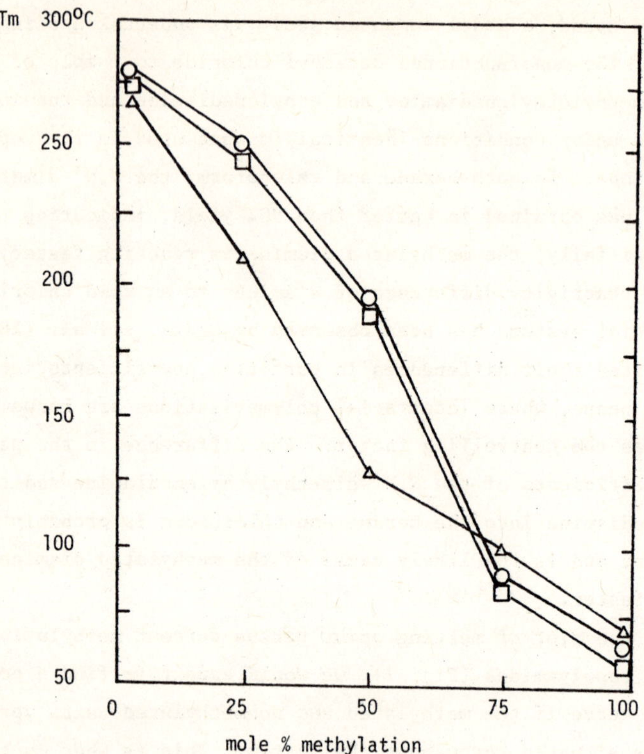


Fig. 1. Effect of methylation on the melting points of copolyamides.
○ Interfacial polymerization in $CHCl_3/H_2O$. □ Melt polymerization.
△ Interfacial polymerization in hexane/$H_2O$.

The formation of blocks in the melt can be explained by the higher condensation rate of the primary amine as compared to the secondary amine (17). This rate difference would result in chains of unsubstituted polyamides being produced in the initial stages of the reaction followed by further condensation of the unused secondary amine to form the block structure.

With the chloroform interfacial system, the formation of blocks is probably a result of the reaction of the secondary diamine in the initial stages of the reaction, due to its greater partition coefficient. With the hexane interfacial system, the reactivity difference between the diamines was not sufficient enough to cause block formation and therefore a random copolyamide resulted.

It is possible, therefore, to synthesize block copolyamides by both the melt and by interfacial polycondensation with the interfacial method in $CHCl_3/H_2O$ giving higher viscosities but lower yields. It is also possible to synthesize random copolyamides by interfacial polymerization in hexane/$H_2O$.

## ACKNOWLEDGEMENT

Support of our research by National Science Foundation (DMR-7813402) is gratefully acknowledged.

## REFERENCES

(1) W. O. Baker and C. S. Fuller, J. Am. Chem. Soc., 65, 1120 (1943).

(2) B. S. Biggs, C. J. Frosch, and R. H. Erickson, Ind. Eng. Chem., 38, 1016 (1964).

(3) R. Hill and E. E. Walker, J. Polymer Sci., 3, 606 (1948).

(4) E. L. Wittbecker, R. C. Houtz, and W. W. Watkins, Ind. Eng. Chem., 40, 875 (1948).

(5) W. H. Carothers, U. S. Patent 2,130,523 (Sept. 20, 1938).

(6) O. Ya. Fedotava and M. Bin-Tsyuan, Vysokomol. soedin., 2(6), 953 (1960).

(7) O. Ya. Fedotova, I. P. Losev, M. A. Askarov, and R. G. Kostina, Zh. obshch. khim., 29, 672 (1959).

(8) K. Saotome and H. Komoto, J. Polymer Sci.: Part A-1, 5, 107 (1967).

(9) V. E. Shashoua and W. M. Eareckson, J. Polymer Sci., 15, 343 (1959).

(10) O. Ya. Fedotova, M. L. Kerber, and I. P. Losev, Vysokomol. soedin., 2(7), 1020 (1960).

(11) O. Ya. Fedotova, M. L. Kerber, and I. P. Losev, Vysokomol. soedin., 5(6), 881 (1963).

(12) O. Ya. Fedotova, I. P. Losev, and N. M. Kozyreva, Vysokomol. soedin., 5(3), 363 (1963).

(13) O. Ya. Fedotova, M. L. Kerber, and I. P. Losev, Vysokomol. soedin., 6(3), 452 (1964).

(14) O. Ya. Fedotova, M. I. Shtil'man, and Kolesnikov, Vysokomol. soedin., A10(2), 278 (1968).

(15) O. Ya. Fedotova, V. I. Gorakhov, and V. V. Korshak, Vysokomol. soedin., A16(6), 1228 (1974).

(16) M. M. Brubaker, U. S. Patent 2,378,977 (June 26, 1945).

(17) L. F. Beste and R. C. Houtz, J. Polymer Sci., 8, 395 (1952).

(18) N. Ogata, K. Sanui, T. Onozaki, and S. Imanishi, Polym. Prepr., 42, 390 (1980).

# PHASE TRANSFER CATALYSIS

# Syntheses of Some Condensation Polymers by Phase-Transfer Catalyzed Polycondensation

Yoshio Imai

Department of Polymer Chemistry
Faculty of Engineering
Yamagata University
Yonezawa, Yamagata 992: Japan

ABSTRACT

Fundamental studies on the two-phase polycondensation using various phase-transfer catalysts such as quaternary ammonium and phosphonium salts, crown ethers, and poly(ethylene glycol)s yielded information useful for the synthesis of a variety of condensation polymers. We successfully synthesized various types of condensation polymers such as aromatic polysulfonates and polyphosphonates, aromatic polyethers, aliphatic and aromatic polysulfides, and carbon-carbon chain polymers of high molecular weights by the phase-transfer catalyzed polycondensation from combinations of aromatic disulfonyl chlorides, phosphonic dichlorides, activated aromatic dichlorides, and aliphatic dihalides, with bisphenols, aliphatic and aromatic dithiols, and active methylene compounds. The two-phase polycondensation was generally carried out in a water-immiscible organic solvent-aqueous alkaline solution system at room temperature or above. In conclusion, this method of polycondensation offers a highly versatile and convenient synthetic method for a variety of condensation polymers.

## INTRODUCTION

Recently phase-transfer catalysts have been widely used for organic synthesis, particularly for nucleophilic substitution reactions [1,2]. In general, those nucleophilic substitution reactions between alkyl halides in an organic phase and anions in an aqueous phase, which do not occur usually because of phase separation, can often be strongly catalyzed by addition of a small amount of quaternary ammonium or phosphonium salts. The catalysis of these onium salts is due to the ability of the organic--soluble cations to repeatedly bring anions into the organic phase in a form suitable for reactions; this effect is termed as "phase transfer catalysis". The outline of the catalysis sequence is expressed by Eq. (1):

$$\begin{array}{llcccccc} \text{Organic phase} & R\text{-}X & + & Q^+Y^- & \longrightarrow & R\text{-}Y & + & Q^+X^- \\ & & & \uparrow\downarrow & & & & \downarrow\uparrow \\ \text{Interface} & \hline & & & & & & \\ \text{Aqueous phase} & M^+X^- & + & Q^+Y^- & \rightleftarrows & M^+Y^- & + & Q^+X^- \end{array} \qquad (1)$$

Quaternary ammonium salt, $Q^+ = R_4N^+$, selected due to its high solubility in the organic phase, transfers anion $Y^-$ into the organic phase as $Q^+Y^-$, which then reacts with alkyl halide RX to yield the substitution product RY. The coproduced $Q^+X^-$ is reconverted rapidly to $Q^+Y^-$ by anion exchange with nucleophile $M^+Y^-$ from the aqueous phase. Crown ethers and poly(ethylene glycol) derivatives are also as useful for the phase-transfer catalyzed reactions as onium salts.

Our attention has been focused on the application of phase transfer catalysts to polymer synthesis, since a few studies on the polycondensation using phase transfer catalysts appeared in the literature. Known examples are the syntheses of aromatic polyesters and polysulfonates by interfacial polycondensation in the presence of onium salt accelerators [3] and polycabonate synthesis by solid-liquid two-phase polycondensation using crown ethers [4,5]. The present paper summarizes our recent works on the syntheses of various condensation polymers by phase-transfer catalyzed polycondensation, where the following phase transfer catalysts were employed: tetrabutylammonium chloride (TBAC), benzyltriethylammonium chloride (BTEAC), cetyltrimethylammonium chloride (CTMAC), benzyltriphenylphosphonium chloride (BTPPC), cetyltributylphosphonium bromide (CTBPB), 15-crown-5 (15-C-5), 18-crown-6 (18-C-6), dibenzo-18-crown-6 (DB-18-C-6), dicyclohexyl--18-crown-6 (DC-18-C-6), dibenzo-24-crown-8 (DB-24-C-8), and dicyclohexyl-24-crown-8 (DC-24-C-8).

## RESULTS AND DISCUSSION

### Syntheses of Aromatic Polysulfonates and Polyphosphonates [6-8]

It is known that aromatic polysulfonates of high molecular weights can be prepared from aromatic disulfonyl chlorides and alkaline salts of bisphenols by the interfacial polycondensation technique using onium salt accelerators [3]. However, the function of these onium salts as phase transfer catalysts had not been fully elucidated till 1965. Therefore, we reinvestigated the synthesis of polysulfonates by two-phase polycondensation using various phase transfer catalysts.

The polycondensation of 4,4'-oxybisbenzenesulfonyl chloride (I) with 2,2-bis(4-hydroxyphenyl)propane (II, bisphenol A) was carried out in a dichloromethane-aqueous alkaline solution system at room temperature under the standard Schotten-Baumann's reaction conditions [Eq. (2)].

The results of the polycondensation are summarized in Table 1. In the absence of the catalyst, only low-molecular polysulfonate III was obtained even though the reaction was continued for 24 hr, whereas the addition of these quaternary ammonium salts and crown

TABLE 1. Synthesis of Aromatic Polysulfonate III with Various Phase Transfer Catalysts in Dichloromethane-Water System[a]

| Catalyst | Reaction time (hr) | $\eta_{inh}$ of polymer (dl/g)[b] | |
|---|---|---|---|
| | | Alkaline component | |
| | | NaOH | KOH |
| None | 24 | 0.20 | - |
| TBAC | 2 | 1.44 | - |
| BTEAC | 2 | 1.02 | - |
| CTMAC | 2 | 1.19 | - |
| 15-C-5 | 3 | 1.31 | 1.28 |
| 18-C-6 | 3 | 0.99 | 1.11 |
| DB-18-C-6 | 3 | 0.95 | 1.02 |
| DC-18-C-6 | 3 | 1.28 | 1.48 |
| DB-24-C-8 | 3 | 1.03 | 1.09 |
| DC-24-C-8 | 3 | 0.80 | 1.13 |

[a]The polymerization was carried out with 2.5 mmol of I and II in the presence of 0.05 mmol of the catalyst in 3.5 ml of dichloromethane and 5 ml of the aqueous alkaline solution (1.01M) at 20°C.

[b]Measured at a concentration of 0.5 g/dl in sym-tetrachloroethane at 30°C.

$$Cl{-}SO_2{-}C_6H_4{-}O{-}C_6H_4{-}SO_2Cl \;(\text{I}) \;+\; HO{-}C_6H_4{-}C(CH_3)_2{-}C_6H_4{-}OH \;(\text{II})$$

$$\longrightarrow \left[ {-}SO_2{-}C_6H_4{-}O{-}C_6H_4{-}SO_2{-}O{-}C_6H_4{-}C(CH_3)_2{-}C_6H_4{-}O{-} \right]_n \;(\text{III}) \quad (2)$$

ethers increased the average molecular weight of the polymer remarkably. These catalysts react with bisphenol II to form such compound that are more soluble in the organic phase than the alkali salts of the bisphenol. Thus, the contact of II with I in the organic phase is accelerated and the polymer of higher molecular weight is obtained.

Among these catalysts, TBAC and DC-18-C-6 were found to be highly efficient, leading to the formation of the polysulfonate with an inherent viscosity of as high as 1.4 dl/g. In the crown ether catalyzed system, potassium hydroxide was generally more effective as alkaline component than sodium hydroxide for preparing the polymer of high molecular weights.

The relationship between inherent viscosities of the resulting polymer and reaction time in the polycondensation with crown ether catalyst is shown in Figure 1. The phase-transfer catalyzed polycondensation proceeded fairly rapidly at 20°C and was essentially completed after 2 hr.

As the next step to the phase-transfer catalyzed synthesis of aromatic polysulfonates, we synthesized structurally related aromatic polyphosphonates. The polycondensation of phenylphosphonic dichloride (IV) with bisphenol A (II) leading to polyphosphonate (V) was conducted with various phase transfer catalysts in a dichloromethane-aqueous alkaline solution system at a temperature

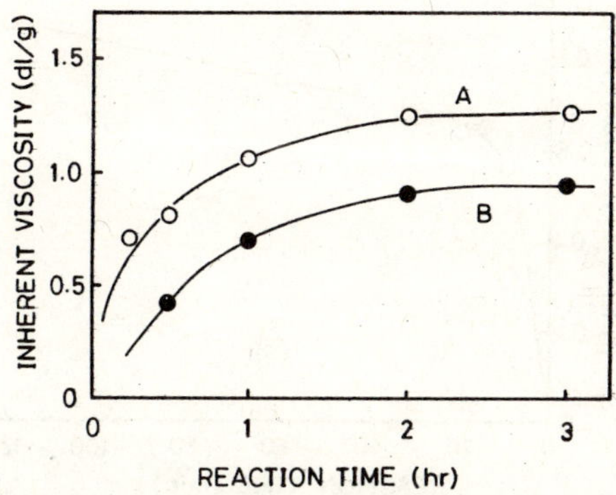


FIG. 1. Polycondensation of I and II in dichloromethane-aqueous KOH system at 20°C with DC-18-C-6 (A) and DB-18-C-6 (B) catalysts. Reaction conditions, see Table 1.

range of lower than 0°C, because IV is known to be highly susceptible to alkaline hydrolysis [Eq. (3)].

$$\underset{\text{IV}}{\text{Cl-}\overset{\text{O}}{\overset{\|}{\text{P}}}\text{-Cl}\ (\text{P-C}_6\text{H}_5)} + \text{II} \longrightarrow \underset{\text{V}}{\left[ -\overset{\text{O}}{\overset{\|}{\text{P}}}(\text{C}_6\text{H}_5)\text{-O-C}_6\text{H}_4\text{-C}(\text{CH}_3)_2\text{-C}_6\text{H}_4\text{-O-} \right]_n} \quad (3)$$

Figure 2 shows the plots of inherent viscosities of the polyphosphonate against reaction time in the polycondensation with BTPPC catalyst. The two-phase polycondensation proceeded very rapidly at -5°C and was almost completed in 20 min. Thus, phenylphosphonic dichloride was found to be more reactive with alkaline salt of bisphenol A than aromatic disulfonyl chlorides.

The results of the polycondensation are summarized in Table 2. In the absence of any catalyst, only low molecular weight polymer V was produced, whereas the addition of the quaternary onium salts

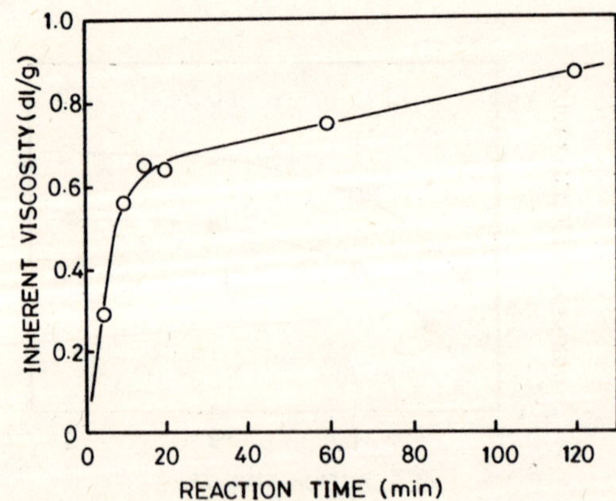


FIG. 2. Polycondensation of II and IV with BTPPC catalyst in dichloromethane-aqueous KOH system at -8∼-5°C. Reaction conditions, see Table 2.

TABLE 2. Synthesis of Aromatic Polyphosphonate V with Various Phase Transfer Catalysts in Dichloromethane-Water System[a]

| Catalyst | $\eta_{inh}$ of polymer (dl/g)[b] | |
|---|---|---|
| | Alkaline component | |
| | NaOH | KOH |
| None | 0.05 | 0.05 |
| TBAC | 0.58 | 0.36 |
| BTEAC | 0.21 | 0.21 |
| CTMAC | 0.60 | 0.70 |
| BTPPC | 0.59 | 0.88 |
| CTBPB | 0.33 | 0.28 |
| 15-C-5 | 0.09 | 0.21 |
| 18-C-6 | 0.14 | 0.30 |
| DB-18-C-6 | 0.09 | 0.59 |
| DC-18-C-6 | 0.19 | 0.64 |
| DB-24-C-8 | 0.10 | 0.17 |
| DC-24-C-8 | 0.24 | 0.50 |

[a]The polymerization was carried out with 2.5 mmol of II and IV in the presence of 0.05 mmol of the catalyst in 1.3 ml of dichloromethane and 5 ml of the aqueous alkaline solution (1.01 M) at -8∼-5°C for 2 hr.

[b]Measured at a concentration of 0.5 g/dl in sym-tetrachloroethane at 30°C.

and crown ethers increased more or less molecular weight of the polymer. Among them, the two-phase polycondensation was strongly catalyzed by some lipophilic phase transfer catalysts such as TBAC, CTMAC, BTPPC, DB-18-C-6, DC-18-C-6, and DC-24-C-8, leading to the formation of the polyphosphonate with high inherent viscosities of more than 0.5 dl/g. In these crown ether catalyzed systems, potassium hydroxide was generally more effective as alkaline component in the aqueous phase than sodium hydroxide for producing the polymer of higher molecular weights. On the other hand, the polycondensation was not affected by the alkaline components in the quaternary onium salt catalyzed systems with regard to inherent viscosity of the resulting polymer.

Synthesis of Aromatic Polyethers [9]

It is known that aromatic polyethers with high molecular weights were generally prepared by the aromatic nucleophilic substitution polymerization of activated aromatic dihalides with alkaline salts of bisphenols in anhydrous polar aprotic solvents at elevated temperatures [10-13]. We successfully synthesized similar aromatic polyether (VII) by the phase-transfer catalyzed polycondensation of bis(4-chloro-3-nitrophenyl) sulfone (VI) and bisphenol A (II) using such catalysts as quaternary ammonium salts, crown ethers, and poly(ethylene glycol)s [Eq. (4)]. Very recently, we have learnt the phase-transfer catalyzed synthesis of another type of polyethers [14].

$$\underset{\text{VI}}{Cl\text{-}C_6H_3(NO_2)\text{-}SO_2\text{-}C_6H_3(NO_2)\text{-}Cl} + \text{II} \longrightarrow \underset{\text{VII}}{\left[ \text{-}C_6H_3(NO_2)\text{-}SO_2\text{-}C_6H_3(NO_2)\text{-}O\text{-}C_6H_4\text{-}C(CH_3)_2\text{-}C_6H_4\text{-}O\text{-} \right]_n} \quad (4)$$

Our polycondensations were carried out in a water-immiscible organic solvent-aqueous alkaline solution system at room temperature. Figure 3 shows the course of the polycondensation with DC-18-C-6 catalyst in terms of inherent viscosity of the resulting polyether. The phase-transfer catalyzed polycondensation proceeded fairly slowly at 20°C in a dichloromethane-water system and went essentially to completion in 24 hr.

The influence of catalysts on the polycondensation in a dichloromethane-water system is given in Table 3. All of these

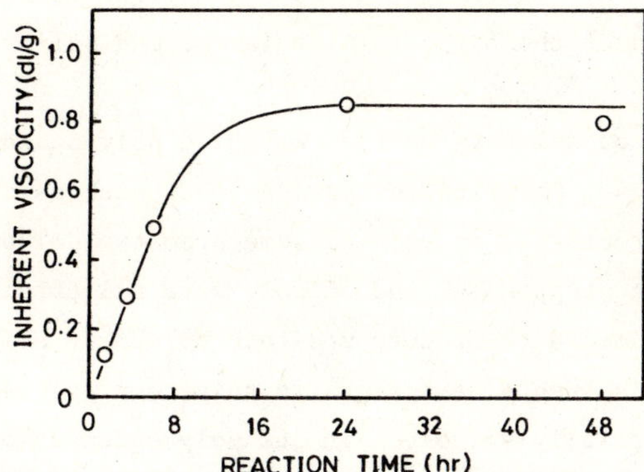


FIG. 3. Polycondensation of II and VI with DC-18-C-6 catalyst in dichloromethane-aqueous KOH system at 20°C. Reaction conditions, see Table 3.

TABLE 3. Synthesis of Aromatic Polyether VII with Various Phase Transfer Catalysts in Dichloromethane-Water System[a]

| Catalyst | Alkaline component | Reaction time (day) | Polymer $\eta_{inh}$ (dl/g)[b] |
|---|---|---|---|
| TBAC | NaOH | 2 | 0.42 |
| CTMAC | NaOH | 6 | 0.26 |
| BTEAC | NaOH | 6 | 0.15 |
| 15-C-5 | KOH | 6 | 0.36 |
| DB-18-C-6 | KOH | 4 | 0.37 |
| DC-18-C-6 | KOH | 1 | 0.84 |
| DC-18-C-6 | NaOH | 6 | 0.31 |

[a]The polymerization was carried out with 2.5 mmol of II and VI in the presence of 0.05 mmol of the catalyst in 3.5 ml of dichloromethane and 5 ml of the aqueous alkaline solution (1.01 M) at 20°C.

[b]Measured at a concentration of 0.5 g/dl in DMF at 30°C.

crown ethers and quaternary ammonium salts except BTEAC were effective for yielding aromatic polyether VII with reasonable molecular weights. Above all, DC-18-C-6 was the most suitable for the preparation of the highest molecular weight polymer under these conditions.

The effect of solvents on the two-phase polycondensation is shown in Table 4. Chlorinated hydrocarbons, such as chloroform and 1,2-dichloroethane, as well as some aromatic solvents like nitrobenzene, acetophenone, and anisole were all effective as the polymerization media to produce moderate molecular weight polymer VII. In these solvents, the aromatic polyether was readily soluble or at least partially soluble. In the polycondensation in a nitrobenzene-water system, an increase in reaction temperature brought about a marked reduction in reaction time; the polycondensation at 80°C for 2 hr afforded the polyether with an inherent viscosity of 0.5 dl/g.

TABLE 4. Synthesis of Aromatic Polyether VII in Various Organic Solvent-Water System with DC-18-C-6 Catalyst[a]

| Solvent | Reaction temperature (°C) | Reaction time (hr) | Polymer $\eta_{inh}$ (dl/g)[b] |
|---|---|---|---|
| Dichloromethane | 20 | 24 | 0.84 |
| Chloroform | 20 | 24 | 0.53 |
| 1,2-Dichloroethane | 20 | 24 | 0.42 |
| Nitrobenzene | 20 | 24 | 0.47 |
| Nitrobenzene | 80 | 2 | 0.51 |
| Nitrobenzene | 100 | 1 | 0.42 |
| Acetophenone | 20 | 24 | 0.33 |
| Anisole | 20 | 24 | 0.31 |

[a]The polymerization was carried out with 2.5 mmol of II and VI in the presence of 0.05 mmol of DC-18-C-6 in 3.5 ml of the organic solvent and 5 ml of 1.01 M aqueous potassium hydroxide.

[b]Measured at a concentration of 0.5 g/dl in DMF at 30°C.

We also found that poly(ethylene glycol)s (PEGs) catalyzed the two-phase polycondensation in a manner similar to those with crown ethers leading to aromatic polyether VII of moderately high molecular weights. As can be seen from Table 5, PEG with molecular weight of 2,000 was found to be the most efficient catalyst among these PEGs. The polycondensation using PEG catalysts in a dichloromethane-water system proceeded much more slowly than the polymerization with crown ether catalysts. This may provide the first successful application of poly(ethylene glycol)s to polymer synthesis, which are both inexpensive and nontoxic compared with crown ethers.

## Syntheses of Aliphatic and Aromatic Polysulfides [15-17]

In the early 1950s, aliphatic polysulfides were synthesized by the aliphatic nucleophilic substitution polymerization of dibromoalkanes with sodium salt of dithiols under anhydrous conditions [18]. Recently, we developed a convenient method for the preparation of polysulfides by the two-phase polycondensation in

TABLE 5. Synthesis of Aromatic Polyether VII with Poly(ethylene glycol) Catalysts in Dichloromethane-Water System[a]

| Catalyst | $\eta_{inh}$ of polymer (dl/g)[b] Reaction time (day) | | | |
|---|---|---|---|---|
| | 1 | 3 | 5 | 7 |
| PEG-600 | 0.10 | 0.28 | 0.27 | 0.29 |
| PEG-1,000 | 0.19 | 0.31 | 0.33 | 0.36 |
| PEG-2,000 | 0.16 | 0.33 | 0.41 | 0.47 |
| PEG-3,000 | 0.18 | 0.33 | 0.29 | 0.32 |
| PEG-20,000 | 0.14 | 0.25 | 0.26 | 0.34 |

[a]The polymerization was carried out with 2.5 mmol of II and VI in the presence of 0.5 g of PEG catalyst in 3.5 ml of dichloromethane and 5 ml of 1.01 M aqueous potassium hydroxide at 20°C.

[b]Measured at a concentration of 0.5 g/dl in DMF at 30°C.

a potassium hydroxide aqueous solution. The polycondensation of 1,4-dibromobutane (VIII) and 1,6-hexanedithiol (IX) leading to polysulfide (X) was first carried out in various organic solvent-water system with DC-18-C-6 catalyst at 80°C [Eq. (5)].

$$\underset{\text{VIII}}{Br(CH_2)_4Br} \quad + \quad \underset{\text{IX}}{HS(CH_2)_6SH} \longrightarrow \underset{\text{X}}{\left[-(CH_2)_4-S-(CH_2)_6-S-\right]_n} \quad (5)$$

The results of the polycondensation are summarized in Table 6. All of these polymerization media employed afforded the polysulfide with moderately high inherent viscosities, whereas the polymer with the highest viscosity was produced in the absence of organic solvents. Thus, the polycondensation proceeded readily in a melt-water two-phase system at 80°C, since polysulfide X has a polymer-melting temperature of 75°C.

Table 7 shows the effect of catalysts on the polycondensation in an aqueous potassium hydroxide. These general phase transfer catalysts including crown ethers and quaternary ammonium salts, as well as sodium dodecyl sulfate (SDS) which serves as an anionic

TABLE 6. Synthesis of Aliphatic Polysulfide X in Various Organic Solvent-Water System with DC-18-C-6 Catalyst[a]

| Solvent | Reaction temperature (°C) | Polymer $\eta_{inh}$ (dl/g)[b] |
|---|---|---|
| Dichloromethane | 20 | 0.30 |
| Chloroform | 80 | 0.31 |
| Benzene | 80 | 0.30 |
| Nitrobenzene | 80 | 0.58 |
| Acetonitrile | 80 | 0.58 |
| None | 80 | 0.73 |

[a]The polymerization was carried out with 2.5 mmol of VIII and IX in the presence of 0.05 mmol of DC-18-C-6 in 2.5 ml of the organic solvent and 5 ml of 1.01 M aqueous potassium hydroxide for 48 hr.

[b]Measured at a concentration of 0.5 g/dl in chloroform at 30°C.

TABLE 7. Synthesis of Aliphatic Polysulfide X with Various Catalysts in Aqueous Potassium Hydroxide[a]

| Catalyst | Polymer $\eta_{inh}$ (dl/g)[b] |
|---|---|
| 15-C-5 | 0.73 |
| 18-C-6 | 0.73 |
| DB-18-C-6 | 0.93 |
| DC-18-C-6 | 0.84 |
| DB-24-C-8 | 0.63 |
| DC-24-C-8 | 0.49 |
| TBAC | 0.83 |
| CTMAC | 0.58 |
| SDS | 0.63 |
| None | 0.51 |

[a]The polymerization was carried out with 2.5 mmol of VIII and IX in the presence of 0.05 mmol of the catalyst in 5 ml of 1.01 M aqueous potassium hydroxide at 80°C for 24 hr.

[b]Measured at a concentration of 0.5 g/dl in chloroform at 30°C.

surface-active agent, were all effective for producing polysulfide X of high molecular weights. It is interesting to note that the polycondensation proceeded without use of any catalyst in this system, leading to the formation of the polymer with moderately high molecular weight.

Figure 4 shows the course of the polycondensation in an aqueous potassium hydroxide at 80°C in terms of inherent viscosity of the resulting polymer. In the presence of DC-18-C-6 catalyst, the maximum inherent viscosity was reached in 24 hr. On the other hand, the maximum was attained after 48 hr in the absence of the catalyst. Thus, the polycondensation reaction was greatly accelerated by the addition of DC-18-C-6 catalyst in the initial stage. It was also observed that a longer reaction time at 80°C resulted in a slight reduction in inherent viscosity; this is assumed to be caused by alkaline hydrolysis of the resulting polysulfide.

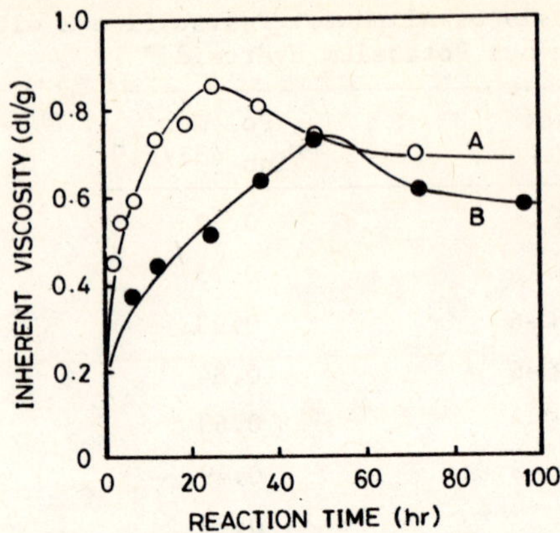


FIG. 4. Polycondensation of VIII and IX in aqueous KOH at 80°C with DC-18-C-6 (A) and without the catalyst (B). Reaction conditions, see Table 7.

Another type of polysulfides (XIII) were synthesized by the two-phase polycondensation of bis(3-chloroacryloyl)benzenes (XIa and XIb), which are one of the most reactive dichlorides, with 4,4'-oxybisbenzenethiol (XII) [Eq. (6)]. The polycondensation was carried out in a chloroform-water system at room temperature with some phase transfer catalysts.

$$\underset{\textbf{XI}}{Cl{-}CH{=}CH{-}\underset{\underset{O}{\|}}{C}{-}Ar{-}\underset{\underset{O}{\|}}{C}{-}CH{=}CH{-}Cl} \quad + \quad \underset{\textbf{XII}}{HS{-}C_6H_4{-}O{-}C_6H_4{-}SH}$$

$$\longrightarrow \underset{\textbf{XIII}}{\left[{-}CH{=}CH{-}\underset{\underset{O}{\|}}{C}{-}Ar{-}\underset{\underset{O}{\|}}{C}{-}CH{=}CH{-}S{-}C_6H_4{-}O{-}C_6H_4{-}S{-}\right]_n} \qquad (6)$$

XIa: Ar = $m$-phenylene, XIb: Ar = $p$-phenylene

The course of the polycondensation in terms of inherent viscosity of polysulfide XIIIa formed is shown in Figure 5. Even

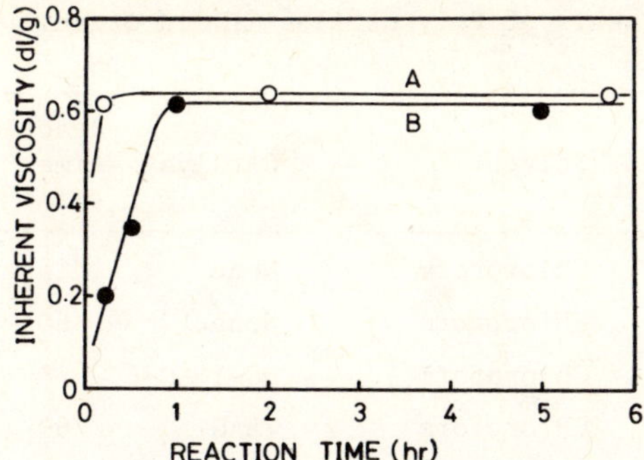


FIG. 5. Polycondensation of XIa and XII in chloroform-aqueous KOH system at 15°C with DC-18-C-6 (A) and without the catalyst (B). Reaction conditions, see Table 8.

without use of any catalyst, the polycondensation proceeded rapidly with precipitation of the polymer and went essentially to completion in 1 hr at room temperature to yield the polysulfide having an inherent viscosity of 0.6 dl/g. The use of DC-18-C-6 increased the initial rate of the polycondensation remarkably and the polymerization was almost completed in 10 min.

Table 8 shows the results of the polycondensation. The polysulfides having inherent viscosities above 0.5 dl/g were readily obtained from two bis(3-chloroacryloyl)benzenes with or without use of phase transfer catalysts. These activated dichlorides are so highly reactive, almost comparable to ordinary dicarboxylic acid chlorides, that the use of catalysts like DC-18-C-6 was not essential to this type of polycondensation for producing high molecular weight polysulfides XIII.

We took a further step to the phase-transfer catalyzed synthesis of aromatic polysulfide (XIV), which is structurally related to the above mentioned aromatic polyether VII. The polycondensation of bis(4-chloro-3-nitrophenyl) sulfone (VI) with 4,4'-oxybisbenzenethiol (XII) was conducted in a nitrobenzene-water system at room temperature by using various phase transfer catalysts [Eq. (7)].

TABLE 8. Synthesis of Polysulfides XIII in Organic Solvent-Water System[a]

| Dichloride | Solvent | Catalyst | Reaction time (min) | Polymer $\eta_{inh}$[b] (dl/g) |
|---|---|---|---|---|
| XIa | Chloroform | None | 15 | 0.21 |
| XIa | Chloroform | None | 60 | 0.61 |
| XIa | Chloroform | DC-18-C-6 | 10 | 0.62 |
| XIa | CHloroform | TBAC | 60 | 0.72 |
| XIa | Dichloromethane | DC-18-C-6 | 60 | 0.55 |
| XIb | Chloroform | None | 60 | 0.55 |
| XIb | Dichloromethane | None | 15 | 0.42 |
| XIb | Dichloromethane | DC-18-C-6 | 15 | 0.51 |

[a]The polymerization was carried out with 2.5 mmol of XI and XII in the presence of 0.05 mmol of the catalyst in 5 ml of the solvent and 5 ml of 1.01 M aqueous potassium hydroxide at 15°C under nitrogen.

[b]Measured at a concentration of 0.5 g/dl in conc. sulfuric acid at 30°C.

$$\text{VI} + \text{XII} \longrightarrow \left[ \overset{O_2N}{-C_6H_3}-SO_2-\overset{NO_2}{C_6H_3}-S-C_6H_4-O-C_6H_4-S- \right]_n \qquad (7)$$

XIV

Plots of inherent viscosities of the resulting polysulfide as a function of reaction time is shown in Figure 6. The polycondensation proceeded fairly rapidly without use of DC-18-C-6 catalyst at 15°C with precipitation of polymer XIV and was almost completed in 4 hr. In the presence of the catalyst, the initial rate of the polycondensation was highly accelerated and a gradual increase in inherent viscosity of the polymer with reaction time was observed in a solid polymer-nitrobenzene-water three-phase system.

Table 9 presents the effect of catalysts on the polycondensation in a nitrobenzene-water system. All of these phase transfer

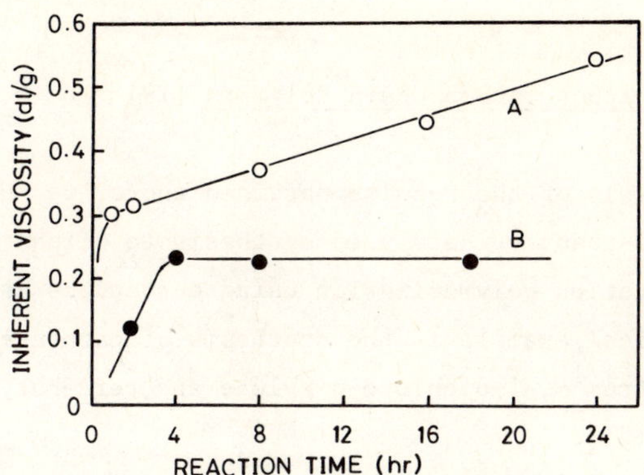


FIG. 6. Polycondensation of VI and XII in nitrobenzene-aqueous KOH system at 15°C with DC-18-C-6 (A) and without the catalyst (B). Reaction conditions, see Table 9.

TABLE 9. Synthesis of Aromatic Polysulfide XIV with Various Catalysts in Nitrobenzene-Water System[a]

| Catalyst | Polymer $\eta_{inh}$ (dl/g)[b] |
|---|---|
| None | 0.16 |
| DC-18-C-6 | 0.54 |
| DB-24-C-8 | 0.51 |
| BTEAC | 0.43 |
| BTPPC | 0.57 |

[a]The polymerization was carried out with 2.5 mmol of VI and XII in the presence of 0.05 mmol of the catalyst in 5 ml of nitrobenzene and 5 ml of 1.01 M aqueous potassium hydroxide at 15°C for 24 hr under nitrogen.

[b]Measured at a concentration of 0.5 g/dl in N-methyl-2-pyrrolidone at 30°C.

catalysts including onium salts and crown ethers were highly effective for producing the aromatic polysulfide with high molecular weights.

Synthesis of Carbon-Carbon Chain Polymers [19]

On the basis of the results obtained above, we successfully extended phase-transfer catalyzed synthesis to aliphatic nucleophilic substitution polymerization using carbanions as a nucleophile. A typical example is the synthesis of carbon-carbon chain polymer (XV) from $\alpha,\alpha'$-dichloro-*p*-xylene and *tert*-butyl cyanoacetate [Eq. (8)].

$$ClCH_2\text{-}C_6H_4\text{-}CH_2Cl \quad + \quad NC\text{-}CH_2\text{-}COO\text{-}C(CH_3)_3$$

$$\longrightarrow \left[ -CH_2\text{-}C_6H_4\text{-}CH_2\text{-}\underset{COOC(CH_3)_3}{\overset{CN}{\underset{|}{\overset{|}{C}}}}- \right]_n \qquad (8)$$

XV

The polycondensation was carried out with a relatively large amount of BTEAC catalyst in a 50% aqueous sodium hydroxide-benzene system at 50°C for 2 hr, and the polymer having an inherent viscosity of 0.7 dl/g was obtained readily. No hydrolysis of *tert*-butyl cyanoacetate or the resulting polymer was observed under these conditions because of a possible steric hindrance of the *tert*-butyl group present. Further studies of this type of polycondensation are now in progress.

## CONCLUSION

The conclusion we reached is that phase-transfer catalyzed polycondensation is broadly applicable to syntheses of various types of condensation polymers by using suitable phase transfer catalysts. A remarkable feature of this method is that it is highly versatile, convenient, and inexpensive (especially in the case of quaternary ammonium salts). Thus, phase-transfer catalyzed synthesis will be extended to much more types of polycondensation and find directly industrial applications in the near future.

## REFERENCES

[1] W. P. Weber and G. W. Gokel, Phase Transfer Catalysis in Organic Synthesis, Springer-Verlag, New York, 1977.

[2] C. M. Starks and C. Liotta, Phase Transfer Catalysis, Principles and Techniques, Academic Press, New York, 1978.

[3] P. W. Morgan, Condensation Polymers by Interfacial and Solution Methods, Interscience, New York, 1965, p. 349.

[4] K. Soga, S. Hosoda, and S. Ikeda, J. Polym. Sci., Polym. Lett. Ed., 15, 611 (1977).

[5] K. Soga, S. Hosoda, and S. Ikeda, J. Polym. Sci., Polym. Chem. Ed., 17, 517 (1979).

[6] Y. Imai, M. Ueda, and M. Ii, Makromol. Chem., 179, 2085 (1978).

[7] Y. Imai, M. Ueda, and M. Ii, Kobunshi Ronbunshu, 35, 807 (1978).

[8] Y. Imai, N. Sato, and M. Ueda, Makromol. Chem., Rapid Commun., 1, 000 (1980).

[9] Y. Imai, M. Ueda, and M. Ii, J. Polym. Sci., Polym. Lett. Ed., 17, 85 (1979).

[10] R. N. Johnson, A. G. Farnham, R. A. Clendinning, W. F. Hale, and C. N. Merriam, J. Polym. Sci., A-1, 5, 2399 (1967).

[11] J. B. Rose, Polymer, 15, 456 (1974).

[12] L. M. Robeson, A. G. Farnham, and J. E. McGrath, Appl. Polym. Symp., 26, 373 (1975).

[13] T. E. Attwood, D. A. Barr, T. King, A. B. Newton, and J. B. Rose, Polymer, 18, 359 (1977).

[14] T. D. N'Guyen and S. Boileau, Polym. Bull., 1, 817 (1979).

[15] Y. Imai, A. Kato, M. Ii, and M. Ueda, J. Polym. Sci., Polym. Lett. Ed., 17, 579 (1979).

[16] Y. Imai, M. Ueda, N. Sakai, and M. Komatsu, Polymer Preprints, Japan, 29, 61 (1980).

[17] Y. Imai, M. Ueda, M. Komatsu, and N. Sakai, Polymer Preprints, Japan, 29, 63 (1980).

[18] C. S. Marvel and A. Kotch, J. Am. Chem. Soc., 73, 481 (1951).

[19] Y. Imai, T. Q. Nguyen, and M. Ueda, to be published.

# Crown Ethers and Phase Transfer Catalysis in Polymer Chemistry

Lon J. Mathias

Department of Chemistry
Auburn University
Auburn University, AL 36830

Abstract: This paper is the first half of a two part review on the applications of crown ethers, cryptands, glymes and various onium ion salts to polymer chemistry. Part 1 surveys the use of these phase transfer catalysts (PTC's) in the synthesis and modification of macromolecules. Applications to ring-opening and vinyl polymerizations generally have involved solubilization and activation by PTC's of organic and inorganic salts which serve as anionic initiators. Change in the polymerization rates, polymer yields and product microstructure often resulted from association of the PTC with the active chain end.

Polymer modification and functionalization reactions have been enhanced by the use of PTC's. Macromolecular substrates have included both soluble and insoluble polymers, although the most popular have been the readily-available Merrifield resins. Generally these two-phase reactions have involved polymer solutions interacting with insoluble reagents or insoluble polymers reacting with the reagent solution. Some of the groups incorporated have been thiols, nitriles, carboxylic acids, esters, and crown ether analogs themselves. Most of this work has aimed at synthesis of polymer derivatives which can serve as chelating agents, catalysts and reagents in subsequent reactions. In almost all cases described, phase transfer catalysis offers definate advantages in selectivity and yield over conventional methods. The outline below gives the general subject areas of both Part 1 and Part 2; the latter will be published shortly.

PART I



PART II



## I. INTRODUCTION

The two parts of this review are intended to provide a reasonably comprehensive coverage of a small part of a field which has been enjoying explosive growth over the last ten years. In general this field includes applications of phase transfer catalysts (PTC's) to synthetic organic chemistry. Of course, other areas investigated include, for example, synthesis of new PTC's; elucidation of fundamentals of metal complexation and phase transfer processes; and comparison of synthetic and natural membrane transport agents. Some of these and other areas of present and potential interest will be described in appropriate sections in the body of the text.

The generally accepted usage of the term "phase transfer catalysis" implies transport of an organic or inorganic salt from a solid or aqueous phase into an organic liquid phase where reaction with an organic-soluble substrate occurs. Often, the reaction could also take place in single-phase polar solvents or solvent mixtures but with less facility or much reduced yields. The activity of the anionic residue (generally in addition or substitution reactions) is enhanced in the nonpolar organic environment. The term "naked anion" describes the reduced or absent solvent sphere around the anion under PTC conditions, a situtation which greatly affects reactivity and selectivity. While the use of the designation "phase transfer catalyst" may not be completely accurate since the PTC's are generally not involved in the actual reaction sequence it is nontheless descriptively appropriate.

There are two basic types of PTC's including crown ethers and analogs on the one hand, and onium ion salts on the other. The former include compounds such as 18-crown-6 1, dibenzo-18-crown-6 2, dicyclohexano-18-crown-6 3; cryptands such as the [2.2.2] compound 4; and terminally capped and uncapped oligooxyethylenes 5. These species function by surrounding a metal cation (usually $K^+$ or $Na^+$) with a number of inward-facing oxygen or nitrogen electron lone pairs while at the same time providing an outer hydrocarbon coating which enhances solubility in a wide range of organic solvents. Furthermore, tight complexation of the cation by the chelating agents promotes ion separation and enhanced anion activity.

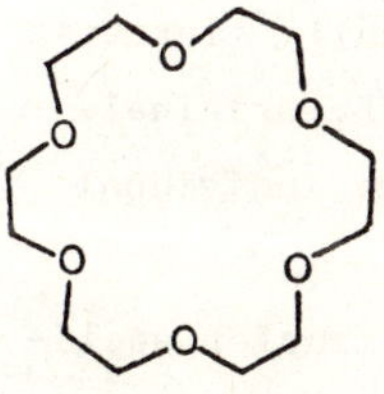

1 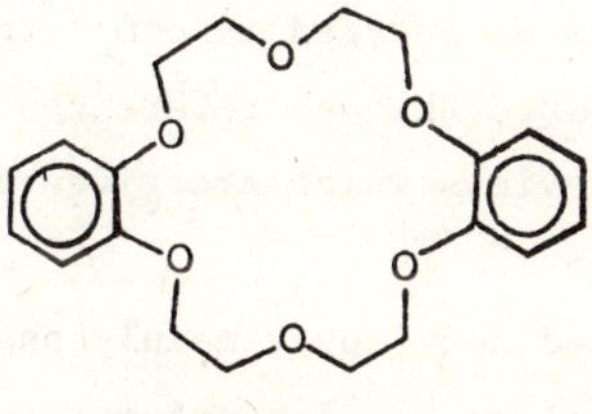 2 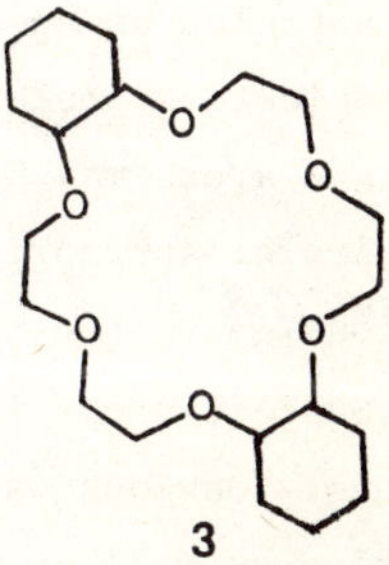 3

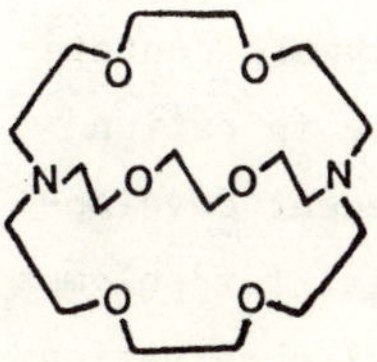

4

$RO(CH_2CH_2O)_nR$

$R = H, CH_3$

5

$(n\text{-butyl})_4N^+X^-$

6

$[CH_3(CH_2)_{7-9}]_3N^+CH_3X^-$

7

$R_4P^+X^-$

8

The onium ion PTC's include the quantenary ammonium salts 6 and 7 (Adogen® 464) as well as a wide variety of phosphonium salts 8. These salts undergo ion exchange with the reagent salt in the aqueous phase or at the solid phase interface. Organic solubility results from medium to long alkyl groups on the onium cation, while the inherent nature of newly formed salts insures the ion separation necessary for increased anion activity.

A great many review articles and books have appeared recently on the synthesis of various types of PTC's [1,2] and their application in organic synthesis. [3,4] Nomenclature and complexation constants have been adequately covered in these and other[5] references and will not be mentioned here. The nature of the activity of these compounds and the advantages of their use in synthesis made it inevitable that they be applied to polymeric systems. In our previous short survey of the use of crown ethers in polymer chemistry[6], we gave representative examples of several important areas. The present review is intended to be as comprehensive as possible and includes work published through the middle of 1980. The breadth of this field, however, and the availability of several review articles on specific aspects of the field requires that not all subjects or references be covered in complete detail. Interest in these areas may be pursued through references to the original articles or appropriate reviews which are given in the individual sections.

Subject areas included here cover mainly phase transfer applications. Obvious uses include PTC for polymer synthesis and modification as well as applying polymer-pendant PTC's to organic synthesis in general. Work is also described which deals with linear polymers (eg., polyoxyethylene) as PTC's and with polymers containing crown ethers and cryptands in the backbone for use in cation chromatography and separation. Additionally, some recent developments in crown ether-related areas which have potential for polymer applications are also included. Finally, some general observations and conclusions are given dealing with the unique properties arising from the combination of PTC's with polymers. Hopefully this review will provide sufficient background for impetus and development of new applications of this increasingly important area.

## II. POLYMER SYNTHESIS WITH PHASE TRANSFER CATALYSTS (PTC's)

### A. Ring-opening and Vinyl Polymerization

In our previous review,[6] we described several applications of crown ethers and cryptands in addition, ring-opening and condensation polymerizations. For chain growth systems, increased and modified anion activity was related to cation complexation by the macrocycles. Solubilization of the polar salts in non-polar solvents and enhanced ion separation were apparently the controlling factors. Examples of ring-opening polymerizations described[6] included β-propiolactone with sodium acetate and 2, and 2-pyrrolidone with 4 and the potassium salt of the monomer. Butadiene homopolymerization and copolymerization with styrene were carried out in hexane with n-butyl sodium and 3 to give higher molecular weight products with higher conversion than without 3.

A more complete examination of the β-propiolactone polymerization has recently appeared.[7] Dichloromethane was used as solvent with an initiator system of dibenzo-18-crown-6 (2) plus potassium acetate. Slomkowski and Penczek found that increasing the crown-to-salt ratio led to a linear increase in the rate of polymerization up to a maximum for a 1:1 ratio indicating essential complete complexation of the potassium acetate. A linear dependence of $\overline{Mn}$ with conversion led to the conclusion that no termination or transfer processes were involved in this polymerization; ie, this is a "living" anionic system. Determination of dissociation constants and extensive kinetic analysis indicated that the propagating carboxylate anion and the crown-complexed potassium form a loose ion pair in $CH_2Cl_2$. The individual rate constant for propagation of the free anion was always greater than that of the ion pair, varying from a ratio of 150 at 35° to 5.6 at -20°. This work clearly indicates the activating influence of the crown ether in promoting loose and free ion pairs of the propagating polymer anion.

In our earlier review we failed to include several pioneering papers on the use of crown ethers and cryptands to initiate anionic alkene polymerizations via complexation with alkalai metals and alkalai salts. In the first of these, for example, sytrene was polymerized quantitatively in benzene or THF using metallic lithium and the [2.1.1] cryptand analogous to 4 [8]. The polymerization was completed in seconds to give $\overline{M}w$ in the millions. Similarly rapid reaction of methyl methacrylate in THF with lithium, sodium or potassium-naphthalene and an appropriate cryptand gave a highly syndiotactic polymer again with $\overline{M}n$ in the millions.[8] Cryptates (cryptand complexes) were also observed to be exceptionally active initiators for the polymerization of isoprene, propylene sulfide and ethylene oxide.[8] A continuation of this work reported on the ready formation of complexes of alkalai sodium-through-cesium metals which were soluble in relatively poor solvents (for these species) such as benzene, toluene, THF, dioxane and tetrahydropyran.[9] These complexes readily formed the dianions of α-methylstyrene and 1,1-diphenylethylene which initiated the generally instantaneous polymerizations of butadiene, isoprene, styrene, methyl methacrylate and 2-vinylpyridine. Even normally inactive alkalai salts were found to be capable of initiating polymerizations when complexed with cryptands.[10] In benzene, for example, the 4 cryptates of KOH and sodium t-amylate respectively polymerized hexamethylcyclotrisiloxane and styrene in good yields. Propylene sulfide and isobutylene sulfide were initiated with the cryptates of KCNS and potassium t-butoxide, respectively.[10] Detailed analysis of the kinetics of the cyrptate-promoted polymerizations of propylene sulfide[11] and ethylene oxide[12] have also been reported by Boileau and coworkers.

Two more recent reports have appeared on the effects of crown ethers on anionic polymerizations. The first of these[13] extends earlier work on the polyaddition of methyl methacrylate initiated with lithium salts in the presence of various cryptands and crown ethers (see references in [13]). The present paper employed other alkalai metals directly as initiators. In the presence of dicyclohexano-18-crown-6 (3) the four metals Na, K, Rb and Cs all readily dissolved in nonpolar solvents such as benzene and toluene. ESR evidence for the potassium solutions at -70° indicated that dissolution actually involved formation of the crown-complexed metal cation and the radical anion of benzene or toluene. Polymerizations were carried out at -78° or from 0° to 25° and generally gave complete conversion in less than a minute. Relatively broad molecular weight distributions were obtained with molecular weights in the range 3-8 x $10^5$ daltons. NMR analysis of the polymer microstructures indicated predominantly syndiotactic products with essentially Bernoullian statistics. Slight increases in the syndiotactic content with the metal used were observed with the order being $Cs^+ < Rb^+ < K^+ < Na^+$. Yields and tacticities were comparable to those obtained with lithium in dioxane and sodium in benzene plus the [2.2.2] cryptand.

A second paper from the same research group[14] described the use of the same four metals for polymerization of butadiene and isoprene. Both the crown ether 3 and the cryptand 4 were employed with benzene and THF as solvents to give homogenous initiator solutions. Polymerizations were carried out at 10° with benzene and -78° with THF. Reactions were complete in one minute with yields of 99-100%. With isoprene in benzene plus 3, small differences in the polymer microstructure were observed with the various metals. The 1,4-, 3,4- and 1,2-contents went from 21:60:19, respectively, for potassium to 42:41:17 for cesium. Much less structural variation was observed in THF although the 1,4-residues decreased significantly with the contents being 12-18:46-50:30-40, respectively.

For butadiene polymerization in benzene with 3, the 1,2-content increased in the order Cs < Rb < K. When the cryptand was employed, however, all four metals gave essentailly the same polymer with a 72-74% 1,2-structure in benzene and 79-86% 1,2-content in THF. While the use of the cryptand rather than the crown ether increased the 1,2-content for all the metals used, the change for potassium was small in both benzene and THF. This plus the leveling effect observed for all four cryptates may be directly attributed to the thermodynamics of metal complexation. The crown ether forms a stronger complex with the potassium cation (which fits exceptionally well into the crown cavity) then with the other larger or smaller alkalai ions. The cryptand, however, forms relatively tight complexes with all four cations mainly because of its bicyclic structure and additional chelating groups.

The butadiene results may be compared to an earlier report[15] in which polymerizations were carried out in hexane, generally at 30°C, with n-butyl sodium and dicyclohexano-18-crown-6 (3) as the cocatalysts. Under these conditions but without the crown ether, butadiene gave a low molecular weight polymer in low conversion. With crown ether present, however, high conversion and high molecular weight were obtained: eg, with inherent viscosities of 1.7-5.2 dl/g. Conversion was somewhat related to the ratio of initiator to crown ether and increased to 95% at higher crown ratios. These changes were attributed to a decrease in the termination reaction brought about through stabilization of the chain end by the crown ether. Of additional importance was a slight increase (from 68-80%) of the vinyl content of the polymer with a decreasing sodium-to-crown ether ratio. The 1,2-percentages were comparable to those obtained in benzene under similar conditions with K, Rb and Cs.

An interesting application of PTC to the synthesis of a block copolymer was based on an initial attempt by Suzuki and coworkers to synthesise block copolymers of oxyethylene and methyl methacrylate.[16] Using the disodium salt of polyoxyethylene (POE) to initiate the anionic polymerization of the vinyl monomer led to products with broad and multimodal molecular weight distributions. Transesterification of the POE terminal anion with the methyl ester was apparantly partially responsible. A more active POE derivative was then synthesized and combined with crown ether and cryptand catalysis.[17] Thus, the POE 1000 (Mn=980) was converted to the corresponding diamine (9):

$$\underset{\overline{M}n=980}{\text{PEO 1000}} \xrightarrow[\text{dry pyridine}]{\text{TsCl}} TsOCH_2CH_2\text{-}(OCH_2CH_2)_n\text{-}OTs \xrightarrow[PhCH_2NH_2]{\text{excess}}$$

several days at 0-10°C      several days at 30°C

$$\underset{\underset{\mathbf{9}}{Mn=1140}}{PhCH_2NHCH_2CH_2\text{-}(OCH_2CH_2)_n\text{-}NHCH_2Ph} \xrightarrow[\text{2 or 4}]{\text{NaNaph.}} \xrightarrow[30^{o}]{\text{MMA}}$$

$$\text{-}[\overset{CH_2Ph}{\overset{|}{N}}\text{-}CH_2CH_2(OCH_2CH_2)_n\text{-}\overset{CH_2Ph}{\overset{|}{N}}]\text{—}[CH_2\text{-}\underset{CO_2CH_3}{\underset{|}{\overset{CH_3}{\overset{|}{C}}}}]_m$$

The secondary amine groups were then converted to the anions with sodium napthalene in tetrahydrofuran. Either dibenzo-18-crown-6 (2) or the cryptand 4 was added followed by the methyl methacrylate and the polymerization then carried out at 30°. Product molecular weights and molecular weight distributions were determined with comparative GPC and $^1$H NMR. Comparison of these values with those of a polymer prepared without a PTC indicated a slight increase in yield with the crown ether and a significant skewing of the MWD to higher values with the cryptand catalyst. The unusual behavior of the polymers in the GPC analysis and somewhat confusing results presented make it difficult to draw any firm conclusions on the effect of the PTC on the course of this reaction. Hopefully, the full paper will clarify these results.

B. Condensation Polymerizations

Several new polycondensations have appeared since the previous survey. A new synthetic route to polycarbonates discribed by Soga and coworkers involves the reaction of α,w-dibromides with potassium carbonate in various organic solvents carried out in the presence of several crown ethers.[18] With p-xylylene dibromide (Figure 1, upper path) the effects of crown ether, solvent, temperature and time on polymer yield and molecular weight were all examined. Diglyme as solvent with 18-crown-6 gave the best yields (50-60%) for high molecular weight products (∿ 12,000 daltons). Polyoxyethylenes of $\overline{DP}$= 4 and 25 were not efficient catalysts. The rate of polymerization increased with temperature to a maximum at 100-120° and then decreased slightly. At 105° the molecular weight gradually increased with time to an apparantly asymptotic value at 48 hours. Surprisingly, polymer prepared at 160° gave evidence of polyether incorporation, presumedly through crown ether ring-opening and incorporation into the backbone. The ratio of catalyst to $K_2CO_3$ was found to be important with a value of 1:10 giving the best yield and molecular weight values.

Similar reaction conditions were applied to a series of α,ω-dibromoalkanes (Figure 1, lower path). For the ethane and propane derivatives, only cyclic carbonates with 5- and 6- membered rings, respectively, were obtained. With several higher alkanes, reasonable yields of high molecular weight (> 20,000 daltons) products were obtained. In genreral, this method provides an

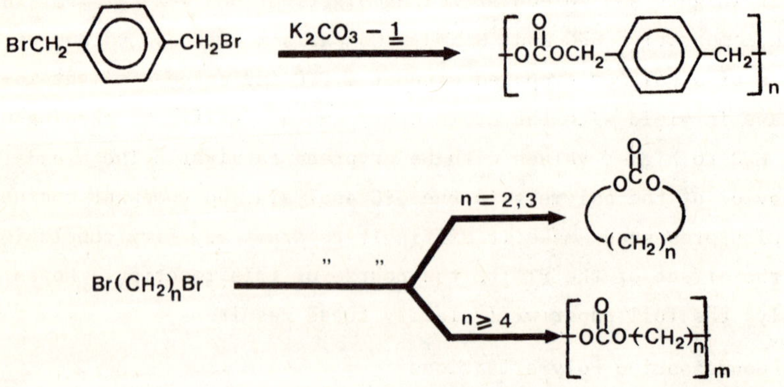

FIG. 1.

excellant alternative to previously avaliable techniques, and complements the use of $CO_2$ rather than $K_2CO_3$ as the carbonate source in the PTC synthesis of polycarbonates described earlier by Soga, et. al.,[19].

Soga's group has also reported a similar procedure for synthesis of a polythiocarbonate.[20] The polycondensation involved α-α'-dibromo-p-xylene with $CS_2$ and the dipotassium salt of 1,4-cyclohexanediol. The reactions were carried out neat or in dioxane, benzene, THF, DMF or acetone solvents with 18-crown-6 as the PTC. At temperatures above 0°, a copolymer was obtained (Figure 2) which resulted from direct attack of the alcoholate anion on the halide along with the desired intermediate formation and attack of the dithiocarbonate. At -78°, however, exclusive formation of the polydithiocarbonate (Figure 2) was observed when $CS_2$ was employed as solvent as well as reagent. Variation in reaction time and PTC content did not markedly affect polymer yield or molecular weight. Values of $\overline{Mn}$ were found to be in the range 1300-2300 daltons. The use of PTC in this case might be best viewed as a means of obtaining oligomers and prepolymers.

Two virtually simultaneous reports have appeared on the use of PTC to synthesize polysulfide elastomers. Imai's group investigated the condensation of, for example, 1,4-dibromobutane with 1,6-hexanedithiol (Figure 3).[21] While polymer was obtained without catalysts, the rate of reaction and inherent viscosities were significantly increased with PTC's like dibenzo-18-crown-6 and tetrabutylammonium chloride (6). The highest viscosities were

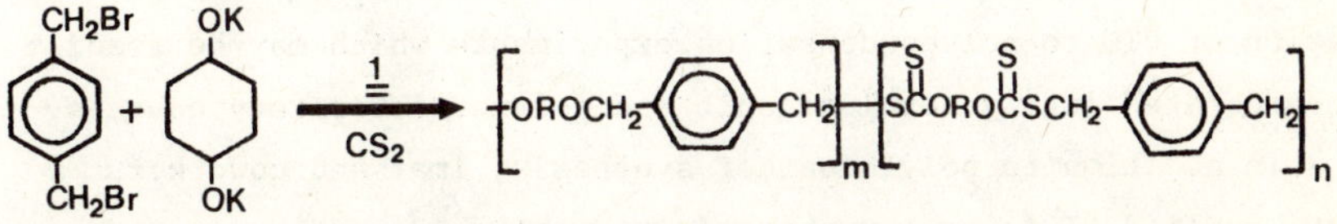

CH₂Br OK
+ —1/CS₂→ [OROCH₂—C₆H₄—CH₂]m [SC(S)OROC(S)SCH₂—C₆H₄—CH₂]n
CH₂Br OK

R = Cyclohexyl; m = 0 at -78°

FIG. 2.

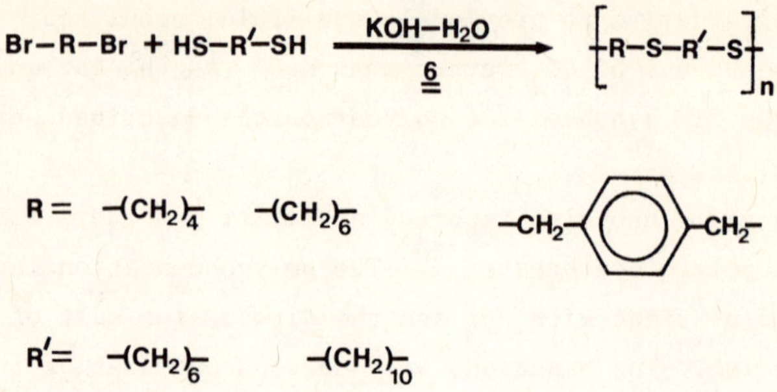

$$\text{Br-R-Br} + \text{HS-R'-SH} \xrightarrow[\underline{\underline{6}}]{\text{KOH-H}_2\text{O}} \left[\text{R-S-R'-S}\right]_n$$

$$\text{R} = \text{-(CH}_2\text{)}_4\text{-} \quad \text{-(CH}_2\text{)}_6\text{-} \quad \text{-CH}_2\text{-C}_6\text{H}_4\text{-CH}_2\text{-}$$

$$\text{R'} = \text{-(CH}_2\text{)}_6\text{-} \quad \text{-(CH}_2\text{)}_{10}\text{-}$$

FIG. 3.

obtained at 80° and in the absence of solvent. Yields were essentially quantitative for this polymer and for the five related polysulfides described.

Petitt has used tetrabutylammonium hydroxide to enhance the rate and yield of the well-known polymer obtained from sodium polysulfide (sulfur dissolved in 10-15% aqueous sodium hydroxide) and 1,2-dichloroethane.[22,23] With or without dichloromethane as solvent for the latter, the polymerization was complete in 5 minutes at reflux temperature with the catalyst, while 45 minutes was required in the absence of the PTC. Similar reaction with 2,3-dibromopropanol and 1,2-dibromostyrene required 12-15 minutes for complete reaction. As expected, reaction of chloroform with the sodium polysulfide (perhaps via the dichlorocarbene) occurred with dichloromethane as the solvent in 30-40 minutes. The yield of polymer was somewhat low and the chlorine content much less than that expected from the suggested reaction pathway. In any event, this procedure provides a convenient demonstration of the application of PTC to polycondensation experiments which may be readily incorporated into general organic and polymer laboratory courses.

In addition to polythioether synthesis, Imai and coworkers have applied PTC's to a number of other polycondensation systems (see reference in this collection). Our earlier survey described aromatic polyether formation via condensation of bis(4-chloro-3-

nitrophenyl)sulfone and 2,2-bis(4-hydroxyphenyl)propane (Figure 4, upper path). The use of crown ethers 2 and 3 allowed essentially quantitative conversion at much lower temperatures (27°) with a $CH_2Cl_2$-KOH-$H_2O$ mixture.[24] This report also contained what is probably the first application of linear polyoxyethylene as a PTC in polymer synthesis, a subject discussed more throughly in a later section of this review.

An earlier paper from this group was concerned with the synthesis of an aromatic polysulfonate from a disulfonyl chloride and bis-phenol A (Figure 4, lower path).[25] Interfacial condensation in the absence of a PTC led to a limiting [n] of 0.2 dl/g. A series of crown ether PTC's including 1 - 3 and various others gave [η] values greater than 0.8 dl/g. With all of the PTC's, KOH gave better results than NaOH. With KOH and 3, for example, polymer with [η] of 1.48 dl/g was obtained in quantitative yield after 3 hours at 20°C.

The above examples along with others in this symposium collection and those appearing continually in the literature indicate the potential that PTC offers in the synthesis of chain- and step-growth polymers. Modification of vinyl and diene polymer microstructure on the one hand and enhancement of the rate and/or yield of ring-opening polymerizations and polycondensations on the other are some of the more obvious advantages. The example included here

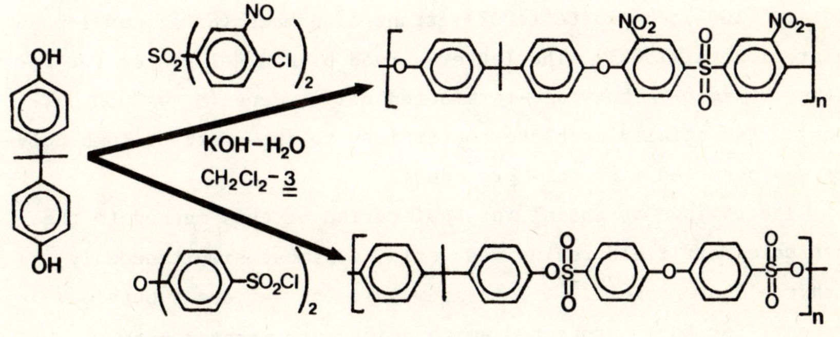

FIG. 4.

of the synthesis of a block copolymer using PTC indicates some further areas of application among many which should be explored; eg, anionic random and graft copolymerizations. This area is one which should enjoy enormous growth over the next few years.

## III. POLYMER MODIFICATION AND FUNCTIONALIZATION

The second main area of PTC applications to polymers is one that has recently enjoyed extensive investigation. It involves the chemical modification of functional polymers using two-and three-phase systems. The former may involve liquid-liquid or liquid-solid reactions while the latter generally involves two liquid phases interacting with an insoluble polymer. This area was not covered in our earlier survey although several reports had appeared in 1976 and 1977.

Roovers published what seems to be the earliest report on the use of PTC in polymer modifications in 1976.[26] The substrate polymer was a soluble, chloromethylated polystyrene with an $\overline{M}n$ of 143,000 daltons. Reactions were generally carried in 50:50 benzene-acetonitrile at 75° with either dicyclohexano-18-crown-6 (3) or dibenzo-18-crown-6 (2) as the PTC. The former was about twice as efficient as the latter under these conditions. The nucleophiles employed included a variety of potassium carboxylates which formed the corresponding bound esters (Figure 5). The following acid anions were used (with yields or % incorporation in parentheses): acetate (100%); benzoate (100%); trans-cinnimate (87%); and 1-phenylpropionate (100%). The latter served as a model system for reaction of mono-carboxylate-terminated polystyrene ($\overline{M}_n$ = 3000 daltons). Essentially complete conversion to the graft or comb-polymer was observed with this procedure.

The obvious potential for application of this method to the insoluble Merrifield resins was verified almost simultaneously with the above report.[27] Potassium salts of various t-butyloxycarbonyl-(or Boc)- protected amino acids were reacted with the resins using 18-crown-6 as the PTC. For example, at 100°

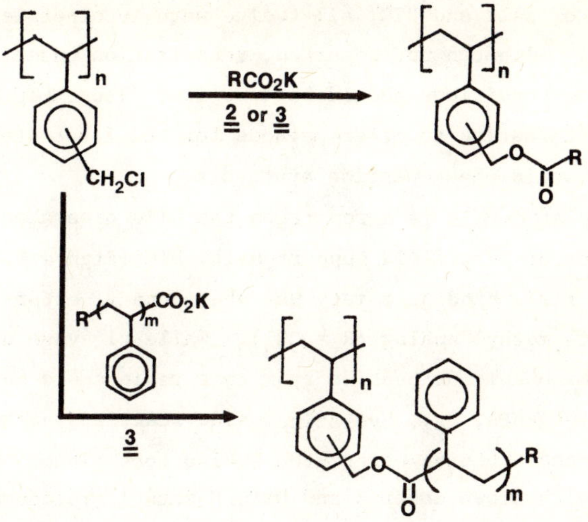

FIG. 5.

and with one equivalent of PTC essentially complete incorporation of Boc-leucine was observed in several solvent systems (Figure 6, upper path). At lower temperatures, however, only dimethylformamide (DMF) gave good results. At room temperature, excess catalyst and amino acid salt were required for complete conversion in 18 hours. With the latter conditions only five of the twelve amino acids examined gave >90% incorporation. At 50° with 1.2

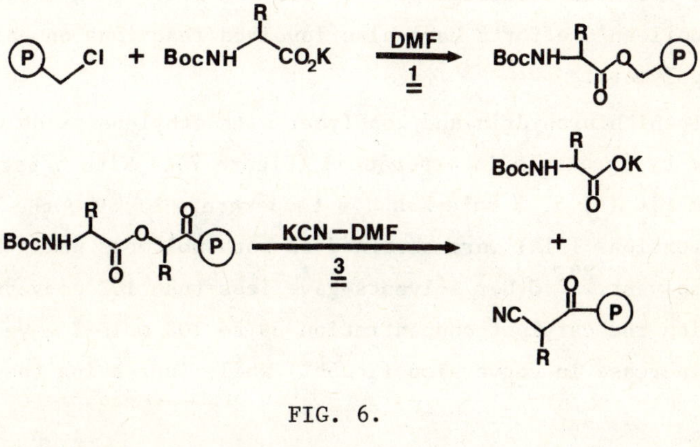

FIG. 6.

equivalents of salt and PTC, all twelve were incorporated at >89% in 18 hours. Although not reported, recemization under these relatively mild conditions should be very low. Thus, PTC offers an excellant alternative to other methods for the first step of the traditional solid-phase peptide synthesis.

In a related vein is a report on the mild cleavage of peptides from a different Merrifield support using PTC (Figure 6, lower path). The resin binding moiety was the phenacyl ester group (R = H) or its methyl analog (R = $CH_3$). While cleavage of the protected amino acid or peptide from this resin could be effected with other reagents, eg., hydrazine, side reactions such as cleavage of side-chain benzyl ester groups also took place. However, use of KCN with crown ether 3 and DMF, N-methylprrolidone or acetonitrile gave clean and high-yield removal of the peptide or amino acid derivative. For exampe, removal of a heptapeptide took place in 86-93% yield. It is somewhat surprising that to the best of our knowledge the use of PTC for solid-phase peptide synthesis and cleavage has not been further investigated.

Two other groups have been investing much effort in the exploration of PTC applications to polymer synthesis and modifications. Boileau in France and Fréchet in Canada along with their respective coworkers have each published several reports in recent years including one in this symposium collection. Again the readily available and well-characterized chloromethylated polystyrene provided the main substrate polymer for the modification work although Boileau's efforts have also involved reactions on epichlorhydrin polymers.

Polyepichlorohydrin and copolymer with ethylene oxide were modified by carbazole incorporation (Figure 7). With a tetrabutylammonium PTC (6) at 5 mole-% and a temperature of 60°, the highest conversions (31%) were obtained on both polymers using DMF as the solvent.[29] Other solvents gave less than 10% conversion. Increasing the catylyst concentration up to 100 mole-% gave only a modest increase in conversion (to 56%) while increasing the re-

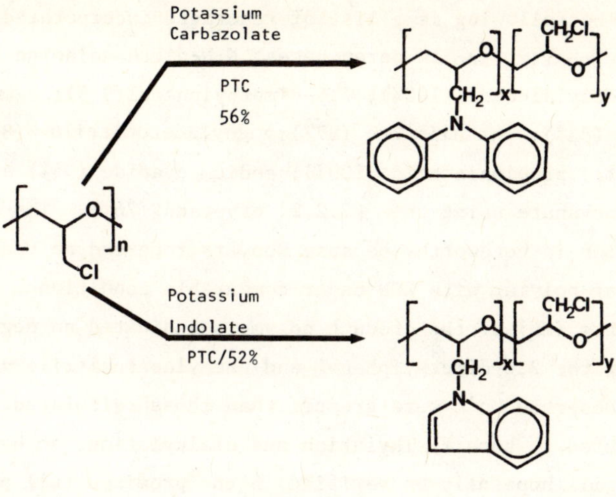

FIG. 7.

action time beyond 5 hours did not markedly effect the incorporation. Comparable degrees of substitution could be obtained with only 5 mole-% of either dicyclohexano-18-crown-6 (3) or the [2.2.2] cryptand 4 indicating a higher catalytic efficiency for these compounds. With a soluble chloromethylated polystyrene, however, essentially complete substitution by carbazole was observed even with the TBA catalyst. A commercial Merrifield resin gave similar results. 2,3-Dimethylindole and a variety of alkyl amines were reacted with the polyepichlorohydrin with poor results (<10% incorporation) for the latter and good incorporation (52%) for the former (Figure 7). With the soluble chloromethylated polystyrene, high conversions (>65%) were obtained with both alkyl and heterocyclic amines.

Several reactions on the soluble chloromethylated polystyrene were further explored subsequently by N'Guyen and Boileau.[30] The polymer (dissolved in DMF) was treated at 60° with 1.5 equivalents of the reactant in aqueous 50% NaOH. The PTC was tetrabutylammonium hydrogensulfate at 5 mole-% and reaction times varied from 0.75 to

3 hours. The following is a list of reactants incorporated along with the % substitution in parentheses: N,N-diethanolamine (75%); p-methoxybenzylalcohol (100%); 2,6-dimethylphenol(93%); 2-mercaptopyridine (82%); phthalimide (87%);phenylacetonitrile (82%); indene (70%); propionic acid (100%); sodium cyanide (95%) and sodium thiocyanate using the [2.2.2] cryptand (76%). The cyanide incorporation is noteworthy because Roovers reported no reaction of a similar polymer with KCN under comparable conditions. Molecular weights ($\overline{Mn}$) of the product polymers indicated no degradation. Indeed, for the 2,6-dimethylphenol and phenylacetonitrile derivatives the observed $\overline{Mn}$'s were greater than those calculated. This was attributed to some C-alkylation and dialkylation, an hypothesis which will hopefully be verified in the promised full paper.

Fréchet's work has been more extensive and has involved some important basic considerations. For example, one of his group's publications in this area dealt with reaction of a difunctional nucleophile, 1,4-butanedithiol, with a lightly-cross-linked chloromethylated polystyrene.[31] In both the presence and absence of a PTC, a high degree of double coupling (ie., further cross-linking) with the dithiol was observed (up to 87%). Only when a large excess of dithiol and base were used (2.6:1.0) with respect to the chloromethyl concentration (0.15 relative ratio) was single coupling obtained in high degree (95%). This work (along with numerous related investigations of site-site interactions in cross-linked polymers) has important consequence to the use of polymers as protecting-groups and anchors for multifunctional compounds.

Three-phase catalysis has been applied to the generation of polymer-bound sulfonium ylids which were then used to form epoxides (Figure 8)[32]. The dimethyl or diethyl sulfonium intermediate was converted to the ylid and used in epoxide synthesis using DMSO as solvent. After two reaction cycles the polymeric reagent had lost all activity. However, when three-phase conditions were used with tetrabutylammonium salts as the PTC, no loss of activity was observed even after five repetitions. The ease of recovery

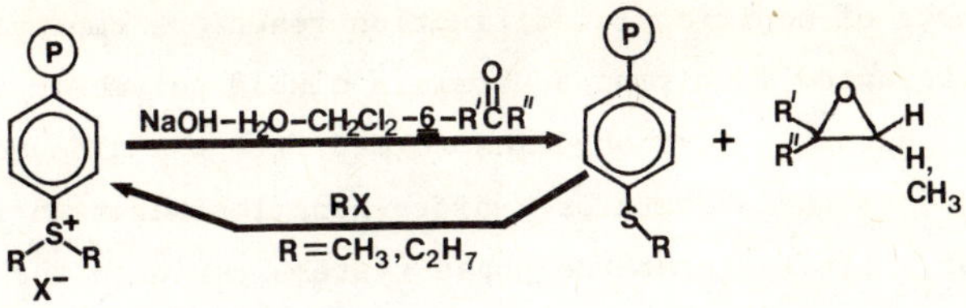

$NaOH-H_2O-CH_2Cl_2-\underline{6}-R'C(O)R''$

RX

$R=CH_3,C_2H_7$

FIG. 8.

and lack of odor of these polymeric reagents coupled with the higher yields of products makes this an excellant system for these and related reactions.

A wide range of nucleophilic substitutions have been reported by Fréchet and coworkers on cross-linked polystyrene derivatives, with the chloromethylated compound being the most heavily investigated.[33] Some of the more valuable conversions are outlined in Figure 9. Solvents (or swelling agents) included o-dichlorobenzene, 1,2-dichloroethane, dichloromethane and benzene while the PTC's were quantenary ammonium salts $\underline{\underline{6}}$ and $\underline{\underline{7}}$. Conversions in most cases were greater than 95%. A three-phase system was also used to convert the chloromethyl groups to thiomethyl groups with excellant conversion, although an inert atmosphere was required to eliminate site-site disulfide formation.

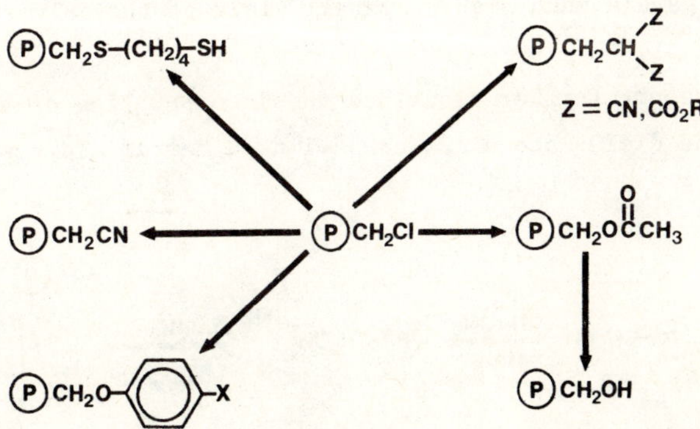


FIG. 9.

A number of additional modification reactions employing PTC's were also reported involving several insoluble polymeric aldehydes, thiols and phenols.[33] Cyanohydrins formed from the aldehyde groups were trapped as the acetate or p-nitro-benzyl ether. An interesting reversal of activity for three-phase systems employed the polymeric thiols and phenols as the nucleophiles. Reaction with various alkyl halides led to ethers and thio ethers in excellant yields.

A subsequent report of the conversion of chloromethylated polystyrene to dinitriles and diamines[34] was later complemented by an extensive investigation of similar compounds by another group.[35] Fréchet and his collegues treated insoluble chloromethylated polystyrene with the anion of malonitrile under three-phase conditions to give the polymer-bound dinitrile in essentially complete conversion (Figure 10). When p-cyanomethylated polystyrene was used as the nucleophile with chloroacetonitrile under similar conditions, conversion to the isomeric dinitrile occurred in 88% yield. Both of these derivatives were quantitatively reduced to the corresponding diamines with $LiAlH_4$.[34]

Ricard and co-workers reported a slightly different procedure for incorporation of the 1,3-diaminopropane group (Figure 11, upper path). Acylation of the polystyrene with chloroacetyl chloride was followed by a three-phase reaction with malononitrile directly or after initial reduction of the ketone group to an alcohol. The former gave a much higher overall yield of the polymer-bound diaminoalcohol.[35]

This report further described the incorporation of various diamines and diols into a 12% cross-linked Merrifield resin. Again

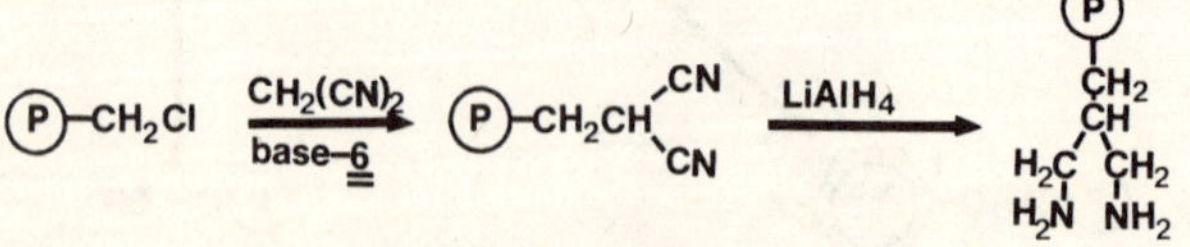

$$\text{(P)}-CH_2Cl \xrightarrow[\text{base-}\underline{\underline{6}}]{CH_2(CN)_2} \text{(P)}-CH_2CH(CN)_2 \xrightarrow{LiAlH_4} \text{(P)}-CH_2CH(CH_2NH_2)_2$$

FIG. 10.

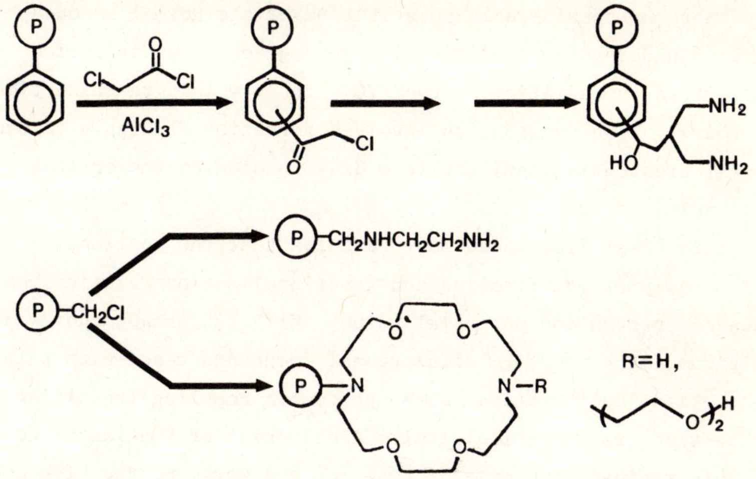

FIG. 11.

using a three-phase system, the chloromethyl groups were converted to various aminomethyl derivatives (Figure 11, middle path). Other diamines examined included 1,2-diaminobenzene and 1,8-diaminonaphthalene. Despite a large excess of diamines, the low yields and the probability of double-coupling indicate a need for more efficient reaction conditions for these modification procedures, perhaps through the use of PTC's other than the benzyl trimethylammonium salt employed here. Nonetheless, the two polymer-bound crown ether derivatives (Figure 11, lower path) did display activity as PTC's and transition metal chelates. This work is reportedly under further development.

An initial communication recently appeared[36] which portends further developments in polymer modification. Poly(vinyl chloride) was allowed to react in THF with solid potassium acetate and 18-crown-6 (1). The reaction was carried out at 46° for varying times and with a range of crown ether concentrations. Up to 7% chlorine replacement by acetoxy groups (measured by quantitative IR) was observed although the "labile" chlorine content of PVC is usually considered to be about 0.1%.[36] Although apparantly there was considerable substitution of the normal secondary chlorines under these conditions, high degrees of substitution were limited by degration of the PVC with very long reaction times (>70 hours). The degradation involved formation of highly colored material presumedly resulting from dehydrohalogenation or loss of acetic acid.

It is clear from these many examples that the application of PTC's to polymer modification and functionalization reactions has tremendous proven and potential value. With the proper choice of conditions, many mono- and difunctional compounds react with relatively clean incorporation, although double coupling (resulting in site-site cross-linking) limits application of the latter to insoluble resins. The ease of reaction and work-up, the high conversions, the apparent generality of these two-phase and three-phase modifications should make this the method of choice for many such reactions.

## REFERENCES

1. C. J. Pederson and K. H. Frensdorff, Angew. Chem. Int. Ed. Eng., 11, 16 (1972).
2. J. J. Christensen, D. J. Eatough and R. M. Izatt, Chem. Rev., 74, 351 (1974).
3. E. V. Dehmlow, Angew. Chem. Int. Ed. Eng., 16, 493 (1977).
4. W. P. Weber and G. W. Gokel, J. Chem. Ed., 55, 350 and 429 (1978).
5. R. M. Izatt and J. J. Christensen, "Progress in Macrocyclic Chemistry," Vol. 1, Wiley-Interscience, New York, N. Y., 1979.
6. L. J. Mathias and K. B. Al-Jumah, Polym. News, 6, 9 (1979).
7. S. Slomkowski and S. Penczek, Macromolecules, 13, 229 (1980).
8. S. Boileau, B. Kaempf, J. M. Lehn and F. Schué, J. Polym. Sci.- Polym. Lett. Ed., 12, 203 (1974).
9. S. Boileau, B. Kaempf, S. Raynal, J. Lacoste, and F. Schué, Ibid., 12, 211 (1974).
10. S. Boileau, P. Hemery, B. Kaempf, F. Schué and M. Viguier, Ibid., 12 217 (1974).
11. P. Hemery, V. Warzelhan and S. Boileau, Polymer, 21,77 (1980).

12. D. Lassalle, S. Boileau and P. Sigwalt, Eur. Polym. J., 13, 587, 599 (1977).

13. S. Alev, A. Collet, M. Viguier and F. Schué, J. Polym. Sci.-Polym. Chem. Ed., 18, 1155 (1980).

14. S. Alev, A. Collet, M. Viguier and F. Schué, Ibid., 18, 1163, (1980).

15. T. C. Cheng and A. F. Halasu, Ibid., 14, 583 (1976).

16. T. Suzuki, Y. Murakami, Y. Tsui, and Y. Takegami, J. Polym. Sci.-Polym. Lett. Ed., 14 ,675 (1926).

17. T. Suzuki, Y. Murakami and Y. Takegami, Ibid., 17, 241 (1979).

18. K. Soga, S. Hosoda and S. Ikeda, J. Polym. Sci.-Polym. Chem. Ed. , 17, 517 (1979).

19. K. Soga, Y. Toshida, S, Hosada and S. Ikeda, Makromol . Chem. 178, 2747 (1977) and 179, 2379 (1978).

20. K. Soga, Y. Toshida, I. Hattori, K. Nagata and S. Ikeda, Ibid., 181, 979 (1980).

21. Y. Imai , A. Kato, M. Ii and M. Udea, J. Polym. Sci.-Polym. Lett. Ed., 17, 579 (1979).

22. G. R. Pettit and G. R. Pettit III, J. Chem. Ed., 55, 472 (1978).

23. G. R. Pettit, J. Polym. Sci. - Polym. Chem. Ed., 18, 345 (1980).

24. Y. Imai, M. Ueda and M. Ii, J. Polym. Sci.-Polym. Lett. Ed., 17, 85 (1979).

25. Y. Imai, M. Ueda and M. Ii., Makromol. Chem., 179, 2085 (1978).

26. J. E. L. Roovers, Polymer, 17, 1107 (1976).

27. R. W. Roeske and P. D. Gesellchen, Tetrahedron Lett., 1976, 3369.

28. J. P. Tam, W. F. Cunningham - Rundles, B. W. Erickson and R.B. Merrifield, Ibid., 1977, 4001.

29. T. D. N'Guyen, A. Deffieux and S. Boileau, Polymer, 19, 423 (1978).

30. T. D. N'Guyen and S. Boileau, Tetrahedron Lett., 1979, 2651.

31. M. J. Farrall and J. M. J. Fréchet, J. Amer. Chem. Soc., 100, 7998 (1978).

32. M. J. Farrall, T. Durst and J. M. J. Fréchet, Tetrahedron Lett., 1979, 203.

33. J. M. J. Fréchet, M. D. deSmet and M. F. Farrall, Polymer, 20, 675 (1979); J. Org. Chem., 44, 1774 (1979)

34. J. M. J. Fréchet, M. deSmet and M. J. Farrall, Tetrahedron Lett. 1979, 137.

35. M. Ricard, D. Villemin and A. Ricard, Ibid., 1980,47.

36. J. Lewis, M. K. Naqvi and G. S. Park, Makromol. Chem.-Rapid Commun., 1, 119 (1980).

# Chemical Modification of Halogenated Polymers Under Phase Transfer Conditions

Jean M. J. Fréchet

Department of Chemistry, University of Ottawa
Ottawa, Ontario, K1N-9B4, CANADA.

ABSTRACT

The chemical modification of several polymers including poly (vinyl bromide), poly (chloromethyl styrene), poly (vinylidene bromide), and poly (epichlorohydrin) was studied using various nucleophiles in multiphase systems with the assistance of a phase transfer catalyst. Satisfactory results were obtained using o-dichlorobenzene as solvent and powerful nucleophiles. The use of dimethylformamide as solvent often resulted in large increases in both reaction rates and conversions but sometimes created other problems. With several polymers dehydrohalogenation or chain cleavage were observed and in general the more basic nucleophiles were more destructive. With poly (vinyl bromide) the reaction was used to produce efficiently polymers containing alkyl sulfide, methyl sulfone, vinyl sulfone, or thioethanol side chains with little apparent polymer degradation.

INTRODUCTION

The preparation of functional polymers by chemical modification[1-3] is an important technique which has been used extensively both industrially to modify the properties of polymers for various technological applications, and in an area such as polymer-supported chemistry[4,5] to prepare chemically reactive polymers to be used as reagents, supports, protecting groups, or catalysts. The polymer which has been used most extensively for this type of application is poly (chloromethyl styrene)[6] which can be modified easily due to the high reactivity of its chloromethyl groups[3]. For example, it can be converted easily and essentially quantitatively to the corresponding quaternary ammonium polymer for use in ion exchange[7], or into polymeric phosphine ligands which are useful in the field of catalysis[8].

While the chemical modification approach is attractive for its apparent simplicity and the fact it can lead to polymers which have structures not readily obtainable through other routes (e.g. poly (vinyl alcohol)), it also suffers from one major drawback: the inability to purify the polymers after modification. If the reactions which are used in chemical modification are not quantitative, the new polymer will necessarily contain some impurities in the form of unreacted groups or other functionalities resulting from side reactions. The purity of the modified polymer will vary with the nature of the starting polymer and with the type of reaction which is used to modify it. In some cases, the presence of just a few impurities will impair the performance of the modified polymer, while in other cases, even appreciable amounts of such impurities will not have a significant influence on the end use or the reactivity of the modified polymer.

## RESULTS AND DISCUSSION

Most of our recent work on the chemical modification of polymers has been centered on the introduction and the modification of various functionalities on crosslinked polystyrene[3], poly (vinyl pyridine)[9], or copoly (styrene acrylonitrile)[10] resins. Over the past few years, phase transfer catalysis has received much attention for its numerous applications in general organic[11] or organometallic[12] chemistry. The first published application[13] of phase transfer catalysis to the chemical modification of a polymer was the preparation of a crosslinked resin containing free thiol groups by reaction of 1,4-butanedithiol with crosslinked chloromethylated polystyrene in the presence of hydroxide and a phase transfer catalyst. The use of this system led to the formation of a product in which most of the dithiol molecules had become attached to the polymer through one extremity only. In contrast, similar reactions carried out in dimethylformamide without phase transfer catalyst led to extensive double binding of the dithiol. Another illustration of the usefulness of phase transfer catalysis for the chemical modification of polymers is afforded by our study of the use of crosslinked polystyrene-based sulfonium salts[14] as alkylidene transfer reagents according to the following scheme.

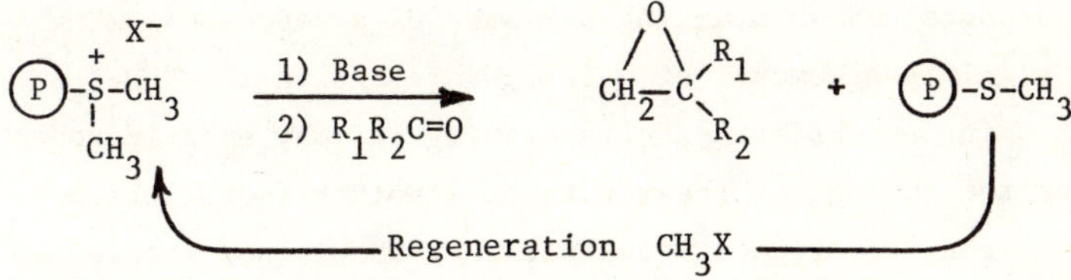


A thorough study of this reaction showed that when a base such as potassium t-butoxide was used to generate the sulfur ylide, the polymer was degraded extensively and could not be regenerated to its initial activity. In contrast, when the reaction was carried out in a three phase system consisting of a suspension of the polymeric sulfonium salt in dichloromethane with aqueous sodium hydroxide containing a small amount of tetrabutyl ammonium hydroxide, the reaction proceeded in higher overall yield and the polymeric sulfide by product could be regenerated and reused repeatedly without apparent loss of activity. These initial results which showed clearly the usefulness of phase transfer catalysis in reactions on polymeric substrates have been extended to numerous other systems[15-17] involving nucleophilic displacements on chloromethylated polystyrene, additions to formylated polystyrene, or reactions in which the polymers themselves acted as the nucleophiles. A similar approach involving the use of phase transfer catalysts to promote the modification of chloromethylated polymers has been described by Boileau and coworkers[18,19]. It is worthwhile noting that in the case of chloromethylated polystyrene most of our reactions were performed in dichloroethane or o-dichlorobenzene which are not miscible with water, while Boileau et al. have used dimethylformamide in most of their phase transfer reactions. The main advantage of using DMF as solvent is that the reaction proceeds at a much higher rate than in hydrocarbons; this is likely due to the higher solubility of the reactive ionic species in the more polar DMF. As DMF has a high affinity for water, it is debatable whether the term "phase transfer catalysis" can be used for reactions such as the ones described by Boileau et al.; the term "ionic assistance" might be more appropriate. Although separate phases can be obtained when mixing DMF and 50% aqueous sodium hydroxide, a significant amount of water is present in the "DMF phase". This in turn affects the reactivity and may even in some cases affect the outcome of the reaction. Another factor which deserves some consideration is the fact that DMF is not stable under strongly alkaline conditions. For example, overnight stirring of a mixture of equal volumes of 50% aqueous sodium hydroxide and DMF containing a little tetrabutyl ammonium hydroxide (a recipe which approximates that used in many phase transfer reactions) produces a two phase system. Analysis of the "DMF phase" shows that it actually contains very little DMF but is rich in dimethylamine and water. Thus the use of DMF as a solvent in phase transfer reactions involving concentrated aqueous hydroxide should be restricted to cases where the dimethylamine which is produced cannot interfere with the transformations being carried out. A second problem with the use of DMF - aqueous hydroxide systems in phase transfer modifications is that unless very large amounts of DMF are used, addition of aqueous hydroxide to the solution of polymer in DMF causes precipitation of such hydrophobic polymers as poly (epichlorohydrin) or chloromethylated polystyrene.

An example of the subtle differences which may result from the choice of one solvent or another in a phase transfer reaction is afforded by the reaction of a solution of chloromethylated polystyrene with an aqueous solution containing both carboxylate and hydroxide ions in the presence of tetrabutylammonium hydroxide. When the reaction is carried out in DMF[19, 20] at 70°, it is complete within a few hours and it produces a soluble hydroxymethyl polymer. In contrast, the same reaction carried out in o-dichlorobenzene is much slower and eventually produces a lightly crosslinked hydroxymethyl polymer.[22] In both cases the reaction fails in the absence of carboxylate ion in the aqueous phase. An explanation for this difference is that in the case of the reaction in o-dichlorobenzene, displacement of chloride by carboxylate produces first a polymer with ester sites. Saponification of these by action of hydroxide produces alkoxide sites (reaction 1 below) which, in turn, can become protonated to yield the hydroxymethyl sites, or can attack directly any neighboring chloromethylated sites to produce a crosslinked material (reaction 2).

Reaction 1:

$$\boxed{P}-CH_2Cl \xrightarrow{RCOO^-} \boxed{P}-CH_2OCOR \xrightarrow{OH^-} \boxed{P}-CH_2O^- + RCOOH$$

Reaction 2:

$$\boxed{P}-CH_2O^- \underset{OH^-}{\overset{H_2O \text{ or } RCOOH}{\rightleftharpoons}} \boxed{P}-CH_2OH$$

$$\boxed{P}-CH_2O^- \xrightarrow{\boxed{P}-CH_2Cl} \boxed{P}-CH_2OCH_2-\boxed{P} \quad \text{Crosslinked Polymer}$$

The concentration of alkoxide in the organic phase is a function of the concentrations of water and alkoxide in the o-dichlorobenzene. Since little water is ever present in this solvent, a relatively high concentration of alkoxide is responsible for the crosslinking which occurs. In the DMF case, the concentration of water in the DMF phase is much higher and thus the importance of alkoxide ions in the overall process becomes insignificant, a small increase in molecular weight may be observed[20] but the polymer remains soluble. A close examination of the analytical data provided by N'Guyen[20, 21] also reveals that in DMF a second reaction may have occurred and remained undetected. While the elemental analysis of the starting chloromethylated polymer is satisfactory, it is less so for the product of the reaction (sum of % C, H, O = 96.4). Since no chlorine remains in the polymer after reaction and the functional yield of the reaction calculated from the Cl analysis of the starting polymer (10.2% for a degree of functionalization of 0.35) and the oxygen analysis of the final polymer (4.1% for a degree of functionalization of 0.29) is only 83%, it can be concluded that a by-product has been formed. The DMF or one of its degradation products may be responsible for this side reaction which does not occur when o-dichlorobenzene is used as solvent in the phase transfer reaction.

In the case of poly (vinyl bromide) the replacement of Br by various nucleophiles under phase transfer conditions occurred very slowly in halogenated hydrocarbons but much more rapidly in DMF. The reactions which were successful are summarized below:

$$-(CH_2-\underset{\underset{SO_2CH_3}{|}}{CH})- \ \text{(I)} \xleftarrow[Q^+]{CH_3SO_2^-} -(CH_2-\underset{\underset{Br}{|}}{CH})- \xrightarrow[Q^+/\,OH^-]{CH_3SH} -(CH_2-\underset{\underset{S-CH_3}{|}}{CH})- \ \text{(II)}$$

$$-(CH_2-\underset{\underset{Br}{|}}{CH})- \xrightarrow[HOCH_2CH_2SH]{OH^-/\,Q^+} -(CH_2-\underset{\underset{S-CH_2CH_2OH}{|}}{CH})- \ \text{(III)}$$

$$\text{(III)} \xrightarrow[\text{2) tosylation-elimination}]{\text{1) oxidation}} -(CH_2-\underset{\underset{SO_2CH=CH_2}{|}}{CH})- \ \text{(IV)}$$

While polymer (I) can be obtained by anionic polymerization of methyl vinyl sulfone[23], polymers such as (III) and (IV) are not readily obtainable through the simple polymerization of vinyl monomers. The pendant vinyl groups of (IV) are very reactive and can be used in further chemical modification reactions, crosslinking, etc.

The reaction of poly (vinyl bromide) with sodium methyl sulfinate under phase transfer conditions gave a product which was darker than the light beige starting material and which contained a large number of sulfone groups as shown by elemental analysis and infrared spetroscopy. Figure 1 shows the IR spectra of poly (vinyl bromide and of polymer (I).

In addition to the strong sulfone stretching bands centered at 1130 and 1300 $cm^{-1}$, the IR spectrum of (I) exhibits weaker bands at 1620 and 3025 $cm^{-1}$ which correspond to the C=C stretch and the C=C-H stretch of unsaturated portions of the polymer backbone. In addition to this spectroscopic data, analysis of the products of several similar reactions carried out under a variety of conditions, indicated that in most cases substitution did not exceed 70% while elimination to form double bonds accounted for 20-30% of the

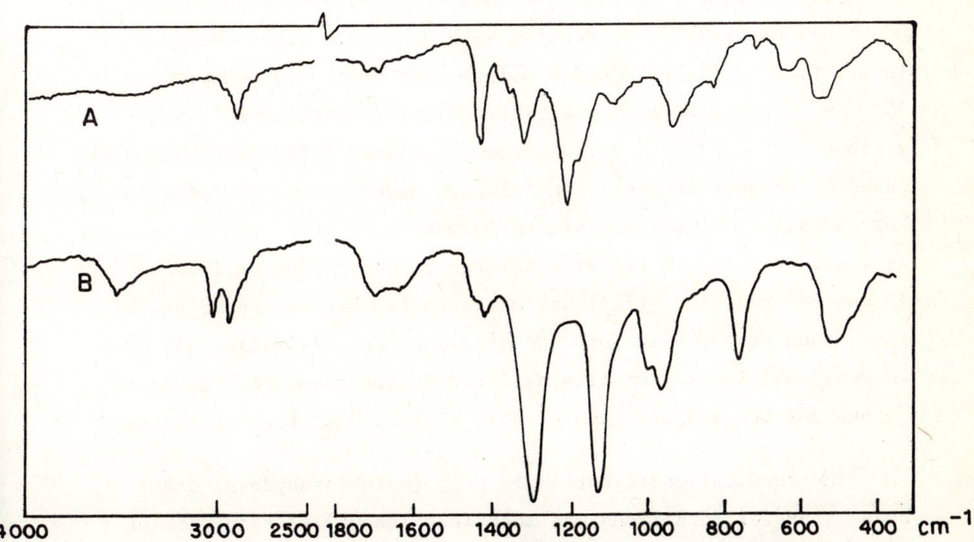


FIG. 1. IR spectrum of poly (vinyl bromide) before (spectrum A) and after (spectrum B) reaction with sodium methyl sulfinate under phase transfer conditions.

sites with usually 10-20% unchanged brominated units. Thus, a more accurate representation of polymer (I) would be that given below for structure (V).

$$\text{(V)}\qquad -(CH_2-\underset{\underset{\underset{CH_3}{|}}{SO_2}}{\underset{|}{CH}})_x-(CH_2-\underset{\underset{Br}{|}}{CH})_y-(CH{=}CH)_z- \qquad \begin{array}{l} x = 0.5\text{-}0.7 \\ y = 0.1\text{-}0.2 \\ z = 0.2\text{-}0.3 \end{array}$$

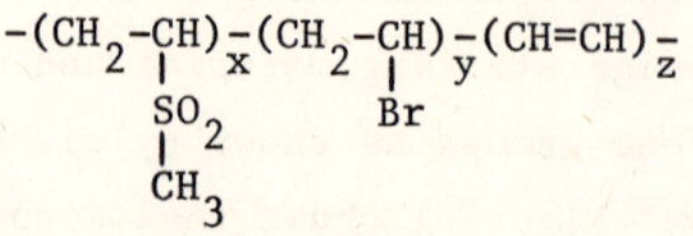

A similar reaction attempted with the less nucleophilic trifluoromethyl sulfinate anion gave only a maximum of 20% substitution while most of the bromine remained on the polymer. In addition, this reaction only occurred when DMF was the solvent. Best results in the substitution reaction were obtained using powerful nucleophiles of low basicity such as the mercaptides of methyl mercaptan or 2-mercaptoethanol with no free hydroxyl present in the reaction mixture. For example, the reaction of poly (vinyl bromide) with an aqueous solution of the anion of 2-mercaptoethanol in the presence of a tetrabutyl ammonium salt acting as phase transfer catalyst, proceeded smoothly to yield a water soluble product which contained essentially no bromine, with up to 90% substitution. The IR spectrum of (III) included a large hydroxyl band centered at 3500 $cm^{-1}$, and elemental analysis indicated a sulfur to oxygen ratio of 1:1. Several derivatives of polymer (III) have been prepared by further chemical modification under classical conditions. For example (III) was acetylated to give a product with no hydroxyl absorption in the IR but with a large carbonyl band at 1730 $cm^{-1}$. In another reaction, (III) was oxidized to the corresponding sulfone, then transformed into its vinyl sulfone derivative (V) by a conventional tosylation-elimination reaction sequence. These reactions and the applications of (IV) will be described elsewhere.

The chemical modification of poly (epichlorohydrin) under phase transfer conditions was much less successful than that of either chloromethylated polystyrene or poly (vinyl bromide). In all cases side reactions including main chain cleavage were observed. In a typical reaction a solution of poly (epichlorohydrin) was treated with a concentrated aqueous solution of acetate and hydroxide ions in the presence of tetrabutyl ammonium hydroxide. In halogenated hydrocarbons the reaction was very slow and only low degrees of substitution (< 10%) could be obtained. In hot DMF high degrees of substitution could be obtained and the final polymer was soluble in water. However, this type of reaction has little value as the polymer chains are cleaved extensively during the reaction. This behavior of poly (epichlorohydrin) has also been observed recently by Boileau et al.[18] for a similar reaction involving the replacement of chloride with carbazole. We are currently studying the reaction of poly (epichlorohydrin) with several softer nucleophiles under various conditions, in all cases our preliminary results indicate incomplete substitution and some chain degradation.

With polyvinylidene bromide the reaction was slightly complicated by the fact that the commercially available starting matereial already contained some unsaturation as evidenced by its elemental analysis and IR spectrum. Reaction of the polymer with aqueous solutions of sodium methylsulfinate or the anions of methyl mercaptan or 2-mercaptoethanol under phase transfer conditions was always accompanied by substantial elimination and, in the most favorable case, the degree of substitution did not exceed 55%. The final polymers were usually very dark in color and had a relatively low solubility. Substitution with sodium methylsulfinate produced a polymer containing 46% of the theoretical amount of sulfur and had an IR spectrum with the expected large sulfone stretching bands at 1135 and 1305 $cm^{-1}$.

## EXPERIMENTAL

Poly (epichlorohydrin) was purchased from Aldrich Chemical Co. (# 18, 186-2). Poly (vinylidene bromide) was purchased from Polysciences Inc. (# 7654), the polymer was light beige-brown in color and contained only 77% bromine (theory 86%); its IR spectrum showed clearly some unsaturation. Poly (vinyl bromide) purchased from Polysciences Inc. (# 4387) was a dark brown-black material partially unsaturated which could not be used in this study, a preparation of poly (vinyl bromide) is given below. Polystyrene was purchased from Polysciences Inc. (# 3432); the white polymer had a nominal molecular weight of 20,000 and an actual molecular weight of Mn = 11,500 and Mw = 56,000 (GPC, Styragel columns in THF). The phase transfer catalysts, Adogen 464 and tetrabutyl ammonium hydroxide were purchased from Aldrich Chemical Co. All other chemicals and solvents were reagent grade, used without further purification except as indicated.

Infrared spectra were recorded on a Pye Unicam SP 1100 or a Perkin Elmer 281 infrared spectrophotometer using potassium bromide pellets. NMR spectra were recorded on a Varian HA 100 or CFT-80 spectrometer. Elemental analyses were performed by MHW laboratories or by Guelph Chemical Laboratories Ltd.

## Preparation of the Polymers

Chloromethylated polystyrene was prepared using a modification of the procedure of Pepper et al.[24] A solution of 50 g polystyrene (Polysciences Cat. # 3432) in 500 mL chloroform and 100 mL chloromethyl ether was treated dropwise and with vigorous stirring with a total of 10 mL anhydrous stannic chloride. After 2 hours at room temperature, 20 mL methanol and 4 mL water were added and most of the solvent was removed under vacuo. The polymer was then precipitated into 1000 mL of methanol. After reprecipitation and drying, a chlorine analysis indicated that the capacity of the polymer was 3.0 mequiv./g (degree of functionalization D.F. = 0.36).

Poly (vinyl bromide) was prepared in a 600 mL stirred pressurized vessel by emulsion polymerization of 50 g vinyl bromide in 70 mL deoxygenated water with 3 drops Anti-foam A, 2 g sodium dodecyl benzenesulfonate, a trace of copper sulfate, and 0.4 g ammonium persulfate. Once the stirring was started, the mixture was heated to 50° and a solution of 0.5 g sodium sulfite in 10 mL water was added over a period of two hours. After four hours the mixture was cooled and the latex was coagulated to yield 29.5 g of dry poly (vinyl bromide). The polymer was slightly colored. Analysis: calculated 74.5% Br; found 74.6% Br.

## Chemical Modification of Chloromethylated Polystyrene

This reaction was carried out as described previously[22] using a solution of 1 g of the polymer in 10 mL O-dichlorobenzene with 0.1 g tetrabutyl ammonium hydroxide and an aqueous phase containing 2 g potassium acetate and 2 g sodium hydroxide in 6 mL distilled water. The mixture was stirred rapidly at 80-85° for two days and formation of a gel was observed. The gel was collected on filter, washed repeatedly with various organic solvents and dioxane-water mixtures, then dried under vacuo to yield 0.63 g of the crosslinked hydroxymethyl polymer. Analysis: Cl, 0.23%; O, 4.74%. The D.F. of the final polymer calculated from the oxygen analysis is 0.34 for a 95% functional yield.

## Chemical Modification of Poly (vinyl bromide)

The general procedure used for these reactions was as follows: The polymer (1 g) was dissolved in 10-25 mL of 1,2-dichloroethane or dimethylformamide and approximately 0.1 g of the phase transfer catalyst was added. The phase transfer catalyst was tetrabutyl ammonium hydroxide for reactions involving aqueous hydroxide and Adogen 464 in all other cases. The aqueous phase (3-10 mL) containing an excess of the nucleophile was added to the polymer solution and the mixture was stirred for 1-4 days at 70-80°. In some cases the nucleophile was added in several portions over the course of the reaction. The polymers were purified by precipitation and reprecipitation, or in the case of water soluble products, by dialysis against distilled water followed by freeze-drying. In all cases the modified polymers were much darker than their off-white precursor.

For example, the reaction with sodium methylsulfinate was carried out in dimethylformamide with 3 mL of a 30% solution of sodium methylsulfinate and with Adogen 464. Three subsequent additions of 0.5 g sodium methylsulfinate were made at 4 hour intervals. After purification and drying the polymer contained 9.8% bromine and 18.9% sulfur.

### Reaction of Poly (vinylidene bromide) with Sodium Methylsulfinate

1 g of poly (vinylidene bromide) (Polysciences Inc., 77% bromine) was dissolved with stirring in 40 mL DMF at 75°. After addition of 0.2 g Adogen 464, 3 g sodium methanesulfinate and 5 mL water the mixture was stirred at 80° overnight. The polymer was precipitated by pouring the reaction mixture in a large amount of distilled water. After filtration and washing, the polymer was reprecipitated, then dried in vacuo. The infrared spectrum of the polymer had large sulfone peaks at 1135 and 1305 $cm^{-1}$ and showed some insaturation. Analysis: Br 28.4%, S 13.95%.

## ACKNOWLEDGEMENTS

Financial support of this work by the Natural Science and Engineering Research Council of Canada is gratefully acknowledged.

## REFERENCES

[1] E.M. Fettes, Ed., "Chemical Reactions of Polymers", Interscience, New York (1964).

[2] J.A. Moore, Ed., "Reactions on Polymers", D. Dreidel Publishing Co., Boston, (1973).

[3] J.M.J. Fréchet and M.J. Farrall, "Chemistry and Properties of Crosslinked Polymers", S.S. Labana, Ed., Academic Press, New York, 1977, pp 59-83.

[4] P. Hodge and D.C. Sherrington, Editors, "Polymer Supported Reactions", J. Wiley, London,(1980). N.K. Mathur, D.C. Norang and R.E. Williams, "Polymers as Aids in Organic Synthesis", Academic Press (1980).

[5] J.M.J. Fréchet, "Synthesis and Applications of Organic Polymers as Supports and Protecting Groups", Tetrahedron Report, in Press (1980).

[6] C. McBurney (Rohm & Haas Co.), U.S. Patent 2591573 (1952); E.L. McMaster (Dow Chemical Co.), U.S. Patent 2616977 (1952); K.W. Pepper, H.M. Paisley and M.A. Young, J. Chem. Soc., 4097 (1953).

[7] K. Dorfner, "Ion Exchangers, Properties and Applications", 3rd Ed., Science, Ann Arbor (1972).

[8] G.O. Evans, C.U. Pittman Jr., R. McMillan, R.T. Beach, and R. Jones, J. Organomet. Chem., 67, 295 (1974); R.H. Grubbs and L.C. Kroll, J. Am. Chem. Soc., 93, 3062 (1971).

[9] J.M.J. Fréchet, G. Darling, and M.J. Farrall, Polym. Prep., 21, (1980) in press; J.M.J. Fréchet, J. Warnock, and M.J. Farrall, J. Org. Chem., 43, 2618 (1978).

[10] J.M.J. Fréchet, M.J. Farrall, C. Benezra, and A. Cheminat, Polym. Prep., 21 (1980) in press.

[11] W.E. Keller, "Compendium of Phase Transfer Reactions and Related Synthetic Methods", Fluka A.G., Buchs (1979); W.P. Weber and G.W. Gokel, "Phase Transfer Catalysis in Organic Synthesis", Springer Verlag, Berlin (1977).

[12] H. Alper, Adv. Organometal. Chem., in press (1980).

[13] M.J. Farrall and J.M.J. Fréchet, J. Am. Chem. Soc., 100, 7998 (1978).

[14] M.J. Farrall and J.M.J. Fréchet, 46th Congress ACFAS, Ottawa (1978); M.J. Farrall, T. Durst, and J.M.J. Fréchet, Tetrahedron Letters, 203 (1979).

[15] M.J. Farrall, Ph.D. Thesis, University of Ottawa, January 1979.

[16] J.M.J. Fréchet, M. de Smet, and M.J. Farrall, Tetrahedron Lett., 137-8 (1979).

[17] J.M.J. Fréchet, M. de Smet, and M.J. Farrall, J. Org. Chem., 44, 1774-9 (1979).

[18] T.D. N'Guyen, A. Deffieux, and S. Boileau, Polymer, 423 (1978).

[19] T.D. N'Guyen and S. Boileau, Tetrahedron Lett., 2651-4 (1979).

[20] T.D. N'Guyen, Thèse de 3ème cycle, Université Pierre et Marie Curie, Paris 6 (June 1978).

[21] See ref. 20 p. 23 for the analysis of the starting chloromethylated polystyrene: C, 82.72%; H, 7.04%; Cl, 10.20%. See page 56 for the analysis of the polymer obtained after reaction with propanoic acid and excess 50% aqueous hydroxide in DMF at 60° for 3 hours in the presence of 10% tetrabutyl ammonium hydroxide: C, 84.42%; H, 7.89%; O, 4.10%; Cl, 0%.

[22] J.M.J. Fréchet, M.D. deSmet and M.J. Farrall, Polymer, 20, 675 (1979).

[23] J.P. Schroeder, D.C. Schroeder, and S. Jotikasthira, J. Polym. Sci., A-1, 10, 2189-95 (1972).

[24] K.W. Pepper, H.M. Paisley, and M.A. Young, J. Chem. Soc., 4097 (1953).

# The Phase Transfer Catalyzed Alkylation of 1-Hydroxybenzotriazole. I. Scope

William A. Feld
Rebecca J. Paessun
M. Paul Serve'

Department of Chemistry
Wright State University
Dayton, Ohio 45435

ABSTRACT

Phase transfer catalyzed alkylation of 1-hydroxybenzotriazole employing a series of alkyl halides was carried out in dichloromethane/water using tetrabutylammonium chloride as the catalyst. Several new 1-alkoxybenzotriazoles were prepared and included the alkyl and benzyl derivatives.

## INTRODUCTION

A frequent problem encountered by synthetic chemists is that of bringing two reagents (e.g. salts and organic substrates) together for the purpose of interaction. The problem is frequently overcome by the use of dipolar aprotic solvents like dimethylsulfoxide or dimethylformamide. The major difficulties with such solvents are difficulty in purification and drying and usually nonexistent recovery. Another possibility is the use of hot methanol or ethanol solutions with the problems of limited salt and organic substrate solubility and the effect of heat on certain heat sensitive reagents or substrates.

An alternate approach for bringing reagents together is to employ phase transfer catalysis (PTC) (1,2). In PTC there are generally two immiscible phases. One phase, usually aqueous, contains the salt which will function as the nucleophile or base and the other is organic and contains the substrate on which the salt will act. The addition of a phase transfer catalyst (e.g. a quaternary ammonium salt) allows for anionic transport from the aqueous phase to the organic phase where reaction takes place. Quaternary ammonium salts like tetrabutylammonium bromide, are good examples of phase transfer catalysts and have been shown to be effective in catalyzing a variety of nucleophilic reactions. Interestingly, however, they are considered soft in the HSAB sense (3) and as such will pair with the softest anion available. This can lead to reduced reaction rates and "poisoned" catalysts in certain circumstances.

Most derivatives of 1-hydroxybenzotriazole (1) have substituents on the benzene ring portion of the molecule and many are used as catalysts in peptide syntheses (4). The hydroxyl group has been subjected to esterification (5) and has been alkylated (6) in alcoholic solutions. The pKa of 1-hydroxybenzotriazole is

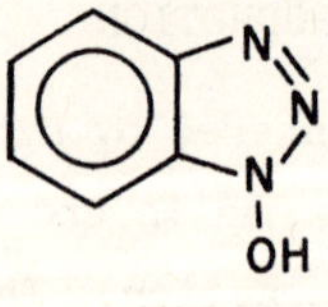

N
N
N
OH

1

about 7.8, which places it between phenols (pKa=9) and carboxylic acids (pKa=5) in acidity, and makes it a good choice for salt formation in an aqueous solution (7).

The objectives of this research were (1) to employ PTC in the alkylation of 1-hydroxybenzotriazole and (2) to extend the alkylation of 1-hydroxybenzotriazole to include previously inaccessible systems.

## RESULTS AND DISCUSSION

A series of phase transfer catalyzed alkylations of 1-hydroxybenzotriazole were carried out employing the alkyl halides listed in Figure 1. Tetrabutylammonium chloride was used as the phase transfer catalyst in all cases and represents a readily available catalyst with an anion similar to the leaving group in most of the alkyl halides employed in this study. The reactions were agitated strongly to provide maximum interfacial contact between the immiscible phases.

The compounds appearing in Table 1 can be compared directly to previously reported 1-alkoxybenzotriazoles as shown. The yields are good but some are low and may represent the "poisoning" effect cited earlier as alkyl iodides were used in some cases. It is also surprising that the cyclopentyl derivative was obtained in such high yield as it represents a secondary alkyl halide.

The examples presented in Table 2 represent new 1-alkoxybenzotriazoles. Of particular interest are the allyl and benzyl derivatives. Each can be uniquivocally identified by NMR absorptions characteristic of venzylic and allylic hydrogen shifted downfield (about 1 ppm) by attachment to oxygen in alkoxybenzotriazoles.

N–OH  →  N–OR

$NaOH / H_2O$

$CH_2Cl_2$ / RX

$(C_4H_9)_4N^+\ Cl^-$

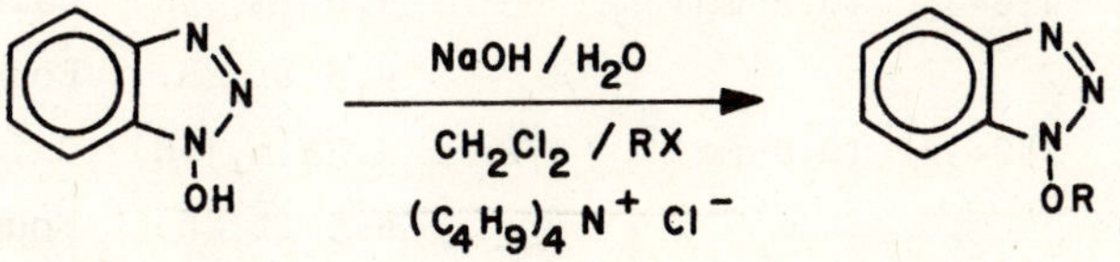

R = $-CH_3$, $-C_2H_5$, $-nC_3H_7$, $-nC_4H_9$, $-nC_5H_{11}$, $-$cyclopentyl, $-nC_7H_{15}$, $-nC_8H_{17}$, $-nC_9H_{19}$, $-$allyl, $-$benzyl

FIG. 1

Table 1

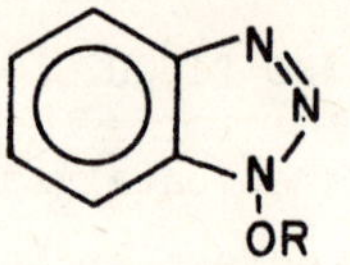

| R | Yield (%) | NMR(ppm) | Reference |
|---|---|---|---|
| $-CH_3$ | 59 | 4.3 (s,3H) | 5,6 |
| $-C_2H_5$ | 76 | 1.45 (t,3H) | 6 |
| | | 4.6 (q,2H) | |
| $-n-C_3H_7$ | 73 | 1.87 (m,5H) | 6 |
| | | 4.5 (t,2H) | |
| $-n-C_4H_9$ | 74 | 1.70 (m,7H) | 6 |
| | | 4.58 (t,2H) | |
| $-n-C_5H_{11}$ | 50 | 1.65 (m,9H) | 6 |
| | | 4.4 (t,2H) | |
| 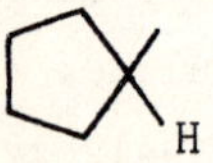 cyclopentyl (H) | 96 | 1.75 (m,8H) | 6 |
| | | 5.2 (m,1H) | |

Table 2

New 1-Alkoxybenzotriazoles

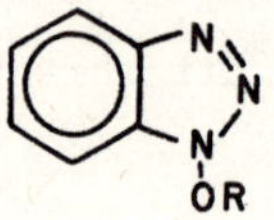

| R | Boiling Point($^{o}$C) | Yield(%) | NMR(ppm) | N% |
|---|---|---|---|---|
| $-n-C_7H_{15}$ | 128-130 (0.25mm) | 63 | 1.6 (m,13H) | Calcd 18.01 |
| | | | 4.5 (t,2H) | Found 17.69 |
| $-n-C_8H_{17}$ | 140-142 (0.05mm) | 72 | 1.3 (m,15H) | Calcd 16.99 |
| | | | 4.5 (t,2H) | Found 16.54 |
| $-n-C_9H_{19}$ | 154-156 (0.05mm) | 65 | 1.4 (m,17H) | Calcd 16.08 |
| | | | 4.55(t,2H) | Found 15.69 |
| Allyl | a | 76 | 5.0 (d,2H) | Calcd 23.99 |
| | | | 5.65(m,3H) | Found 23.57 |
| Benzyl | 138-140 (0.05mm) | 52 | 5.45(s,2H) | Calcd 18.66 |
| | | | | Found18.42 |

a) decomposed on distillation

## CONCLUSIONS

Phase transfer catalyzed alkylation of 1-hydroxybenzotriazole proceeds smoothly and in good yield. A variety of functional groups can be introduced which include the thermally sensitive allyl group.

## EXPERIMENTAL

### Instrumentation and Starting Materials

Infrared spectra were obtained on a Perkin Elmer model 735B spectrometer. Nuclear magnetic resonance spectra were obtained on a Varian model EM-360 spectrometer employing 10-30% w/v solutions in deuteriochloroform with 1% TMS as an internal standard. Alkyl halides and 1-hydroxybenzotriazole were used as received.

### General Phase Transfer Catalyzed Alkylation Procedure

A solution of 0.02 mol 1-hydroxybenzotriazole and 0.02 mol NaOH in 75 mL of water was mixed with a solution of 0.06 mol of the alkyl halide and 0.002 mol of tetrabutylammonium chloride in 75 mL of dichloromethane. The mixture was agitated for 36 hr. The organic phase was then separated and evaporated. The residue was dissolved in ether and passed through a short (1.5 X 3 cm) column of neutral alumina. The residue obtained after evaporation of the ether was subjected to NMR and IR spectral analysis.

## REFERENCES

[1] Weber, W.P. and Gokel, G.W., "Phase Transfer Catalysis in Organic Synthesis," Springer-Verlag, New York, N.Y., 1977.

[2] Starks, C.M. and Liotta, C., "Phase Transfer Catalysis, "Academic Press, New York, N.Y., 1978.

[3] Pearson, R.G., "Hard Acids and Bases," Dowden, Hutchinson and Ross, Inc., 1973.

[4] Konig, W. and Geiger, R., <u>Ber</u>., 103, 788 (1970).

[5] Brady, O.L. and Reynolds, C.V., <u>J</u>. <u>Chem</u>. <u>Soc</u>., 1928, 193.

[6] Serve', M.P., Seybold, P.G., Feld, W.A. and Chao, M.A., *J. Hetero. Chem.*, 13, 509 (1976).

[7] Boyle, F.T. and Jones, R.A.Y., *J. Chem. Soc. Perkin II*, 1973, 160.

# Homogenizing Metal Oxide Catalysts

Charles U. Pittman, Jr., Edwin A. Lewis
and Mohammad Habib

Department of Chemistry
University of Alabama
University, AL 35486

ABSTRACT

Metal oxide particles have been prepared which are "soluble" in hydrocarbon solvents. The reaction of either a metal (or its metal oxide) with water, mineral spirits, and a mixture of carboxylic acids can result in hydrocarbon-miscible metal oxide core particles of 20-1000Å diameter sizes which are coated with a layer of hydrophobic carboxylates. This will only occur if the metal:acid ratio exceeds one. The particle size increases as the metal:acid ratio increases. The surface carboxylate groups are in equilibrium, preventing the use of colligative property measurements in determining molecular weights. Sedimentation velocity studies confirmed that particle aggregation equilibria occur which are solvent dependent. Aggregation is more serious in polar solvents. The distribution of carboxylic acids used did not effect the particle size. Cobalt and manganese oxide particles have been used in oxidations of cyclohexane, toluene, and xylenes. Cobalt oxides were used in hydroformylations of 1-pentene. Nickel oxide catalysts were tested in hydrogenations of 1-pentene, polybutadiene, polyisoprene, and aromatic compounds. The nickel oxides were much more active upon addition of $Et_3Al$. Cobalt oxide systems were active in the syndiotactic 1,2-polymerization of butadiene. The oxide particles could be coated onto supports such as silica, alumina, and kieselguhr followed by oxidation to give supported metal oxides. Reduction with hydrogen or metal hydrides or $Et_3Al$ gave supported metal catalysts. The activity of these supported catalysts were investigated.

## INTRODUCTION

The area of heterogenizing homogeneous catalyst has been an active research topic for the past few years. Homogeneous transition metal complexes, for example, have been immobilized onto crosslinked polymers and silica [1-3]. These heterogenized systems can then be used in fixed catalyst beds. However, the idea of homogenizing a heterogeneous catalyst has largely been overlooked. Consider metal oxide catalysts. How could one "homogenize" a metal oxide? Could a metal oxide be prepared that was "soluble" in a hydrocarbon solvent? It was this goal which was the basis of the work described in this paper.

Soluble metal carboxylate salts (soaps) have been used as catalysts for paint drying, oxidation of p-xylene to teraphthalic acid, and combustion. They have been prepared from metals or metal oxides in the presence of an equivalent of carboxylic acid [4]. However, what would happen if a metal oxide was reacted with a limited amount of carboxylic acid in a similar fashion (ie, if the metal/acid equivalent ratio is kept greater than one)? If an insufficient quantities of acid were available to stoichiometrically form the carboxylate salts, would a mixture of metal oxides and metal carboxylates result? Anthony Alkaitis and Paul Cells, working at Mooney Chemical Co., Inc. [5], pioneered such reactions in 1977. The procedure is illustrated in Figure 1.

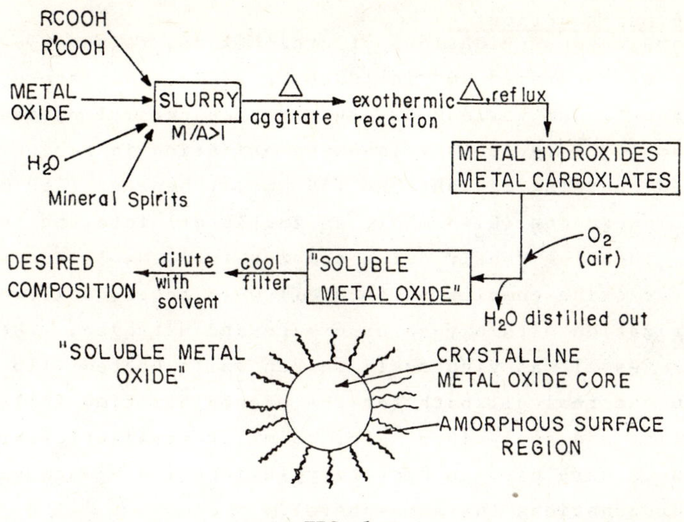


FIG. 1.

Heating metal oxides, water and a mixture of carboxylic acids, which were slurried in mineral spirits (or other solvents), led to an exothermic reaction giving metal hydroxides and carboxylates [5]. Air oxidation and distillation of water gave "soluble" metal oxides [5]. For example, stable hydrocarbon "solutions" of $Mn_3O_4$, CoO, or NiO could be made that were 60% by weight metal. Other successful preparations of $Fe_2O_3$ and CuO, as well as bimetallic systems such as $Mn_3O_4$/CoO, $Mn_3O_4$/BaO and $Mn_3O_4$/ZnO were made [5].

## EXPERIMENTAL

The detailed synthesis of the "soluble" metal oxide particles is described by Alkaitis and Cells in their 1979 patent [5]. Thus, the syntheses will not be further described here. Sedimentation velcocity and equilibrium experiments will be described in the following paper.

Homogeneous Reactions

Homogeneous oxidations of cyclohexane, toluene, or xylenes were carried out in 100cc or 250cc stainless steel autoclaves under various oxygen/nitrogen partial pressures. For example, a typical cyclohexane oxidation in benzene involved charging cyclohexane (59 mmol), benzene (250 mmol), and cyclohexanone (0.48 mmol, an initiator) into the autoclave. Then the cobalt oxide complex (0.13 mmol) and the manganese oxide complex (0.44 mmol) were added, followed by pressurization with oxygen or oxygen and nitrogen. The reactor vessel was vigorously shaken while heated (115-160$^0$) with an external oil bath for the prescribed time followed by cooling and analysis of glc. Similar straightforward procedures were used in hydroformylations ($CO/H_2$ atmosphere), and hydrogenations ($H_2$ atmosphere).

Supported Catalysts

The preparation of the supported heterogeneous catalysts from the "soluble" metal oxides followed the following general method. The support ($Al_2O_3$, $SiO_2$, kieselguhr) was thoroughly dried by heating at 500-600$^0$C under a continuous nitrogen flow. Then under nitrogen or *in vacuo*, the support was treated with a "solution" of the overbased Mooney complexes to form a slurry which was thoroughly mixed. The solvent was removed *in vacuo*. The supported catalysts were then heated to 400-500$^0$ under a flow of oxygen to insure the oxidation of any organic compounds on the surface and to effect surface binding. In many cases reduction was then effected using $NaBH_4$ or $AlEt_3$ or by heating in a hydrogen atmosphere.

As a general example, the preparation and use of alumina-supported NiO and Ni for studies of the demethylation of 1-methylnaphthalene is given. Alumina tablets (1/8") from Calcicat, having a pore volume of 0.5cc/g and a surface area of 200m$^2$/g were dried at 500$^0$C for 3 hrs. under a continuous nitrogen purge. After cooling, 106g of the support was treated with a hexane solution (200ml) of the nickel oxide particles (122Å diameter in mineral spirits, Ni/acid = 33, acid distribution: 20%$C_3$, 50%$C_8$, 15%neo$C_{10}$ and 15%$C_{18}$) containing 21.2g of the nickel complex (to give theoretically a 9% wt. Ni catalyst). The nickel complex was injected; then nitrogen was introducted at 1 atm. and the slurry was mixed for 10 hrs. The solvent was evaporated to give dark tablets which were dried in a steel tube at 400°C for 1 hr. under nitrogen and then an additional 2 hrs. at 400°C under oxygen. This catalyst was designated $Al_2O_3$-NiO.

The supported nickel oxide (just described) (60g) was reduced with a 0.5M aqueous solution of $NaBH_4$ (3.2g, 0.085 mol). After the evolution of gas stopped (~8hrs.) the catalyst had turned green. The catalyst was washed with distilled water three times and dried *in vacuo*. The tablets were heated under a nitrogen flow for 2 hrs. at 400°C and then designated $Al_2O_3$-Ni.

*Testing $Al_2O_3$-NiO and $Al_2O_3$-Ni*

Dried $Al_2O_3$-Ni (10g), cyclohexane (30ml), and 1-methylnaphthalene (5g; 35.1mmol) were charged into a 300ml autoclave which was pressurized to 850psi with hydrogen. The reactor was heated to 320° (pressure increased to 1400psi) and held for 30min. Glc analysis (Hewlett Packard Model 5710A with a model 3380A recorder integrator using internal standard techniques) showed 1-methylnaphthalene 93.6%, naphthalene 3.8%, *trans*-1-methyldecalin 1.6%, *trans*-decalin 0.5%, and *cis*-decalin 0.51%. After 17hrs at 320° the product distribution was, respectively 0%, 1.9%, 11.6%, 11.0%, and 73.7%.

The $Al_2O_3$-NiO catalyst (20g) and 1-methylnaphthalene (10g, 70.3mmol) were charged to an autoclave and reacted as described above (740psi initial hydrogen pressure and 1250psi at 320°). After 18hrs, glc analysis showed 1-methylnaphthalene 28.4%, naphthalene 14.2% *cis*-1-methyldecalin 5.7%, *trans*-1-methyldecalin 7.8%, *cis*-decalin 10.7%, and *trans*-decalin 33.1%

## RESULTS AND DISCUSSION

"Soluble" metal oxide systems were prepared in a variety of concentrations. To characterize the particles, the nature of the metal oxide core was studied by X-ray analysis. A crystalline core domain was surrounded by an amorphous region. The core phases were found to be $Mn_3O_4$ (spinel), CoO (cubic), $Fe_3O_4$ (spinel) and CuO (triclinic). Where two different metals were used in the preparations (ie $Mn_3O_4$/CoO etc.) the atoms of both were present in the core. The amorphous outer region resulted from the dissimilarily of the acid moieties and this confers the high solubility in aromatic and aliphatic hydrocarbon solvents. Electron microscopy ($2x10^6$M) of dried samples revealed small spherical crystalline "ultimate core particles" (30-80Å diameter) which were aggregated into 200-300Å clusters. Infrared and chemical analysis further revealed a structure generalized in Figure 2. Carboxylic acids are attached to the metal oxide particle surfaces. Both carboxylate salt sites and hydrogen-bonded carboxylic acid sites are present. Thus, during the preparation of the core particles, metal oxide cores are formed and surrounded by an amorphous carboxylic acid layer which renders

METAL OXIDE CARBOXYLATE PARTICLE STRUCTURE

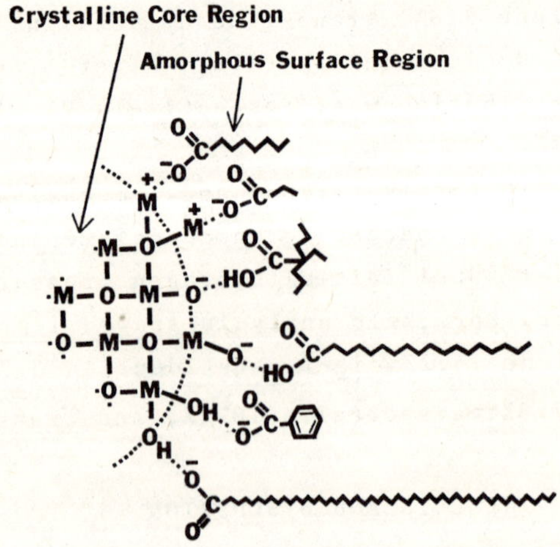

FIG. 2.

the core oxide particles "soluble" in such solvents as mineral spirits, hexane, benzene, etc.

The solubility of these particles in a particular solvent depends upon the distribution of carboxylic acids used in the original preparation. Attempts to characterize the solution behavior and structure of these particles caused much initial confusion. The particles' molecular weights could not be obtained by gel permeation chromatography due to association with the columns and vapor pressure osmometry (VPO) gave molecular weights far to low. This resulted because surface-attached carboxylic acids were in equilibrium with the solution and the VPO technique measures a colligative property. Figure 3 illustrates this principle. Here a core particle, with a molecular weight of $10^6$ is in equilibrium with six dissociated moleucles of carboxylic acid. Thus it will only appear to have a molecular weight of $10^6/7$. Therefore, ultracentrifugation was a perferable method to study the solution behavior. Molecular weights from sedimentation velocity experiments were in agreement with those from sedimentation equilibrium experiments. However, the dynamic equilibria which these core particles experience in solution prevented the general application of sedimentation equilibrium experiments.

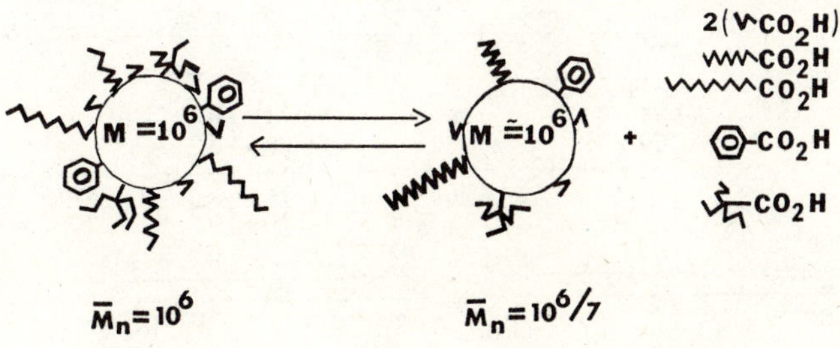


FIG. 3.

Sedimentation velocity studies revealed the observed molecular weight of a given preparation was a function of the solvent. This is the result of a solution aggregation equilibrium which is solvent dependent. Poor solvents gave the highest molecular weights as can be seen in Table 1. For example, a $Mn_3O_4$ sample, made with a metal/acid ratio of 18, was present as "ultimate particles " (48Å diameter) in the good solvent mineral spirits but aggregation of these core particles occured in poor solvents. In THF an average of 20 ultimate particles was present per aggregate. This nonspecific particle aggregation equilibrium is pictured in Figure 4. Thus, to measure the core particle's actual size requires that measurement be made in a good solvent. In a poor solvent the ultimate particles aggregate. By adding a good solvent, the equilibrium is reversed and the "ultimate particles" are again observed.

TABLE 1

Sedimentation Velocity Studies Confirm Aggregation Equlibra Is Solvent Dependent

| Sample | M/A | Solvent | Particle Dia -(A) | Molec. Weight | Ave. No. Of Ultimate Particles |
|---|---|---|---|---|---|
| 40%Mn[A] ($Mn_3O_4$) | 18 | Min. Sp. | 48 | $1.5x10^5$ | 1.0 |
| | | Cyclohex. | 72 | $5.0x10^5$ | 3.3 |
| | | $CCl_4$ | 103 | $1.5x10^6$ | 10.0 |
| | | Bz | 111 | $1.9x10^6$ | 13.0 |
| | | THF | 128 | $3.0x10^6$ | 20.0 |
| 50%Ni[B] (NiO) | 33 | Min. Sp. | 122 | $3.1x10^6$ | |
| | | Isooct. | 132 | $4.1x10^6$ | |
| | | $CCl_4$ | 286 | $40.0x10^6$ | |
| 40%Co[C] (CoO) | 24 | Isooct. | 86 | $9.0x10^5$ | |
| | | $CCl_4$ | 160 | $5.7x10^6$ | |

[A] 20% $C_3$, 40% $C_8$, 20% neo $C_{10}$, 20% $C_{18}$

[B] 20% $C_3$, 50% $C_8$, 15% neo $C_{10}$, 15% $C_{18}$

[C] 30% $C_3$, 55% neo $C_{10}$, 15% $C_{18}$

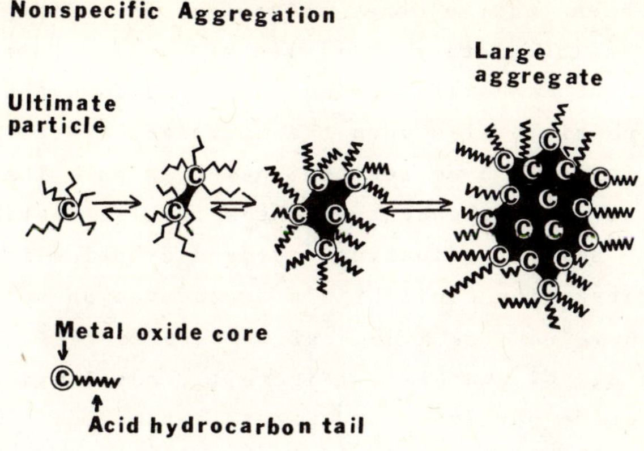


FIG. 4.

The size distribution of the core particles might be an important feature in their use as catalysts. Can a core particle's size be varied by changing the metal/acid ratio during preparation? The answer appears to be yes. In Table 2 one observes this trend. Other similar observa-

TABLE 2

Effect Of Metal/Acid Ratio On Particle Size

| Sample | Metal Acid | Particle Dia., Å | Molecular Weight | Solvent |
|---|---|---|---|---|
| Mn | $6^A$ | 35 | $6.3x10^4$ | Min. Sp. |
| ($Mn_3O_4$) | $12^A$ | 43 | $1.2x10^5$ | Min. Sp. |
| | $18^A$ | 48 | $1.5x10^5$ | Min. Sp. |
| Co | $24^B$ | 179 | $7.4x10^6$ | Min. Sp. |
| (CoO) | $42^B$ | 206 | $1.1x10^7$ | Min. Sp. |

$^A$ 25%$C_3$, 25%$C_{10}$, 15%$C_{18}$, 10%Benzoic

$^B$ 30%$C_3$, 55%neo$C_{10}$, 15%$C_{18}$

<u>Conclusion</u>. PARTICLE SIZE INCREASES AS M/A INCREASES

tions have been made in other solvents. By raising the metal/acid ratios, larger particles are made. Does the distribution of carboxylic acids, employed in synthesis, change the particle size when the metal/acid ratio is constant? In Table 3 we see the answer is no. The ultimate particle size is independent of the acid distribution. However, the acid distribution employed drastically effects the solution aggregation equilibria's dependence on solvent.

From these results a general picture emerges. During the oxidation stage of synthesis (Figure 1), metal hydroxides condense into metal oxide particles and water is removed. The particles grow until all available water is lost (complete M-O-M bond formation) and their surfaces are covered by carboxylate functions.

## Use of "Soluble Metal Oxides" as Homogeneous Catalysts

Metal carboxylates, $(RCO_2^-)_yM^{+y}$, have long been used as soluble combustion catalysts for fuels and as driers in paints and varnishes. Since the surfaces of the soluble Mooney metal oxides have metal carboxylate sites, it was of interest to see if these oxide complexes could function homogeneously like metal carboxylates. Indeed,

TABLE 3

Effect Of Acid Distribution On Particle Size at Constant Metal/Acid Ratio (In Mineral Spirits)

| Metal (Oxide) | Metal / Acid | S | Diameter Å | Molec. Weight |
|---|---|---|---|---|
| | $18^A$ | 52.4 | 48 | $1.51\times10^5$ |
| Mn | $18^B$ | 51.3 | 47 | $1.47\times10^5$ |
| $(Mn_3O_4)$ | $18^C$ | 49.5 | 47 | $1.40\times10^5$ |
| | $18^D$ | 52.5 | 48 | $1.53\times10^5$ |

| A | B | C | D |
|---|---|---|---|
| 25% $C_3$ | 35% $C_3$ | 50% $C_4$ | 40% $C_4$ |
| 25% $C_8$ | 35% $C_8$ | 40% $C_8$ | 40% $C_8$ |
| 25% $C_{10}$ | 10% $C_{10}$ | 10% $C_{18}$ | 20% $C_{18}$ |
| 15% $C_{18}$ | 5% Benzoic acid | | |

Conclusion Size is independent of acid distribution at constant M/A if solvent is a good one.

these complexes were successfully used as combustion catalysts for heavy fuel oils.[6] The complexes, unlike metal oxide powders which have also been used as fuel oil combustion catalysts, are readily dispersed into the oil. Furthermore, they do not agglomerate as metal oxide powders do. Thus, they do not plug up furnace spray nozzles. We report, here, several other attempts to use these complexes homogeneously.

Cyclohexane Oxidation

Cobalt and manganese carboxylates are used commerically to catalyze the oxidation of cyclohexane with oxygen at 115-160$^0$.[7] The special soluble particulate catalysts of $Mn_3O_4$ and CoO also catalyze this reaction. The rates of this oxidation when catalyzed by a soluble CoO complex (Co/acid = 24, 86Å diameter in isooctane) were compared to that when the catalyst was cobalt naphthenate at 115$^0$ and 300psi (1/9, $O_2/N_2$). The rates were approximately equal. Similar comparisons were made of the $Mn_3O_4$ complex (Mn/acid =18, 48Å diameter in mineral spirits) to manganese naphthanate at 140$^0$ and 170psi. Again the rates were approximately equal. The reactions were carried out in autoclaves shaken with a wristaction shaker; thus mass transport (gas/liquid) considerations might be limiting.

The major drawback in these oxidations was the lack of chemical stability of the soluble oxide complexes. As the yield of cyclohexanol and cyclohexanone increased, the catalysts precipitated as a black powder. Possibly, some of the cores' surface acid groups are esterified, thus promoting aggregation and precipitation. Alternatively, the keto and hydroxyl groups could compete with carboxylic acids for surface sites on the core particles.

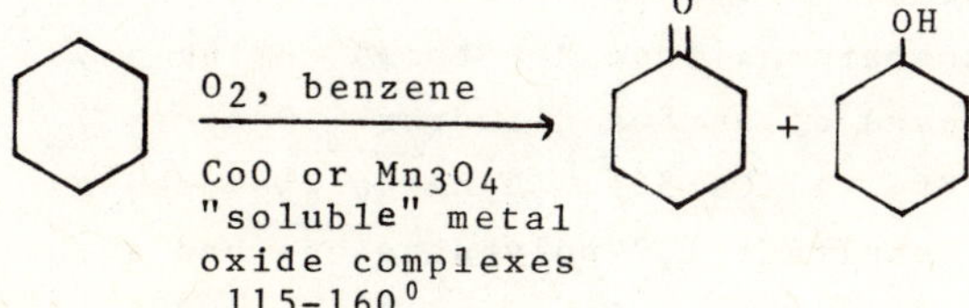

Oxidations of Toluene and p-Xylene

The oxidations of toluene to benzoic acid and p-xylene to p-toluic and terephthalic acids with $O_2/N_2$ mixtures were successfully catalyzed by the $Mn_3O_4$ and CoO soluble Mooney complexes at 160° and 120-200psi in benzene. Since surface sites must be involved in catalysis, it was not surprising that the core complexes were only about 1/10 as active per metal atom as the corresponding metal carboxylate salts. It would be interesting to see if the use of smaller CoO core particles would lead to an increase in rate.

$$C_6H_5CH_3 \xrightarrow[\text{CoO or } Mn_3O_4 \text{ complexes, } 160^0, \text{ 120-200psi}]{O_2/N_2} C_6H_5CO_2H$$

Hydroformylation.

The CoO complexes catalyzed 1-pentene hydroformylations. The soluble core complexes decomposed, in part, to $Co_2(CO)_8$ in these reactions. This was shown by demonstrating $Co_2(CO)_8$ and the CoO complex gave the same normal/branched aldehyde selectivities at a variety of temperatures. Thus cobalt carbonyl served as the true homogeneous catalyst in this case and one cannot view the soluble metal oxide as the catalyst.

$$\text{1-pentene} \xrightarrow[\text{800-1300psi, } 160^0, \text{ CoO complex}]{H_2/CO} \text{hexanal (CHO) + isomers}$$

Polymerizations

Butadiene was selectively polymerized to 1,2-polybutadiene at temperatures from 20° to 70° using a catalyst system composed of an 80Å diameter CoO core complex, $CS_2$, and $AlEt_3$ (Al/Co:50). Benzene was used as the solvent and the resultant 1,2-polybutadiene had a high crystallinity and melting points ranged from 203-

$213^0$. The polymer was dissolved into hot o-dichlorobenzene and precipateted into methanol A strong IR absorption at $660m^{-1}$ characterized the 1,2-mode of polymerization. No polymer was obtained in the absence of $CS_2$. Typical reactions employed butadiene (5g), benzene (30ml), cobalt complex (0.05mmol of Co), $CS_2$ (0.42mmol) and $AlEt_3$ (2.5mmol). Under these conditions the polymer yield was 40% at $20^0$ after 24 hr. and 19% at $60^0$ after 3.2 hr.

Apparently, Zeigler-Natta type catalyst sites are formed on the surfaces of the soluble metal oxide complexes. Similar catalyst systems prepared from homogeneous cobalt octoacte were 10 to 20 times more active per cobalt atom added. This is due to the fact that only metal atoms at the surface of the particles are available as catalyst sites whereas all the metal are available catalyst sites with cobalt octoate.

$$\text{butadiene} \xrightarrow[\text{Cs}_2,\ \text{AlEt}_3]{\text{CoO complex},\ 20\text{-}70^0} \text{1,2-polybutadiene}_n$$

Hydrogenations

The soluble nickel oxide core particles (Ni/acid = 33, diameter in mineral spirits = 122Å) were active catalysts in the hydrogenation of olefins and polybutadiene if first complexed with $AlEt_3$. Neither the NiO complex nor nickel octoate was active without the presence of $AlEt_3$ unless the reaction temperature exceeded $130^0$. In example reactions carried out at $100^0C$ and using $AlEt_3/Ni = 3.7$, 1-pentene was rapidly hydrogenated and isomerized in cyclohexane using either the nickel oxide core particles or nickel octoate. The activities were comparable. Table 4 summarizes a few representative reactions.

*Cis*-polybutadiene was hydrogenated using catalyst systems composed of the NiO particles reduced with $AlEt_3$ or nickel

TABLE 4

Hydrogenation of 1-Pentene Catalyzed by Soluble NiO-Complexes Versus Nickel Octoate [a]

| Catalyst | Temp °C | $H_2$ Pressure psig | Time hr. | n-Pentane % | 2-Pentene %(*cis*/*trans*) |
|---|---|---|---|---|---|
| Nickel octoate[b] | 100 | 450 | 3 | 0 | 0 |
| NiO-particles[b] | 98 | 450 | 16 | 0 | 0 |
| NiO-particles[b] | 160 | 475 | 110 | 100 | |
| Nickel octoate | 100 | 200 | 0.5 | 66 | 8/26 |
| " " | " | " | 12.0 | 100 | 0 |
| NiO-particles | " | " | .33 | 52 | 6/21 |
| " " | " | " | 12 | 100 | 0 |
| Nickel octoate | 60 | 110 | .25 | 24 | 57/17 |
| NiO-particles | 72 | 110 | .25 | 51 | 40/9 |

[a] Reactions run in cyclohexane (15ml) using 1-pentene (2.0ml, 18.2 mmol), nickel species (0.98 mmol), and $AlEt_3$ (3.6mmol) unless noted otherwise.
[b] No $AlEt_3$ added.

octoate/$AlEt_3$ (Al/Ni=3.5). Using 1.0mmol of Ni, the hydrogenations were complete after 10hrs. at 120° and 110psi. Nickel octoate/n-BuLi and nickel octoate/n-BuLi/phenol systems were also active but n-BuLi or n-BuLi/phenol was unable to activate the NiO core particles. Unlike *cis*-polybutadiene, neither the NiO particle nor nickel octoate systems were able to catalyze the hydrogenation of polyisoprene.

$H_2$ 60-100°, 500psi

NiO Complex or
Ni Octoate
$AlEt_3$

NiO particles
$AlEt_3$
n + $H_2$
120°, 110psi
n

## Preparation and Use of Supported Heterogeneous Catalysts from "Soluble" Metal Oxide Complexes

Supported metal oxide catalysts are usually prepared by impregnating support materials ($Al_2O_3$, $SiO_2$, Kieselguhr etc.) with aqueous solutions of metal salts followed by drying and oxidation (with $O_2$ at 400-800°) to the oxide. Reduction with hydrogen (200-500°), or with metal hydrides, give the supported metals. We have shown that the soluble metal oxide particles can be used to prepare supported metal oxide or metal catalysts. The route is illustrated in Figure 5. One might be able to tailor the particle size of the supported metal oxide (or metal) by varying the size of the core complex employed or by using the aggregation equilibria (illustrated in Figure 4).

The metal oxide complexes were deposited on support surfaces from an organic solvent. After removal of the solvent, oxidation with $O_2$ at 400-700° gave the supported metal oxide. Reduction with $H_2$, $NaBH_4$ or $AlEt_3$ gave the

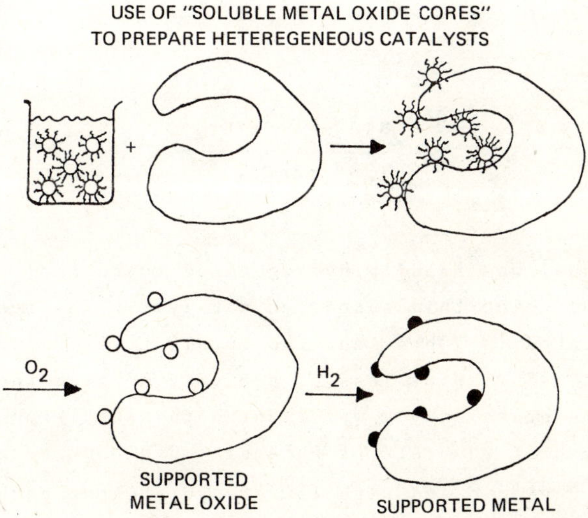

FIG. 5.

supported metal. Supported oxides (CoO, NiO, $Fe_2O_3$, and $CoO-Fe_2O_3$) were prepared on alumina and kieselguhr. After reduction with $AlEt_3$ all these systems were very active hydrogenation catalysts for aromatic systems. These are described in the next chapter in more detail.

Hydrogenation of Aldehydes to Alcohols

A reduced $SiO_2$-Ni-Cu-Cr- catalyst was prepared and used in the hydrogenation of aliphatic and aromatic aldehydes to alcohols. These reactions were patterned after those described in US Patent 3,803,055 which used a silica-supported catalyst prepared by the classic salt impregnation techniques. Silica (8-12mesh, $300m^2/g$, pore volume = 1cc/g) was treated with aqueous NaOH to bring its pH to 8.2. After washing and drying ($120^0$ *in vacuo* and $500^0$, 4hrs., under nitrogen flow), the silica was treated with pentane solutions of both the soluble NiO (Ni/acid = 33, 122Å diameter) and CuO (Cu/acid = 19, 280Å diameter in $CCl_4$) and chromium octoate, followed by drying in vacumn. The resulting solid was heated for 2 hrs. at $500^0$ in oxygen to give a supported metal oxide catalyst analyzing for 8%Cu, 2.7%Ni and 0.4%Cr. The oxide catalyst was reduced by heating in a hydrogen atmosphere at $200^0$ (1hr.) and $300^0$ (1.5hr.)

$$H_2 + \text{butanal (CHO)} \xrightarrow[160^0,\,200psi]{SiO_2\text{-Ni-Cu-Cr}} \text{butanol (OH)}$$

Butanal was readily hydrogenated to butanol at $160^0$ and 200psi using this supported catalyst. The reaction was completed in 20hrs. and the catalyst was recycled several times without loss of activity. Under the same conditions benzaldehyde was reduced to benzylalcohol at about one half the rate of butanal. Cyclohexanecarboxaldehyde, contaminated with 15% of cyclohexane carboxylic acid, was reduced to cyclohexylcarbinol at a rate comparable to the benzaldehyde reductions.

Demethylation of 1-Methylnaphthalene

The alumina-supported Ni and NiO catalysts (described in the experimental section) were prepared by deposition of soluble NiO particles on $Al_2O_3$ followed by oxidation ($O_2$, 400°) and reaction with $NaBH_4$. They were active in the demethylation of 1-methylnaphthalene at 320° and 1400psi. Not only did demethylation occur, but hydrogenation of the naphthalene ring system occurred in competition with demethylation. Product distributions as a function of time showed that the hydrogenation was independent of demethylation and that naphthalene produced by demethylation was sequentially reduced to *cis* and *trans* decalin. The $Al_2O_3$-Ni catalyst was substantially more active than $Al_2O_3$-NiO.

$CH_3$ — $H_2$, 320°, 1400psi / $Al_2O_3$-Ni, batch, 17hr. → 2% + $CH_3$ 13% ; 74% + 11%

## CONCLUSIONS

Metal oxide particles with surfaces containing bound carboxylates have been prepared by reacting metal oxides (or metals) with a deficiency of carboxylic acids. The outer amorphous layer renders these metal oxides soluble in hydrocarbon solvents. Aggregation equilibria were discovered and aggregation is promoted by polar solvents. These particles serve as catalysts in "homogeneous reactions" where catalysis actually takes place at the solution-particle interface. The unique feature here is that a liquid-solid interface is the site of catalysis, yet the solid is "solublized" in the liquid to give a fluid-phase reaction which has many features of homogeneous catalysis. This represents a new type of interfacial catalysis.

Several advantages may result from the application of this concept. First, these particles will be compatible with the use of hydrocarbon solvents. In exothermic reactions, the solvent may be used to carry heat away from the particles' surfaces and "hot-spots" should not become a problem. Reactions that are promoted by the solvent can now be carried out, if that solvent is also matched with the particles' carboxylic acid distribution to insure solubility. These particles will resist agglomeration, precipitation, and the clogging of equipment which can occur if suspended metal oxides were employed instead. By the use of centrifugation gradient separations, in good solvents, narrow particle size distributions can be isolated. These might be used to prepare supported (heterogeneous) metal oxide catalysts with defined metal oxide particle sizes.

## REFERENCES

[1] C.U. Pittman, Jr., Chemtech., 560 (1973).

[2] R.H. Grubbs, Chemtech., 512 (1977).

[3] D.D. Whitehurst, Chemtech., 44 (1980).

[4] H.M. Olson, US Patent 2,573,049 (1951).

[5] A. Alkaitis and P.L. Cells, US Patent 4,163,986 (1979).

[6] For further information contact Mooney Chemicals, Inc. (814-432-2125).

[7] US Patent 3,671,588 (1972) to Allied Chemical Corporation.

# Characterization of Hydrocarbon Soluble Metal Oxides. Precursors to Supported Catalysts

Edwin A. Lewis , Mohammed Habib, and Charles U. Pittman, Jr.

Department of Chemistry
University of Alabama
University, Alabama 35486

ABSTRACT

The metal oxide carboxylate complexes described in the previous chapter have been characterized by infrared spectroscopy, x-ray diffraction, electron microscopy, and analytical ultracentrifugation. The molecular weights and solution particle diameters have been determined for a number of the hydrocarbon soluble particles by analytical ultracentrifugation methods, and the use of a spherical model consistent with the particulate shape observed by electron microscopy. The size of the soluble complexes has been studied as a function of: metal, metal/acid ratio, acid composition, and solvent. The molecular weights for the ultimate particles are reasonably independent of the metal employed in the synthesis and are relatively constant for materials with similar metal/acid equivalents ratios. The single particle molecular weights for the complexes studied ranged from approximately $5x10^4$ to $1.5x10^6$g mole$^{-1}$. The solution size distribution was polydisperse in all cases, with aggregates of the ultimate particles prevalent. Weight average molecular weights in excess of $10^9$g mole$^{-1}$ have been observed. The aggregation is dependent on the surface acid composition and on the solvent in which the soluble complex is dispersed. Solubility and stability of these materials have been examined in a number of solvents. The metal oxide particles are initially soluble in octane, iso-octane, cyclohexane, mineral spirits, carbon tetrachloride, benzene, and tetrahydrofuran. However, most of the complexes eventually precipitate from dilute solutions in carbon tetrachloride, benzene, and tetrahydrofuran. The stability of the particles is decreased in the presence of oxygen, or when carboxylic acids, alcohols, or ketones are present even in small amounts, and is decreased even further at temperatures above 100°C. Heterogeneous catalysts have been prepared by deposition of several of the soluble metal oxides onto supports such as alumina, silica, or kieselguhr followed by oxidation to yield supported metal oxides or reduction to yield supported metal. The application of few supported metals and metal oxides in hydrogenations of olefins, benzene, and naphthalene is described.

## INTRODUCTION

In the preceding chapter entitled "Homogenizing Metal Oxide Catalysts" we described the synthesis and some potential applications of a new type of high molecular weight soluble metal oxide carboxylate complex. Any catalytic activity of the soluble metal oxide particles must depend to a large degree on the availability of the metallic core region to the reactants in solution, and therefore both the surface to volume ratio and the surface structure in terms of carboxylate coating density must be extremely important. A schematic diagram of the particulate structure is presented in Figure 1. In looking at this structure it should be obvious that an

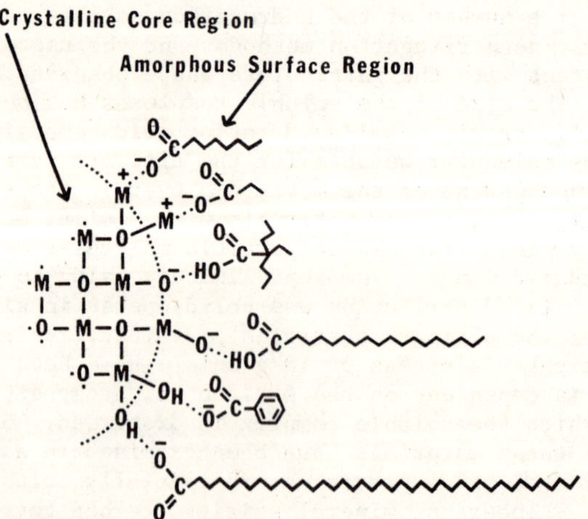


Figure 1. Schematic drawing of the crystalline core region of a metal-carboxylate complex.

increase in particle diameter would place a smaller fraction of the metal ions at the core surface, but it is more difficult to draw any conclusions about the dependence of reagent accessibility of the core on particle size. Nevertheless, one of the more important characteristics of these materials must be their molecular size, with the largest complexes expected to more closely resemble heterogeneous catalysts while the smallest complexes should look more like homogeneous catalysts. A very attractive feature of these oil soluble metal oxide particles is that the conditions of synthesis and the composition of the complex can apparently be manipulated to produce particles with a range of sizes. This ability to control the solution size of catalytically active particles could lead to the tailoring of the catalytic properties of these materials to specific applications. In addition the variation in solution size could presumably be exploited in the preparation of the supported metal and metal oxide catalysts, for example in controlling the catalytic site density or supported metal cluster size distribution.

The initial tasks of the work presented here were to elucidate the chemical structure of the novel hydrocarbon soluble metal oxide carboxylates and to determine the factors that influence size and solubility of the particles. In this regard the choice of metal, the metal/acid equivalents ratio, and the acid composition have been examined as synthesis variables, while solvent, polar solvent additives, oxidizing and reducing conditions, and temperature have been examined as post synthetic variables affecting size, solubility and structural stability. The final task of these investigations was to explore the application of the hydrocarbon-soluble metal oxides to the preparation of supported metal and metal oxide catalysts. In this regard supported metal and metal oxide materials were prepared from their hydrocarbon soluble complexes and tested in hydrogenation reactions.

## EXPERIMENTAL

Preparation of Hydrocarbon Soluble Metal Oxides. The detailed synthesis of the soluble metal oxide particles has been described in detail by Alkaitis and Cells in their 1979 patent (1), and was schematically diagrammed in the previous chapter (see figure 1, previous chapter). The synthesis of one specific complex of CoO is described below and some of the metals, carboxylic acids, and metal/acid equivalents ratios that have been utilized in the preparation of other complexes are listed in Table 1. It should be pointed out, that not all possible combinations implied in Table 1 have been prepared or examined; however, the information in this table does serve to indicate the range of variation possible in the complex preparation. The synthesis can be accomplished from either the metal or metal oxide and a mixture of two or more acids (usually three) mixed in the desired equivalents ratio.

The specific procedure for preparation of a cobalt carboxylate material is as follows. A mixture of 3 parts of propionic acid, 23 parts of neodecanoic acid, 49 parts $C_{9-13}$ acid (neo-isomers), 2 parts polypropylene glycol, 0.2 parts of an antioxidant, 32 parts of mineral spirits, 12 parts cobalt metal, and 5 parts water was charged to a reaction vessel. The mixture was agitated and heated

Table 1

Metal Oxide Carboxylate Complex Component Variations.[a]

| | | | | |
|---|---|---|---|---|
| Metal or Metal Oxide Employed: | Cr, Mn, Fe, Co, Ni, Cu, Mn-Co | | | |
| Carboxylic Acids Employed: | $C_3$, $C_8$, neo $C_{10}$, $C_{18}$, $C_{40}$, Benzoic | | | |
| Typical Acid | 15% $C_3$<br>23% $C_8$<br>62% neo $C_{10}$ | 20% $C_3$<br>70% neo $C_{10}$<br>10% $C_{18}$ | 20% $C_3$, 50% $C_8$<br>15% neo $C_{10}$<br>15% $C_{18}$ | 25% $C_3$, 20% $C_8$<br>25% $C_{10}$, 15% $C_{18}$<br>15% $C_{40}$ |
| Metal/Acid Equivalents Ratio: | 6, 12, 18, 24, 30, 36, 42 | | | |

a. All possible variations have not been prepared or investigated.

b. Several acid mixtures not listed here have been employed including many with aromatic acid moieties comprising from 5 to 15% of the mixture.

to 75°C. Air was then purged through the batch as an oxygen source to promote the reaction. Upon completion of the reaction the material was heated in the absence of air to 150°C to remove water. The product was cooled, filtered, and diluted with mineral spirits. The resulting soluble CoO carboxylate had a metal/acid equivalents ratio of approximately 6.0.

Structural Studies. Infrared studies were done on $CCl_4$ and hydrocarbon solutions, and on films of the hydrocarbon soluble metal oxide samples dried on salt plates. The infrared spectra of the complexes were compared to reference spectra for carboxylic acids and carboxylate salts, with particular attention to a comparison of the bands arising from both monomeric and dimerized carboxylic acid groups.

X-ray diffraction experiments were done on $CCl_4$ and hydrocarbon solutions of the metal oxide complexes. Diffraction patterns were obtained using a powder type camera with the liquid samples held in a glass capillary, which was noncrystalline and highly transparent to x-rays. The crystal phase identification was made in reference to the diffraction patterns of known materials.

Transmission electron microscopy studies of samples prepared by evaporation of $CCl_4$ solutions on Formvar film were made along with electron diffraction studies. The micrographs were normally taken at 2,000,000 x magnification and the diffraction patterns were recorded with a sample to photographic plate distance of 50 cm. The crystalline phase identification was again made in reference to the electron diffraction patterns of known materials.

Solution Molecular Weight and Particle Size Measurements. The metal oxide particle size and molecular weight in solution was examined by the use of sedimentation velocity techniques employing an analytical ultracentrifuge. These experiments were the method of choice since colligative property measurements would severely underestimate the complex size as the metal oxide particle solutions contain significant amounts of dissociated carboxylic acid and perhaps even small amounts of simple metal ion carboxylates, and because gel permeation chromatography experiments would also suffer from such association equilibria and from the expected interactions between the gel matrix and the particulate surfaces. Initial attempts to determine particle molecular weights by sedimentation equilibrium methods also failed, as particle aggregation occurring during the time required to reach an equilibrium concentration distribution in the centrifuge cell resulted in precipitation of aggregates rather than achieving a distribution of the ultimate particles. This aggregation process has been examined further and will be discussed later in this chapter. Subsequent sedimentation equilibrium measurements on a few samples in good solvents have corraborated the velocity results.

All of the ultracentrifuge measurements were done using a Beckman Model E analytical ultracentrifuge equipped with a photoelectric scanner. All experiments employed an AnD rotor run at speeds of from 5,000 to 20,000 rpm and the temperature was controlled at 20°C. The cells employed double sector aluminum centerpieces with Kel-F gaskets and the scanner was operated in the split beam absorbance mode. Sample solutions had absorbances of approximately 0.5 in the wavelength range of 280 to 450 nm. The sedimentation coefficients, S, were calculated from the least squares slope of the logarithim of the boundary radius plotted as a function of time. The partial specific volumes, $\bar{v}$, for the soluble complexes were calculated from the densities of concentrated solutions of known composition. In all cases the $\bar{v}$ values were close to the weighted average of the reciprocal of the metal oxide and carboxylic acid densities, thus the assumption of limited (i.e. negliglble) preferential solvation appeared to be a reasonable one.

The use of sedimentation velocity data along with a spherical model for the solution complex shape has allowed the determination of molecular weights and particle diameters by combination of the following equations:

$$S = \frac{M(1-v\rho)}{Nf} \quad (1)$$

$$f = 6\pi\eta R_s \quad (2)$$

$$R_s = \frac{3\bar{v}M}{4\pi N}^{1/3} \quad (3)$$

where S is the sedimentation coefficient, M is the molecular weight, $\bar{v}$ is the partial specific volume, $\rho$ is the solution density, N is Avogadros number, f is the fricitional coefficient for a sphere, $\eta$ is the solvent viscosity, and $R_s$ is the spherical particle radius. The use of a spherical model seemed to be a reasonable assumption and electron micrographs have shown both the ultimate particles and the aggregates to be spherical when solvent is evaporated to leave a particulate film.

Supported Catalysts Prepared from Soluble Metal Oxides. The preceeding chapter described the deposition of several soluble metal oxide preparations onto supports such as alumina, Silica, or Kieselguhr followed by oxidation with oxygen, to give supported metal exides, and reduction with hydrogen or other reagents ($AlEt_3$ or $NaBH_4$), to give supported metals. The preparation of several other supported heterogeneous catalysts from the soluble metal oxides is now described.

Scheme I, A. shows the preparation of an alumina-supported cobalt oxide and cobalt from soluble cobalt oxide particles (Co/acid = 42). These catalysts were employed in olefin, benzene, and naphthalene hydrogenations. Two types of Kieselguhr-supported nickel catalysts were prepared as shown in Scheme I, B. The first was reduced with hydrogen and the second with $AlEt_3$. These were used in the hydrogenation of benzene. An alumina supported iron and an alumina-supported mixed cobalt iron catalysts were prepared as shown in Scheme I,C. and Scheme I,D. respectively. These catalysts were also tested in the hydrogenation of benzene.

Scheme I.

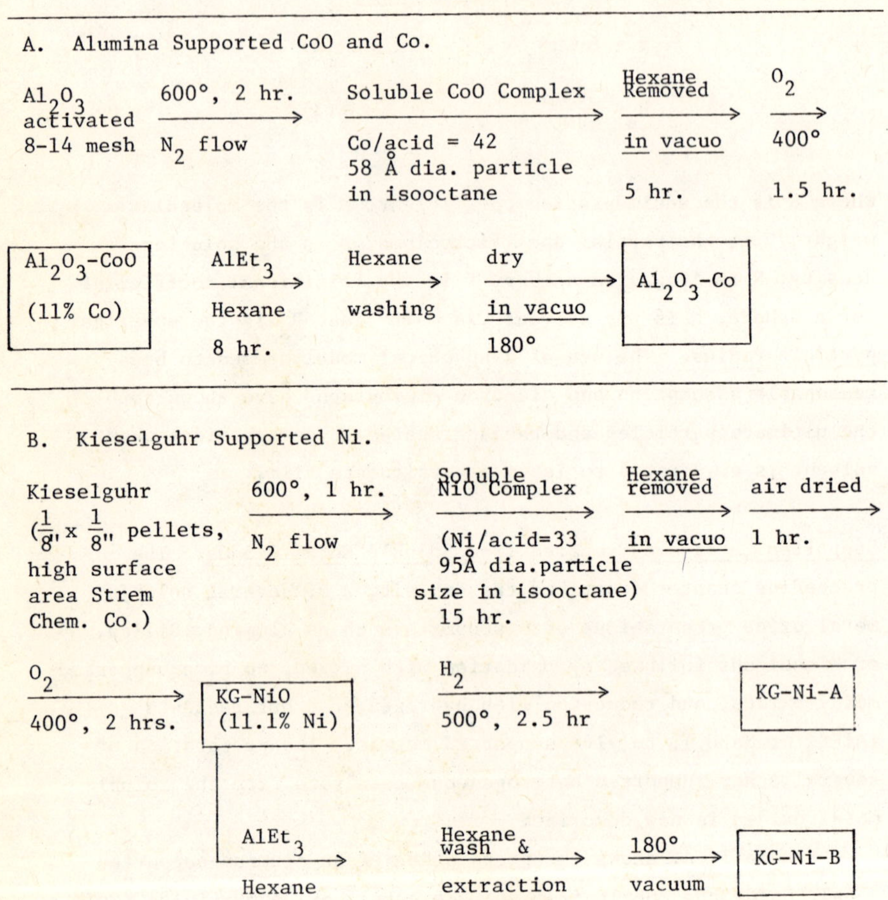


## RESULTS AND DISCUSSION

Soluble Metal Oxide Carboxylate Structure. The formation of a metal carboxylate salt is characterized in the IR by the formation of a strong asymmetric vibrational band in the region from 1650-1540 $cm^{-1}$ and a somewhat weaker symmetric stretch at 1450-1360 $cm^{-1}$. The former replaces the corresponding acid C=O band at 1740-1680

---

C. Alumina Supported Fe.

$Al_2O$ Activated 8-14 mesh Fisher $\xrightarrow[\text{in vacuo}]{200°,\ 2\ hr.}$ $\xrightarrow[\text{Fe/acid = 9, 5 hr.}]{\text{Soluble } Fe_2O_3 \text{ Complex}}$ $\xrightarrow[\text{in vacuo}]{\text{pentane removed}}$ $\xrightarrow[\text{1 hr.}]{\text{air dried}}$

$\xrightarrow[500°,\ 2\ hr.]{O_2}$ $Al_2O_3$-$Fe_2O_3$ 10% Fe $\xrightarrow[12\ hrs.,\ 25°]{AlEt_3/\text{hexane}}$

$\xrightarrow[\text{wash}]{\text{Hexane}}$ $\xrightarrow[\text{in vacuo}]{180°,\ 2\ hr.}$ $Al_2O_3$-Fe

---

D. Alumina Supported Co/Fe.

$Al_2O_3$ activated 8-14 mesh Fisher $\xrightarrow[\text{in vacuo}]{200°}$ $\xrightarrow[\text{3 hr.}]{\text{Mixture of Soluble CoO and } Fe_2O_3 \text{ Complexes}}$ $\xrightarrow[\text{in vacuo}]{\text{pentane removed}}$ $\xrightarrow[\text{1 hr.}]{\text{air dried}}$

$\xrightarrow[\text{in vacuo}]{200°,\ 1\ hr.}$ $\xrightarrow[500°,\ 2\ hr.]{O_2}$ $Al_2O_3$-CoO-$Fe_2O_3$ 5% Co, 5% Fe

$\xrightarrow{AlEt_3 \text{ (Hexane) } 25°}$ $\xrightarrow[\text{2) 180°, 2 hr. in vacuo}]{\text{1) washed with Hexane}}$ $Al_2O_3$-Co-Fe

---

$cm^{-1}$ and the characteristic acid dimer band at 960-875 $cm^{-1}$ disappears. The spectra of all of the soluble metal oxide carboxylates are indicative of carboxylate salts, $RCO_2M$.

In addition to indicating the type of acid-metal bonding, the infrared spectra in the region of 3600-3000 $cm^{-1}$ reveal a characteristic difference in the -OH absorption between the high molecular weight soluble metal oxides and simple metal carboxylate soaps. The presence of extensive hydrogen bonding of the -OH is evident in the spectra of the soluble complexes while only minor if any -OH and H-bonding is observed for the soaps. The large amount of -OH and H-bonding seen in the complexes is logically associated with the inorganic or metal portion of the particle. The infrared studies are consistent with the particulate structure shown in Figure 1 and would seem to corraborate the hypothesis of core (inorganic) and surface (carboxylate salt) regions.

When solutions of the soluble metal oxides were subjected to FeK radiation both crystalline and amorphous scattering patterns were observed. The sharpness of the lines in the diffraction patterns obtained on these solutions was indicative of crystallites in the 100-300 Å size range. The diffraction from the crystalline phase was sufficiently distinctive that the crystalline domains could be identified by reference to the diffraction patterns of known metal oxides. Crystallites of $Mn_3O_4$, CoO, $Fe_3O_4$, and CuO were observed in complex particulate solutions while the amorphous scattering was attributed to the metal carboxylate surface region surrounding the metal oxide crystalline core region. Of course the amorphous patterns were not sufficiently distinctive for phase identification, however, the mere demonstration of crystalline regions is sufficient to distinguish between low metal/acid equivalents ratio complexes (soaps) and the high metal/acid equivalents ratio materials we were investigating. Again the X-ray diffraction results are in accord with the structural model given in Figure 1.

Electron microscopy and electron diffraction studies on dried samples of the soluble metal oxide complexes demonstrated the particulate nature of these materials and corraborated the x-ray identification of the crystalline core regions of the complexes. Figure 2 shows a typical electron micrograph for a complex preparation. The material shown here is a $Mn_3O_4$ complex (Mn/acid = 18) and the ultimate particle size, about 3 mm on the print, corresponds

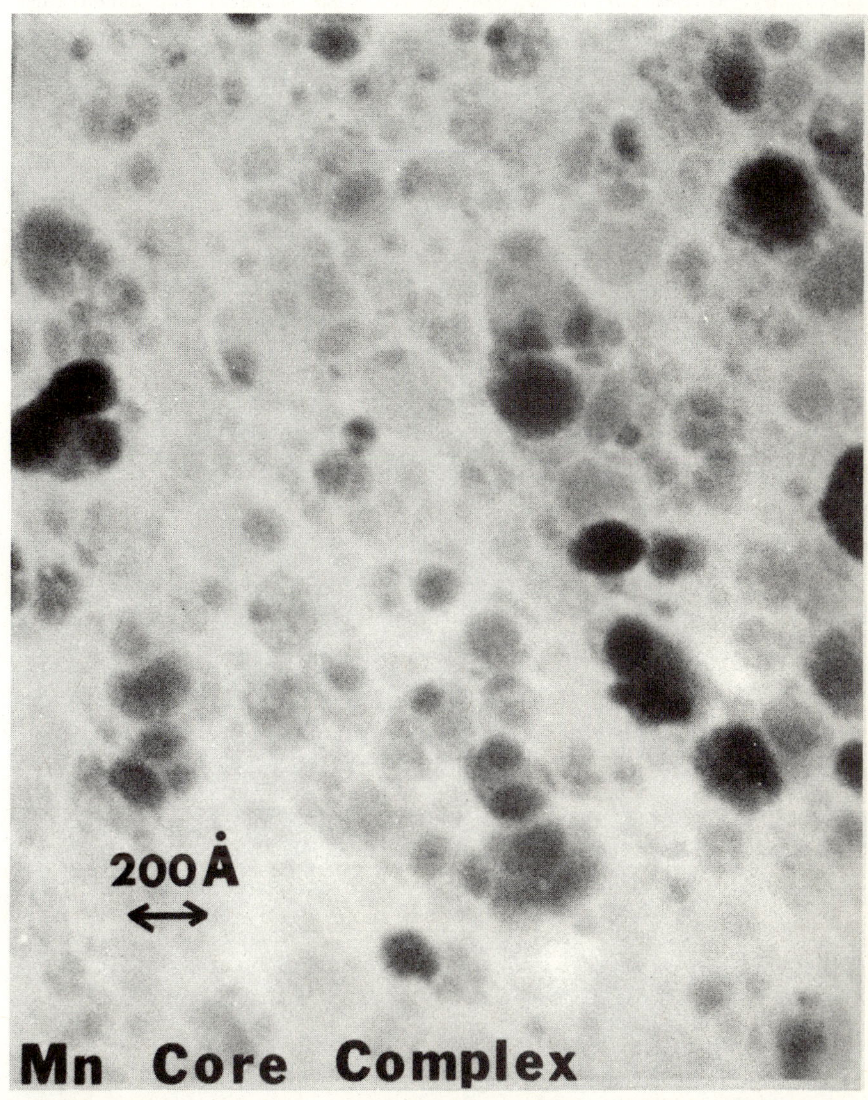


Figure 2. Electron micrograph of a carbontetrachloride evaporated sample of a soluble $Mn_3O_4$ complex with a metal/acid ratio of 18.

to a diameter of approximately 50 Å. In addition to the ultimate particles, larger particles, 200 to 300 Å in diameter and appearing to be aggregates of the smaller units, are prevalent. Both the solution particle size distribution and the aggregation of particles in solution as opposed to the dried film particulate dimensions will be discussed below. It should also be obvious from examination of Figure 2 that both the ultimate particles and aggregates are well described as spheres. Again the electron microscopy studies substantiate the hypothetical structure we presented in Figure 1.

Solution Size, Size Distrubution, and Controlling Factors. In the preceding chapter we presented three conclusions about the soluble metal oxide solution size; 1) the ultimate particle size tended to increase as the metal/acid equivalents ratio increased, 2) the particule size was independent of surface acid composition in a good solvent, and 3) particle aggregation was quite solvent dependent. Actually the situation is even somewhat more complicated.

The very first sedimentation experiments we conducted gave an indication that the soluble metal oxide particles were present in a distribution of sizes. Even for a single sample the size distribution was apparently dependent on the solvent and to a lesser degree on variations in the sedimentation velocity experiment itself, e.g. rotor speed.

The solution size measurements were complicated to an extreme degree by both surface equilibria and ultimate particle aggregation, even in good solvents. These complications foiled completely our early attempts at determining the molecular weight distributions by sedimentation equilibrium methods. We then decided to get estimates of the particle's solution diameters from the sedimentation coefficients measured in sedimentation velocity experiments combined with the use of a spherical model to calculate the frictional coefficients. The particle solution dimensions obtained from the sedimentation velocity measurements were in excellent agreement with the size estimates made from the electron micrographs of dried solutions. In addition all of the hydrodynamic experiments corraborated the complex model given in Figure 1.

Figure 3 illustrates three different types of sedimenting boundaries that were observed in various ultracentrifuge experiments on the soluble metal oxides. Single sharp boundaries were only rarely seen, while both single broad boundaries (indicative of polydisperse samples) and multiple boundaries (indicative of mixtures of discrete particles) were quite commonly observed. We have labeled samples exhibiting broad boundaries as containing non-specific aggregates of ultimate particles while solutions showing multiple boundaries are labeled as containing specific aggregates. These two aggregation models are shown schematically in Figure 4. The type of aggregation a particular soluble complex participates in must be determined by the particulate surface topography (dependent on acid distribution and solvent) and by packing considerations.

Prior to continuing our discussion of the aggregation equilibria and the concommittant solubility effects we would like

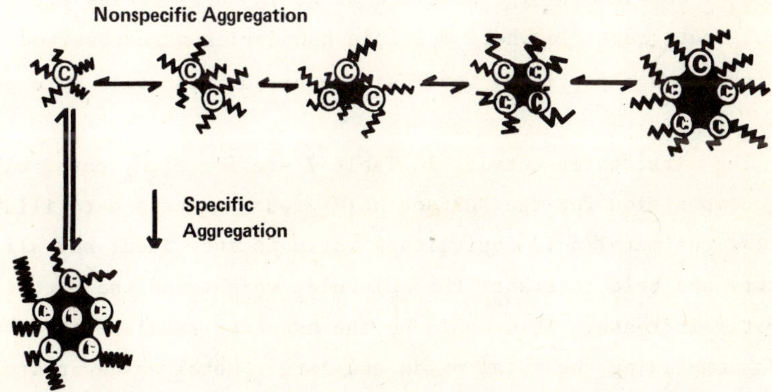


Figure 3. Schematic drawing of nonspecific and specific aggregation models. The nonspecific case yields a continuum of particle sizes while the specific yields only discrete particle sizes.

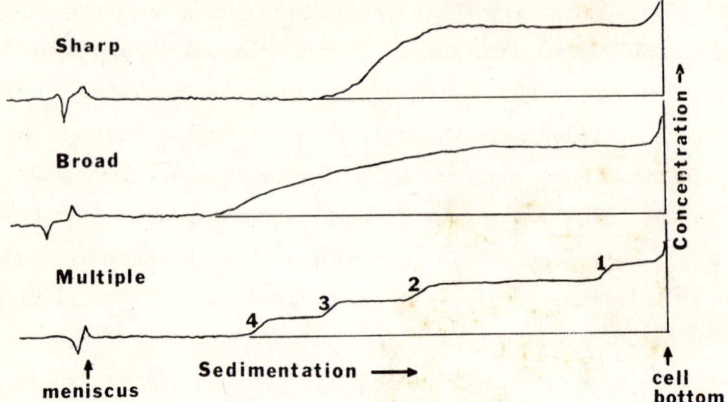


Figure 4. Boundary morphology as traced from the photoelectric scanner recordings of three different velocity experiments

to briefly review the solution size effects described in the previous chapter. Tables 2 and 3 summarize representative sedimentation velocity work and report the molecular weights and particle diameters for a number of complex metal oxide carboxylates in various solvents. Table 2 presents the data on a number of different $Mu_3O_4$ core complexes, while Table 3 presents the data on CoO, NiO, CuO, and mixed $CoO/Mu_3O_4$ complexes. The particle sizes in these tables are for the ultimate particle where multiple boundaries were observed but are for average particles where single broad boundaries were present.

The first three entries in Table 2 are for $Mn_3O_4$ cores with the same composition for the surface acid groups. These data illustrate that as the metal/acid equivalents ratio is increased; and all other factors are held constant, the molecular weight and the particle diameter increase. This would be the expected result with fewer acids complexing the metal oxide and larger metal oxide crystalline domains prevailing. It seems obvious that at some point the solubility of the particle surface would no longer be sufficient to allow dispersion of the particle, but this limit is difficult to determine experimentally. In the cases observed in this study

Table 2

Selected Hydrodynamic Data for Soluble Manganese Oxide Carboxylates

| Metal | M/A (equivalent ratio) | Sedimentation coefficient(s) | Molecular Weight | Diameter | Solvent |
|---|---|---|---|---|---|
| Mn ($Mn_3O_4$) | 6[a] | 29 | $6.3X10^4$ | 35 | Mineral Spirits |
| Mn ($Mn_3O_4$) | 12[a] | 38 | $1.2X10^5$ | 43 | " |
| Mn ($Mn_3O_4$) | 18[a] | 52 | $1.5X10^5$ | 48 | " |
| | 18[b] | 51 | $1.5X10^5$ | 47 | " |
| | 18[c] | 49 | $1.4X10^5$ | 47 | " |
| | 18[d] | 52 | $1.5X10^5$ | 48 | " |
| Mn ($Mn_3O_4$) | 18[d] | 104 | $0.5X10^6$ | 72 | Cyclohexane |
| | | 184 | $1.2X10^6$ | 80 | Octane |
| | | 214 | $1.2X10^6$ | 95 | Isooctane |
| | | 220 | $1.5X10^6$ | 103 | $CCl_4$ |
| | | 281 | $1.9X10^6$ | 111 | Benzene |
| | | 369 | $2.9X10^6$ | 128 | THF |

[a]Acid composition 25% $C_3$, 25% $C_8$, 25% $C_{10}$, 15% $C_{18}$, and 10% benzoic acid

[b]Acid composition 35% $C_3$, 35% $C_8$, 10% $C_{10}$, 15% $C_{18}$, and 5% benzoic acid

[c]Acid composition 50% $C_4$, 40% $C_8$, 10% $C_{18}$

[d]Acid composition 40% $C_4$, 40% $C_8$, 20% $C_{18}$

it appears that the solubility is decreased mainly by particle aggregation, and that the maximum ultimate particle has not been observed for any of the metal/acid ratios achieved so far. Entries three through six in Table 2 illustrate the apparent independence of particle size on acid composition, while entries seven through twelve show an apparent dependence on solution particle size on the solvent in which the metal oxide carboxylates are dispersed. The increase in particle size in $CCl_4$, benzene, and THF is thought to be due to the aggregation of the ultimate particles by one of the schemes illustrated in Figure 4. The hypothesis is that

Table 3

Selected Hydrodynamic Data for Soluble Metal Oxide Carboxylates.

| Metal | M/A (equivalent ratio) | Molecular Weight | Diameter Å | Solvent |
|---|---|---|---|---|
| Co(CoO) | 24[a] | $8.9X10^5$ | 86 | Isooctane |
| | | $4.6X10^6$ | 148 | Benzene |
| | | $5.7X10^6$ | 160 | $CCl_4$ |
| | | $7.4X10^6$ | 180 | Min. Spirits |
| Co(CoO) | 42[a] | $5.3X10^5$ | 74 | Isooctane |
| | | $8.9X10^6$ | 180 | Benzene |
| | | $1.1X10^7$ | 206 | Min. Spirits |
| | | $3.3X10^9$ | 1270 | $CCl_4$ |
| Ni(NiO) | 33[b] | $9.0X10^5$ | 90 | Isooctane |
| | | $3.1X10^6$ | 122 | Min. Spirits |
| | | $6.2X10^6$ | 154 | THF |
| | | $7.9X10^6$ | 167 | Benzene |
| | | $4.0X10^7$ | 286 | $CCl_4$ |
| CoMn | 20[c] | $1.7X10^5$ | 56 | Isooctane |
| | | $1.9X10^5$ | 60 | Benzene |
| | | $5.2X10^5$ | 80 | Min. Spirits |
| | | $3.4X10^7$ | 320 | $CCl_4$ |
| Cu(CuO) | 19[d] | $4.4X10^6$ | 121 | $CCl_4$ |
| | | $4.3X10^8$ | 540 | Isooctane |
| | | $5.5X10^8$ | 608 | Benzene |
| | | $7.3X10^8$ | 668 | Min. Spirits |

a. Acid Composition 50% $C_4$, 40% $C_8$, 10% $C_{18}$

b. Acid Composition 20% $C_3$, 50% $C_8$, 15% neo $C_{10}$, 15% $C_{18}$

c. Acid Composition 20% $C_3$, 30% $C_8$, 30% neo $C_{10}$, 20% $C_{18}$

d. Acid Composition 25% $C_3$, 20% $C_8$, 25% $C_{10}$, 15% $C_{18}$, 15% $C_{40}$

entanglement of the longer carboxylic acid hydrocarbon chains on the surface of one particle with surface groups on other particles depends on the ability of the solvent to solvate these groups relative to the ability of surface groups to solvate one another. Further evidence to support this general mechanism of aggregation can be taken from the fact that the oil-like solvents, mineral spirits and isooctane, are the best solvents for the most complexes, i.e. the average particle size is smallest in these solutions.

In addition the extremely large particles formed by the Cu complex even in a rather good oil-like solvent such as isooctane seems consistent with the proposed aggregation mechanism when one considers just how sticky a surface coated with $C_{40}$ acids might be toward other $C_{40}$ coated particles. The apparent inversion of "goodness" of solvent seen for the Cu complex in Table 3 ($CCl_4$ giving the smallest particles) is a anomaly arising from the precipitation of most of the complex from $CCl_4$ solutions leaving the smaller aggregates as the average solution particles.

Specific versus non-specific aggregation has already been defined (see Figure 4.) and the data in Table 4 show specific aggregation of a $Mn_3O_4$ core complex in mineral spirits. Even in a good solvent like mineral spirits, aggregation was commonly encountered with the better solvents showing specific aggregation most often and the poorer solvents almost always showing non-specific aggregation.

By definition, a good solvent should yield solutions with a significant concentration of ultimate particles and small aggregates. As stated previously, the average particle size in a good

Table 4

Multiple Boundary Evidence for Specific Aggregation of a $Mn_3O_4$ Core Complex in Mineral Spirits

| Boundary Number[a] | Sedimentation Coefficient | Molecular Weight | Number of Ultimate in the Aggregate[b] |
|---|---|---|---|
| 1 | 120.2 | $1.5X10^6$ | 94 (93.8) |
| 2 | 83.9 | $6.3X10^5$ | 40 (39.4) |
| 3 | 41.7 | $2.2X10^5$ | 14 (13.8) |
| 4 | 33.6 | $8.0X10^4$ | 5 (5.00) |
| 5 | 29.1 | $6.3X10^4$ | 4 (3.93) |

a. see Figure 3.

b. smallest observed particle assumed to be the ultimate particles had $M_w$ estimated to be $1.6X10^4$.

solvent is almost independent of the surface acid composition. This is reasonable since the "goodness" of the solvent would be a measure of its ability to solvate the various acid hydrocarbon tails projecting from the complexed metal oxide surface. Table 5 presents data on the particle size of some $Mn_3O_4$ complexes in a somewhat poorer solvent, isooctane (for comparison see Table 3 in the preceding chapter). These data indicate that the compatibility of the surface acids with the solvent is the most important factor governing the average solution particle size.

Solubility and Stability Studies. From the above discussions of particle structure (see Figure 1) and particle aggregation, it is obvious that the carboxylate composition at the surface should have a profound influence on solubility. All of the materials

Table 5

Effect of Acid Distribution on Particle Size in a Poorer Solvent

| $Mn_2O_3$ Complexes in Iso-octane<br>Metal/Acid | Diameter, Å | Molecular Weight |
|---|---|---|
| 6[a] | 128 | $1.3X10^6$ |
| 6[b] | 55 | $2.4X10^5$ |
| 18[a] | 76 | $4.9X10^5$ |
| 18[c] | 72 | $4.4X10^5$ |
| 18[d] | 97 | $1.3X10^6$ |
| 18[e] | 142 | $4.8X10^6$ |

[a]25% $C_3$, 25% $C_8$, 25% $C_{10}$, 15% $C_{18}$, 10% Benzoic

[b]20% $C_3$, 70% neo $C_{10}$, 10% $C_{18}$

[c]35% $C_3$, 35% $C_8$, 10% $C_{10}$, 15% $C_{18}$, 5% Benzoic

[d]20% $C_3$, 40% $C_8$, 20% $C_{18}$, 20% neo $C_{10}$

[e]50% $C_8$, 40% $C_8$, 10% $C_{18}$

prepared have shown good initial solubility in a variety of solvents including octane, iso-octane, cyclohexane, mineral spirits, mineral oil, carbon tetrachloride, benzene, and tetrahydrofuran. Most of the complexes eventually precipitate from $CCl_4$, benzene, and THF. The precipitation from dilute solutions is slow requiring from 24 hours to one week and is thought to reflect the time for maximum surface entanglement of separated particles. Perhaps carboxylate groups with halogenated alkyl moieties would enhance the solubility in $CCl_4$, while inclusion of more aromatic carboxylic acids in the surface layer might be expected to improve solubility in benzene.

The very nature of the weak ionic and H-bonding interactions holding the non-polar surface coating to the metal oxide core accounts for one of the drawbacks to the use of these materials as catalysts in reactions producing polar products. As the solvent polarity is increased, the surface groups are released from the core and precipitation of the metal oxide results. The stability of cobalt catalyst to polar solvent additives and an oxidizing agent is given in Table 6. It is clear that particle decomposition of the materials studied under mild conditions is serious and at elevated temperatures (above 100°C) and at $O_2$ pressures of 10 atmosphere the particle stability is even further decreased. The CoO and $Mn_3O_4$ soluble particles are readily destroyed by peroxides in solution and they are less stable when heated above 100° in alcohols or carboxylic acids. The cobalt particles go, in part, to $Co(OAc)_3$ when heated to 150°-200° in acetic acid.

Selected stability studies of a $Mn_3O_4$ core catalyst material under the more rigorous conditions are presented in Table 7. The main conclusions of the particle stability studies can be summarized as follows: 1) the core catalysts show good stability in inert solvents, e.g. cyclohexane, and in inert atmospheres ($N_2$) even at temperatures as high as 200°C and pressures up to 200 psi. 2) the catalyst stability is unaffected by changing the

Table 6

Catalyst Particle Stability at 25°C and Ambient Pressure With Ketone, Alcohol, and Peroxide.[a]

| Sample | Additive | Observation |
|---|---|---|
| 0.012 g CoO Catalyst 1 ml benzene | none | no ppt. or other changes after 18 hr. |
| " | Cyclohexanone (0.05ml) | no precipitation after 8 hr. |
| " | Cyclohexanol (0.05ml) | turned cloudy and some brown ppt. observed after 6 hr. |
| " | Cyclohexanone (0.05ml) Cyclohexanol (0.05ml) | turned cloudy and brown ppt. was observed after 5 hr. |
| " | t-butylhydro-peroxide (0.05ml) | darkened and ppt. appeared in 2 hr. The sample was completely precipitated in 12 hr. |

[a]A similar series of studies using 0.01g of a $Mn_3O_4$ core catalyst in 1 ml of benzene gave the following stability order as a function of additive: no additive >>cyclohexanone >cyclohexanol >t-butyl-hydroperoxide. In 15 hrs. the complex was completely precipitated with perioxide.

atmosphere to a reducing one, e.g. $H_2$, CO, or a mixture of $H_2$ and CO, 3) the stability is decreased in an oxidizing atmosphere e.g. $O_2$ or $O_2/N_2$ mixture, 4) the catalysts are also much less stable when water, cyclohexanol, or cyclohexanone are added in small amounts and even more sensitive to the addition of carboxylic acids, and 5) exposure to peroxide causes decomposition of the metal oxides out of the core particles even at room temperature and at atmospheric pressure.

### Evaluation of Soluble Metal Oxides as Precursors to Supported Catalysts

Several applications of the soluble metal oxides were described in the previous chapter. The soluble complexes had rather low catalytic

Table 7

Stability of $Mn_3O_4$ Core Catalyst to Various Reagents and Conditions.

| Solvent & Reagent | Gas | Pressure Psig | Temp °C | Time hr. | Stability Observations and Comments |
|---|---|---|---|---|---|
| Cyclohexane/Benzene | $N_2$ | 120 | 195 | 35 | no change, no ppt formed |
| Cyclohexane | $H_2$ | 130 | 178 | 24 | no change, no ppt formed |
| Cyclohexane | CO | 130 | 180 | 12 | no change, no ppt formed |
| Cyclohexane | $H_2$/CO 1:1 | 135 | 177 | 24 | no change, no ppt formed |
| Benzene | $O_2/N_2$ | 235 | 188 | 9-17 | catalyst precipitates from solution |
| Cyclohexane/Benzene Cyclohexanone (0.2%) | $N_2$ | 195 | 190 | 8 | Catalyst precipitated, solution becomes light yellow instead of normal red. |
| Benzene/Cyclohexane | $N_2$ | atm. | 146 | 12 | most of catalyst ppt as a brown precipitate and solution become yellow. |
| Benzene/Cyclohexane | $N_2$ | 205 | 195 | -- | large amount of ppt and solution became orange. |

activities in many applications but their use in preparing supported heterogenous catalysts looks quite promising. Preparation of these supported catalysts could in principle make use of the size controlling factors to achieve a particular catalytic site density or to control the size of the supported metal cluster after reduction. With this potential for surface tailoring of the support in mind, we prepared a number of catalysts. The use of the Scheme I catalysts in hydrogenations of olefins, benzene and naphthalene is described below.

Hydrogenations Over Alumina--Supported Co. 1,5-Cyclooctadiene hydrogenation was studied at 80° and 200 psi ($H_2$) using the $AlEt_3$-Co (Scheme I, A.) The same reaction was studied at 120° and 300 psi using the oxidized form of the catalyst ($Al_2O_3$-CoO). This reaction was selected because it is analogous to the hydrogenation of olefins over supported cobalt catalysts reported in U.S. Patent 3,855,324 (2). The reactions were followed by observing the hydrogen pressure drop and terminated within 1 to 2 hrs. after no further pressure drop was observed. The solvent was hexane.

$$2H_2 + \text{[cyclooctadiene]} \xrightarrow[\text{120°, 300 psi, } Al_2O_3\text{-CoO; Hexane}]{\text{80°, 200 psi, } Al_2O_3\text{-Co; or}} \text{[cyclooctane]}$$

Reactions using $Al_2O_3$-Co at 80° were complete in about 24 hr. when 2.6 g (24.5 mmol) of diene, 4-10 ml. of hexane and 10 g of catalyst were employed. The catalyst was recycled five times without appreciable loss of activity. Hydrogenations employing the cobalt oxide catalyst $Al_2O_3$-CoO were much slower. When 6 g of catalyst was used with 0.9 g (8.1 mmol) of diene in 5 ml of hexane at 120°, it took 25 hrs. to achieve complete hydrogenation at 300 psi. Thus, to achieve similar hydrogenation rates, the cobalt oxide catalyst requires a temperature 40° greater than the reduced form of the catalyst.

The hydrogenation of trans, trans, cis-1,5,9-cyclododecatriene, catalyzed by the reduced $Al_2O_3$-Co catalyst, proceeded in the same fashion as the hydrogenations of 1,5-cyclooctadiene. The first reaction (80°, 200 psi, 15 ml hexane, 3 g triene, 10 g catalyst) gave only 37% conversion (to cyclododecane) in 48 hrs and 100% conversion in 120 hr. In each subsequent recycle reaction (same conditions except only 4 ml of hexane was used), however, the reduction was complete in 24 hr. until the fifth recycle where it took 36 hrs. to achieve complete reduction. Apparently, as was the case in 1,5-cyclooctadiene hydrogenations, the first cycle involves some reaction process at the surface to activate the catalyst. To achieve approximately equal rates, the cobalt oxide catalyst, $Al_2O_3$-CoO, had to be used at 120°, some 40° higher than that of the reduced form of the catalyst, and at higher pressures (340 versus 200 psi).

$3H_2$ + [1,5-cyclooctadiene] —(80°, 200 psi, $Al_2O_3$-Co / 120°, 340 psi, $Al_2O_3$-CoO)→ [cyclooctane]

Benzene was reduced to cyclohexane over the reduced ($Al_2O_3$-Co) form of the supported cobalt at 80° and 200 psi of hydrogen. Remarkably, the hydrogenation of benzene occurred at a rate similar to that of the 1,5-cyclooctadiene hydrogenations. This is suprising under such mild conditions. At 80° and 200 psi, a 97% conversion to cyclohexane was achieved in 45 hrs. using 6 g of benzene and 10 g of catalyst. The catalyst used had previously been used 5 times in 1,5-cyclooctadiene hydrogenations.

$3H_2$ + [benzene] —(80°, 200 psi, $Al_2O_3$-Co)→ [cyclohexane]

Kieselguhr-Supported Nickel Catalysts in Benzene Hydrogenations. Both the Kieselguhr-supported nickel catalyst, KG-Ni-A (reduced by hydrogen) and KG-Ni-B (reduced by $AlEt_3$), were active in promoting the hydrogenation of benzene to cyclohexane. At 80° and 200 psi hydrogen pressure the KG-Ni-B catalyst was more active. In 6 hrs., complete conversion of benzene to cyclohexane was observed using 8.8 g benzene and 5 g of catalyst. Using KG-Ni-A in an identical reaction, 77% conversion was observed in 7.5 hrs. and complete conversion was noted in 14 hrs.

$H_2$ + [benzene] —(KG-Ni-A, 80°, 200 psi / KG-Ni-B, 80°, 200)→ [cyclohexane]

Benzene Hydrogenation Catalyzed by $Al_2O_3$-Fe. Benzene was readily hydrogenated to cyclohexane at 120° and 150 psi in the presence of $Al_2O_3$-Fe (Scheme I, C). In 14 hrs. the conversion was quantitative. It is interesting to note only traces of cyclohexane was obtained at 130° and 25-45 psi after 48 hrs. These pressure effects may signify mass transport problems. Whatever the case may be, the reductions did proceed at rather low temperatures (120°).

$H_2$ + benzene → cyclohexane

$Al_2O_3$-Fe, 150 psi., 120°

$Al_2O_3$-Co-Fe, 150 psi., 120°

(Catalysts described in Scheme I, 7 and 8)

Benzene Hydrogenation Catalyzed by $Al_2O_3$-Co-Fe. Benzene was hydrogenated to cyclohexane using $Al_2O_3$-Co-Fe (Scheme I, D) at 120° and 150 psi of hydrogen. Using 8.8 g of benzene and 8 g of catalyst, the conversion was 94% in 15 hrs. However, hydrogenation was very slow at 25-45 psi even at 160°.

## CONCLUSION

The influence of metal/acid equivalents ratio on the catalyst particle size has been demonstrated and the influence of surface carboxylate structure and solvent on particle aggregation and solubility explained in these studies. The particle stability along with the solution size and solubility of these materials are consistent with the particulate structure represented in Figure 1 and this structure, which might be described as resembling an inverted micelle, explains the rather low catalytic activity of these materials in some of the applications reported in the previous chapter. The direct use of the soluble metal oxide materials as precursors to supported catalysts looks very good.

## REFERENCES

1) A. Alkaitis and P. L. Cells, US Patent 4,163,986, Mooney Chemical Co.) 1979.

2) J. K. Mertzweiler and H. M. Tenney, US Patent 3,855,324, (Esso Research and Engineering Co.) 1971.

# NEW POLYCONDENSATION SYSTEMS

# Interfacial Synthesis Part I: Phase-Transfer Catalyzed Synthesis of Polyhydroxy Ether

Ajit K. Banthia, David Lunsford, Dean C. Webster and J.E. McGrath

Chemistry Dept. and Polymer Materials and Interfaces Laboratory
Virginia Polytechnic Institute and State University, Blacksburg, Virginia 24061

## ABSTRACT

A convenient, rapid and efficient method for the synthesis of linear, moderate molecular weight polyhydroxy ethers has been developed. Based on the principle of "phase-transfer catalysis" the method involves the reaction of quaternized bis-phenate ion-pair with epichlorohydrin in aqueous alkali-dioxane system, at the reflux temperature of ≈ 86°C, using a quaternary ammonium salt as an effective reagent for the transportation of the bis-phenate-A ion to the organic phase in a highly reactive form. In contrast to several other processes where phase-transfer catalysts are involved, the present procedure appears to be truly catalytic with respect to quaternary ammonium salt. The successful use of polyethylene glycols (MW 300-14,000) as phase-transfer catalysts is also of great importance due to their low toxicity and relatively low cost. The role of phase-transfer catalysts in biphasic systems for preparative macromolecular systhesis has also been briefly outlined.

## INTRODUCTION

The term "phase-transfer catalysis" (PTC) was coined by Starks [1] to describe reactions which are accelerated by small quantities of quaternary ammonium or phosphonium cations ($Q^+$) which are usually visualized as assisting in the transfer of one reactant across an interface [1,2]. Subsequently, macrocyclic crown ethers, cryptates, etc. (C) were used extensively for the same purpose [3-6]. Due to the extreme usefulness of PTC reactions in preparative organic chemistry, synthetic exploitation of PTC to effect reactions in biphasic systems has been recently surveyed [7-9].

A typical phase-transfer reaction is the cyanide displacement on 1-halooctane [1,4,10] (Fig. 1):
Here, the organic soluble $Q^+$ (or C) transports the anion from the aqueous phase to the organic phase. Moreover, linear or cyclic ethers serve the same purpose [4,10]. Starks and Owens [2] collected evidence that such processes occur in the organic phase and do not involve micelles. Arguments against micellar catalysis rested on the following factors: the observed linearity of reaction rate as a function of transfer agent concentrations in the organic phase, the consideration of the catalytic efficiency as a function of transfer agent aggregation in the organic phase, the effectiveness of the poor surfactants like tetradodecyl ammonium salts as catalysts and the ability of these salts to

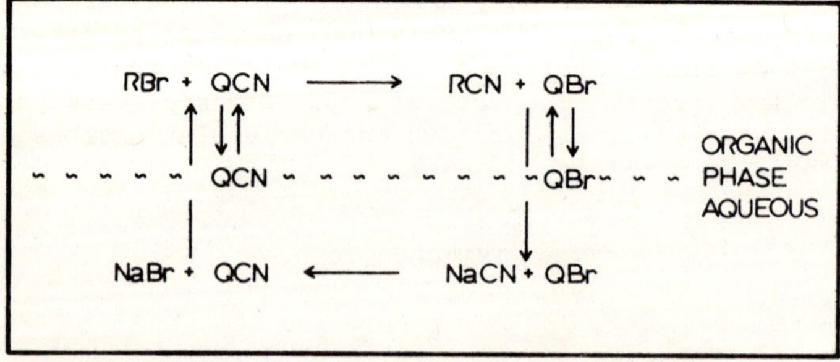


FIG. 1. Generalized Phase - Transfer Reaction Scheme.

participate in the displacement reacitons at high rates even in non-polar media.

Phase transfer catalysis (PTC) could in principle offer several advantages over solution conducted nucleophilic displacement step-growth polymerizations. These would include the substitution of inexpensive solvents for anhydrous aprotic solvents such as DMSO, DMAC, etc., lower reaction temperature, and shorter reaction times. This process is, of course, strikingly analogous to what macromolecular chemists have called "Interfacial Polymerization" for some time [11-13].

Indeed, Morgan utilized tetraethylammoniumchloride effectively as an accelerator for the synthesis of poly(arylester) homopolymers [14] under typical interfacial conditions [12,13].

The utilization of quaternary ammonium or phosphonium salts (or any suitable PTC) for the polymer synthesis from bifunctional phenols can be viewed by considering the PTC cycle (Fig. 2) shown here.

Thus, it is logical to correlate the migrational mobility of the newly formed ion-pairs to and across the phase boundary with enhanced rate in these biphasic polymerization systems.

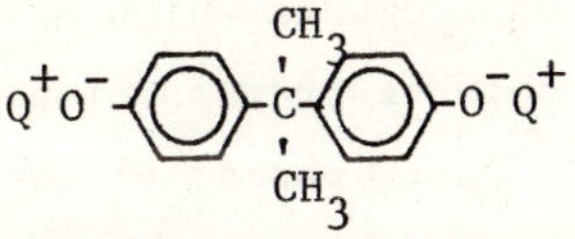

$$Q^{+}O^{-}-C_6H_4-C(CH_3)_2-C_6H_4-O^{-}Q^{+}$$

Both the lyophilic nature of the quaternary ammonium or phosphonium cations (or any PTC), the extractibility of the phenate ion by them and the polarity of the organic phase seems to have a significant influence on the catalytic effect. Moreover, the availability of the transported reagent in a highly reactive form will be more advantageous. Furthermore, a continuous cycle is allowed by the fact that the quaternary salt migrates back and forth across the interface, thus providing a certain neces-

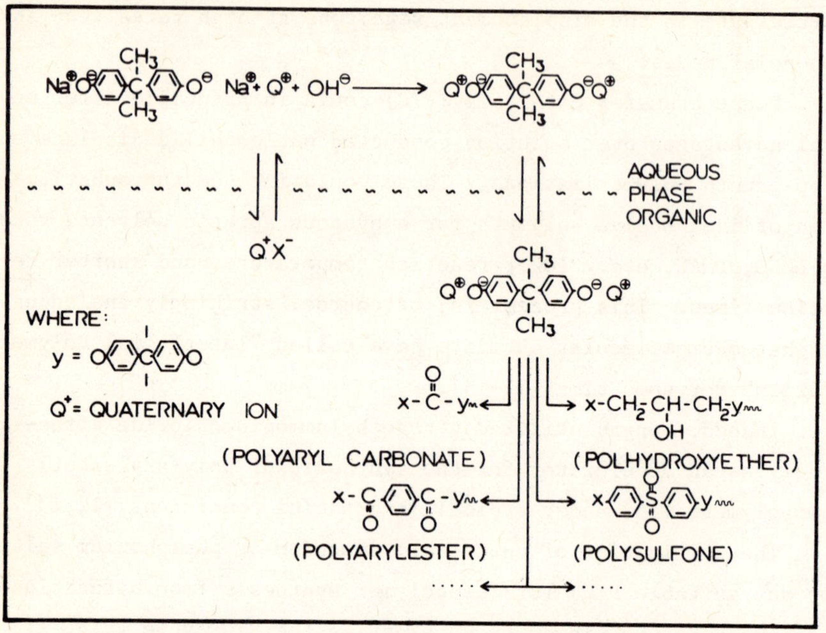

FIG. 2. Biphasic Reaction Scheme of Polymer Synthesis From Bifunctional Phenols.

sary concentration in each phase. Under these conditions the basic and essential requirements which are fulfilled are: a) the PTC should be able to transport one reactant from its phase to the other reactant's phase; b) the transported reactant should be available in a highly reactive form; and c) the PTC should be recycled.

Thus, despite the apparent similarity of these PTC reactions and prior interfacial polymer syntheses, there may be significant mechanistic difference between them, and there may also be ample opportunities for new synthetic applications in preparative polymer chemistry. Mechanistically, in the former case, the locus of the polymerization may be considered to be more in the organic phase than at the interface.

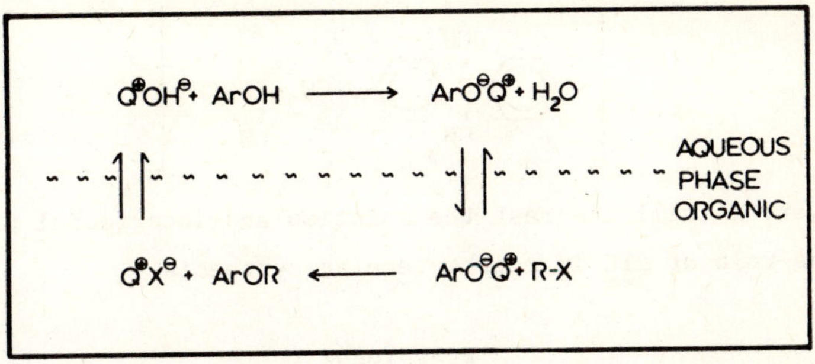

$Q^{\oplus}OH^{\ominus} + ArOH \longrightarrow ArO^{\ominus}Q^{\oplus} + H_2O$

AQUEOUS
PHASE
ORGANIC

$Q^{\oplus}X^{\ominus} + ArOR \longleftarrow ArO^{\ominus}Q^{\oplus} + R\text{-}X$

FIG. 3. Model Reaction Scheme For Hindered Phenols in Biphasic Medium.

Intriguing uses of PTC model ether syntheses have been given by Freedman et al [15] and McKillop et al [16]. In both cases the phenol is added as a solution in concentrated sodium hydroxide along with the quaternary ammonium salt and mixed with a methylene chloride solution of an alkylating agent. The phenol is converted into the corresponding quaternary ammonium phenoxide in the aqueous phase, which due to discrete solubility in the organic phase will migrate into the methylene chloride solution. The migration is followed by rapid alkylation and the formation of the phenol ether. Even highly hindered phenols gave high yields.

These model experiments [1,2,14-18] suggest a possible route for the preparation of high molecular weight polyhydroxyether or, even, poly(arylene ether sulfones) and their block copolymers under mild reaction conditions in the presence of a good PTC. The polymerization might proceed even without utilizing a dipolar aprotic solvent.

This paper will describe our preliminary results on the PTC synthesis of polyhydroxyethers (PHE).

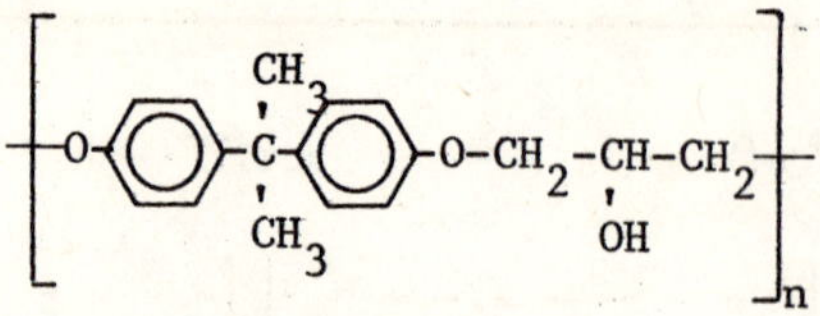

$$\left[ -O-C_6H_4-C(CH_3)_2-C_6H_4-O-CH_2-CH(OH)-CH_2- \right]_n$$

Moreover, it will contrast the solution and interfacial processes and the role of PTC in macromolecular synthesis.

## EXPERIMENTAL

2,2-bis(p-hydroxyphenyl) propane (Bisphenol-A) purified polymerization grade from Union Carbide Corporation was used. Epichlorohydrin was obtained from Aldrich Chemical Company and was distilled before use. All PTC's were pure grade commercial materials and were used without further purifications. Spectrograde dioxane and tetrahydrofuran were used as received.

A typical synthesis of bisphenol A derived polyhydroxyether by interfacial PTC polycondensation is given below.

To a 200 ml. three-necked flask equipped with a stirrer, thermometer and condenser is added 4.2 gm. (0.105 mol) of NaOH in 30 ml. of water followed by 11.415 gm. (0.05 mol) of bisphenol-A, 4.625 gm. (0.05 mol) of epichlorohydrin in 15 ml. of dioxane and 0.228 (0.001 mol) of benzyltriethylammonium chloride. The mixture is rapidly stirred and heated to reflux which occurs at a temperature of about 86°C. Heating is continued for six hours. The mixture is cooled. The aqueous phase separated by decantation and the organic phase diluted with 75 ml. of dioxane, or viceversa. The organic rich phase is poured into about 100 times its volume of acidified cold water yielding a relatively solvent-free polymer in the form of shreds. The polymer is dried overnight in a vacuum oven at 50°C. Intrinsic viscosity [η] of the polymer in THF at 25°C typically is 0.24 dl/gm. The yield is 8.9 gm (≈55.5%).

## RESULTS AND DISCUSSIONS

### Chemistry of Polyhydroxyether (PHE) Synthesis

Polyhydroxyethers (PHE) are the linear analogs of three-dimensional epoxy resin networks. They can be synthesized by the reaction of bifunctional phenols such as bisphenol-A with either (i) epichlorohydrin or, alternatively with (ii) its preformed monomeric diglycidyl ether using suitable basic catalysis [19, 20]. However, due to the complexity and commercial interest in these systems relatively little systematic details of the syntheses are available in the published literature.

These classes of polymer are different from the well known epoxy resins in two aspects: first, they are much higher in molecular weight, and, secondly, they do not contain any terminal epoxide functionality.

Theoretically, the equimolar amounts of bifunctional phenols, epihalohydrin and base should provide high molecular weight PHE as evident from the following equation (1).

$$\mathrm{nHO{-}R{-}OH} + \mathrm{n}\underset{\diagdown O \diagup}{\mathrm{CH_2{-}CH}}\mathrm{{-}CH_2Cl} + \mathrm{nMOH} \longrightarrow \left[ \mathrm{{-}R{-}O{-}CH_2{-}\underset{OH}{CH}{-}CH_2{-}O{-}} \right]_n + \mathrm{nMCl} + \mathrm{H_2O} \qquad \ldots (1)$$

where R - any alkyl or, aryl unit

M - alkali metal

The reaction proceeds through the formation of both chlorohydrin (Eq. 3) and monoepoxy intermediates (Eq's. 6,8,9). The following reaction scheme provides an insight into the mechanistic aspects of PHE synthesis.

$$\mathrm{C_6H_5{-}OH} + \mathrm{OH^{\ominus}} \rightleftharpoons \mathrm{C_6H_5{-}O^{\ominus}} + \mathrm{H_2O} \qquad \ldots (2)$$

$$C_6H_5\text{-}O^{\ominus} + \underset{\diagdown O \diagup}{CH_2\text{-}CH}\text{-}CH_2\text{-}Cl \longrightarrow C_6H_5\text{-}O\text{-}CH_2\text{-}\underset{O^{\ominus}}{\underset{|}{CH}}\text{-}CH_2\text{-}Cl \qquad \ldots (3)$$

$$C_6H_5\text{-}O^{\ominus} + \underset{\diagdown O \diagup}{CH_2\text{-}CH}\text{-}CH_2\text{-}Cl \longrightarrow C_6H_5\text{-}O\text{-}\underset{CH_2O^{\ominus}}{\underset{|}{CH}}\text{-}CH_2\text{-}Cl \qquad \ldots (4)$$

$$C_6H_5\text{-}O\text{-}\underset{CH_2O^{\ominus}}{\underset{|}{CH}}\text{-}CH_2\text{-}Cl \longrightarrow C_6H_5\text{-}O\text{-}CH\begin{matrix} \diagup CH_2 \diagdown \\ \quad\quad O \\ \diagdown CH_2 \diagup \end{matrix} + Cl^{\ominus} \qquad \ldots (5)$$

$$C_6H_5\text{-}O\text{-}CH_2\text{-}\underset{O^{\ominus}}{\underset{|}{CH}}\text{-}CH_2\text{-}Cl + M^{\oplus} \longrightarrow C_6H_5\text{-}O\text{-}CH_2\text{-}\underset{\diagdown O \diagup}{CH\text{-}CH_2} + MCl \qquad \ldots (6)$$

$$C_6H_5\text{-}O\text{-}CH_2\text{-}\underset{O^{\ominus}}{\underset{|}{CH}}\text{-}CH_2\text{-}Cl + C_6H_5\text{-}OH \longrightarrow$$

$$C_6H_5\text{-}O\text{-}CH_2\text{-}\underset{OH}{\underset{|}{CH}}\text{-}CH_2\text{-}Cl + C_6H_5\text{-}O^{\ominus} \qquad \ldots (7)$$

$$C_6H_5\text{-}O\text{-}CH_2\text{-}\underset{OH}{\underset{|}{CH}}\text{-}CH_2\text{-}Cl + M^{\oplus} + OH^{\ominus} \longrightarrow$$

$$C_6H_5\text{-}O\text{-}CH_2\text{-}\underset{\diagdown O \diagup}{CH\text{-}CH_2} + MCl + H_2O \qquad \ldots (8)$$

$$C_6H_5\text{-}O\text{-}CH_2\text{-}\underset{OH}{\underset{|}{CH}}\text{-}CH_2\text{-}Cl + C_6H_5\text{-}O^{\ominus} \longrightarrow$$

$$C_6H_5\text{-}O\text{-}CH_2\text{-}\underset{\diagdown O \diagup}{CH\text{-}CH_2} + C_6H_5\text{-}OH + Cl^{\ominus} \qquad \ldots (9)$$

$$C_6H_5\text{-}O\text{-}CH_2\text{-}\underset{\diagdown O \diagup}{CH\text{-}CH_2} + C_6H_5\text{-}O^{\ominus} \longrightarrow$$

$$C_6H_5\text{-}O\text{-}CH_2\text{-}\underset{O^{\ominus}}{\underset{|}{CH}}\text{-}CH_2\text{-}O\text{-}C_6H_5 \qquad \ldots (10)$$

$$C_6H_5\text{-O-CH}_2\text{-}\underset{\diagdown O \diagup}{\text{CH-CH}_2} + C_6H_5\text{-O}^{\ominus} \longrightarrow C_6H_5\text{-O-}\underset{\substack{| \\ CH_2\text{-O-}C_6H_5}}{\text{CH}}\text{-CH}_2\text{O}^{\ominus} \qquad \ldots(11)$$

$$C_6H_5\text{-O-CH}_2\text{-}\underset{\substack{| \\ O^{\ominus}}}{\text{CH}}\text{-CH}_2\text{-O-}C_6H_5 + C_6H_5\text{-OH} \longrightarrow C_6H_5\text{-O-CH}_2\text{-}\underset{\substack{| \\ OH}}{\text{CH}}\text{-CH}_2\text{-O-}C_6H_5 + C_6H_5\text{-O}^{\ominus} \qquad \ldots(12)$$

The reactions [3,10] represent the "normal" addition of phenoxide anion to the least substituted carbon atom of the epoxide(oxirane) ring [21,22]. In contrast, the "abnormal" addition of the phenoxide anion to the most substituted carbon-atom of the oxirane ring[23-25] which may also occur will lead to the formation of either 1,3-chlorohydrin (Eq. 4) or 1,3-hydroxy ether (Eq. 11). The former will thus act as a chain-terminating step (Eq. 4), due to the relative inertness of 1,3-chlorohydrin to any further nuleophilic substitution. Moreover, subsequent ring closure of the same by dehydrohalogenation will result in the formation of a four-membered oxetane ring, which is also unreactive. However, it may be pointed out that unlike reaction (4) reaction (11) should not effect the polymer molecular weight.

Subsequent studies with model compounds [26] confirmed that there is a negligible amount of 1,3-chlorohydrin, formed. Formation of the intermediate monoglycidylether and the subsequent chain extension step are outlined in reactions (6-9) and (10) respectively.

One possible mode of branching which may occur in the presence of any unreacted epichlorohydrin is shown below.

$$\text{C}_6\text{H}_5\text{-O-CH}_2\text{-}\underset{\underset{\text{O}^{\ominus}}{|}}{\text{CH}}\text{-CH}_2\text{-O-C}_6\text{H}_5 + \underset{\diagdown\text{O}\diagup}{\text{CH}_2\text{-CH}}\text{-CH}_2\text{-Cl} \longrightarrow \text{C}_6\text{H}_5\text{-O-CH}_2\text{-}\underset{\underset{\text{O-CH}_2\text{-}\underset{\underset{\text{O}^{\ominus}}{|}}{\text{CH}}\text{-CH}_2\text{-Cl}}{|}}{\text{CH}}\text{-CH}_2\text{-O-C}_6\text{H}_5 \quad \ldots(13)$$

However, one should consider that for the base catalyzed reaction of diglycidylether of bisphenol-A and bisphenol-A, it has been shown that [27] as long as any phenolic groups is present, the epoxide will react with the phenolic functionality with the exclusion of any epoxide-(pendant secondary) alcohol reaction. This will certainly rule out any possibility of extensive branching, at least in the early stages of polymerization.

The alternative synthetic procedure is based on the reaction of diepoxides with equimolar quantities of dihydricphenols, in presence of a suitable base catalyst [28,29,30]. However, there is very little published literature describing the actual synthetic procedure. Nevertheless, several workers have studied the reaction of dihydricphenol with diepoxides [27,31,32] and pointed out that only a few selective catalysts can possibly lead to the formation of linear polyhydroxyether.

Table I summarizes the results of the synthesis of polyhydroxyether from bisphenol-A and epichlorohydrin (EPI) in presence of various phase-transfer catalysts. A comparison of several quaternary ammonium bromide salts bearing different alkyl groups indicates that as the length of the longest chain increases, the catalytic efficiency of the quaternary ion ($Q^+$) levels off. Even quaternary ammonium salts bearing lower alkyl groups such as $(C_2H_5)_4N^+$ act as an efficient catalyst. This is not surprising if one considers the fact that the polarity of dioxane allows it to be a good solvent for such small quaternary ions ($Q^+$) as $(C_2H_5)_4N^+$, $(C_3H_7)_4N^+$ etc. The polarity of this solvent is also

TABLE I. Interfacial Synthesis of Polyhydroxyether: with Various Phase-Transfer Catalyst

[Bisphenol-A] = [EPI] - 0.02 mole;
Solvent composition - $H_2O$ (40)/dioxane (20);
Temperature = 86°C;
Time = 6 hrs.;
[C] - 0.02 mole per mole bisphenol-A;
[Base] - 3.00 mole per mole bisphenol-A.

| CATALYST, C | BASE | $[n]^{THF}_{25°C}$ | % Yield |
|---|---|---|---|
| $(C_2H_5)_4NBr$ | NaOH | 0.17 | 54.6 |
| $[CH_3(CH_2)_3]_4NBr$ | NaOH | 0.17 | 49.2 |
| $[CH_3(CH_2)_6]_4NBr$ | NaOH | 0.10 | 38.7 |
| $[CH_3(CH_2)_3]_4PBr$ | NaOH | 0.12 | 47.5 |
| DC-18-C-6 | NaOH | 0.14 | 38.7 |
| $C_6H_5CH_2P(C_6H_5)_3(Cl)$ | NaOH | 0.12 | 47.5 |
| $C_6H_5CH_2N(C_2H_5)_3(Cl)$ | NaOH | 0.25 | 47.5 |
| $C_6H_5CH_2N(C_2H_5)_3(Cl)$ | KOH* | 0.34 | 70.0 |

*[C] - 2.55 mole per mole bisphenol-A

likely to affect the intrinsic reactivity of various $Q^+$ due to significant decreases in cation/anion interaction energy so that the exact choice of the catalyst structure may not be a very important factor in determining high reactivity [33]. In addition, the presence of concentrated alkali further influences the solubility so that even lower alkyl groups can often be used as the quaternary onium salt PTC's.

Contrary to literature reports, ammonium ions were found to be as effective as phosphonium ions bearing the same alkyl substituents. This may also relate to the fact that the polarity of the organic phase is of great importance in the present case.

Moreover, the presence of concentrated alkali can also considerably reduce the differences in the intrinsic reactivity of both tetraalkyl ammonium and phosphonium catalysts [7].

It may also be significant that the phosphonium cations have been reported to undergo a facile decomposition reaction around 50°C in the presence of hydroxide ion [34-37] and with strong sodium hydroxide (50% wt.) this decomposition occurs even at room temperature

$$R_4P^+OH^- \longrightarrow R_3PO + RH \qquad \ldots(19)$$

to give trialkyl phosphonium oxide which can not be recycled back to $R_4P^+X^-$. Clearly, the catalytic activity will be irreversibly lost under these conditions. Furthermore, this reaction is thought to be more rapid than the Hoffman elimination of the corresponding ammonium analog. In fact, the quaternary ammonium salts are stable in dilute sodium hydroxide (ca. 2-3M) around 80°C for 48 hrs. Thus, it is reasonable to assume that they will be fairly stable in the reaction conditions utilized for the present work. Thus, under strong alkaline conditions the quaternary ammonium salt should be the preferred catalyst over the corresponding phosphonium salt.

Surprisingly, benzyl triethyl ammonium ion was a very efficient catalyst compared to the several other PTC's which had been used, even considering a representative crown ether like dicyclohexyl-18-crown-6. In this context, it may be pointed out that for the synthesis of model phenolic ethers under typical PTC conditions, benzyl trialkyl ammonium ion was found to be very effective as PTC [34-37].

All these results strongly support the idea that the effectiveness of various PTC's must be related to the specific different system under investigation. The effectiveness will be governed for any individual system on the basis of several factors, some of which are given below:

* the polarity of the organic phase
* concentration of the inorganic salts in the aqueous phase
* the presence of solvating compounds for the particular anion
* reaction in presence of concentrated alkali
* solubility of the resulting macromolecule
* consideration of which step(s) are rate determining

Here, one should also consider whether the rate determining step of the PTC cycle or sequence is in the organic phase reaction or in the phase-transfer step i.e., the PTC regeneration step. The latter will also be an important factor affecting the relative activities of catalysts having different structures.

Thus, one should be cautious about generalizations on the role of PTC in macromolecular synthesis, at least at the present level of understanding

We noted that a change from $Na^+$ to $K^+$ counterion for the bisphenate anion in the aqueous phase increased the polymer yield significantly and produced polymer having significantly higher intrinsic viscosity values. The result seems logical if we consider the probability of the presence of the anion as an ion-pair aggregate in relatively low-dielectric constant solvents like tetrahydrofuran or dioxane. In the case of small cations such as $Na^+$, these ion-pairs should be relatively "tight" (as compared to $K^+$ or $Q_4N^+$) and the electrostatic restraint will be considerably larger for the subsequent ether forming reaction. Thus, it should be desirable to utilize a loose ion-pair complex, where the positive charge will be spread over a large volume. This would decrease the electrostatic barrier to ether formation from $Li^+ > Na^+ > K^+ > Q_4N^+$.

A comparison of the cation radii shows only a moderate increase from tetramethyl to tetrabutyl ammonium cation [$(CH_3)_4N^+ < (C_2H_5)_4N^+ < (n\text{-}C_3H_7)_4N^+ < (n\text{-}C_4H_9)_4N^+ \simeq 2.85<3.48<3.98<4.37$Å] [38] and the calculated cation-anion interaction energies for the corresponding bromide salts [$Br^- = 1.95$Å] [39] will be 3.1<2.8<2.5<2.4 K Cal/mole respectively in dioxane ($\varepsilon = 2.2$). Thus, the relatively similar effectiveness of these quaternary ammonium salts as PTC is not at all surprising.

The base concentration seems to play an important role. Surprisingly, even less than a stoichiometric amount of base (< 2 mole of base per mole of bisphenol-A) yields a polymer of reasonable molecular weight. It is reasonable to presume that the presence of partially aqueous alkali soluble organic phase tends to increase the migration and the solubility of phenoxide

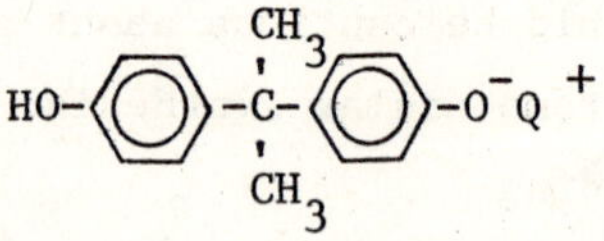

$$HO-C_6H_4-C(CH_3)_2-C_6H_4-O^-\,Q^+$$

ion-pair and the reaction then proceeds through the formation of the monoepoxyintermediate. In fact, a base concentration in excess of 1 mole per mole of bisphenol-A is sufficient to yield a polymer (Table II).

Even the increased solubility of the phenoxide ion-pair in water, or better aqueous alkali - miscible solvent does not effectively enhance the rate of phenyl ether formation in the absence of any PTC (Table III). This is not surprising if one considers the fact that the rate of similar reactions in homogeneous systems (eg. dioxane) is higher by factor of $3 \times 10^4$ when the phenyl counter-ion is either quaternary ammonium or phosphonium than when it is an alkali metal (eg., $k_{Bu_4N^+}/k_{K^+} = 3.3 \times 10^4$ for dioxane) [18]. A tenfold increase in the catalyst, $(C_2H_5)_4N^+Cl^-$, concentration from $1.79 \times 10^{-3}$ moles to $1.79 \times 10^{-2}$ moles does not affect the intrinsic viscosity of the resulting polymer.

TABLE II Interfacial Synthesis of Polyhydroxyether: Effect of Base Concentration

[Bisphenol-A] = [EPI] = 0.02 mole;
[$C_6H_5CH_2N(C_2H_5)_3(Cl)$ - 0.0004 mole;
Solvent - $H_2O$ (40)/Dioxane (20);
Temperature - 86°C;
Time - 6 hrs.

| [BASE] Per Mole Bisphenol-A | $[\eta]^{THF}_{25°C}$ | % Yield |
|---|---|---|
| KOH | | |
| 2.55 | 0.34 | 70% |
| 2.00 | 0.28 | 76% |
| 1.70 | 0.30 | 82% |
| 1.40 | 0.20 | 93% |
| 0.85 | Sticky Paste | --- |
| NaOH | | |
| 2.1 | 0.24 | 55.5% |
| 1.1 | 0.14 | 50.0% |

TABLE III Interfacial Synthesis of Polyhydroxyether: Effect of Catalyst Concentration

[Bisphenol-A] - 0.050 mole; [EPI] - 0.050 mole;
[NaOH] - 0.1575 mole; $H_2O$/Dioxane - 120 ml./50 ml.;
Temperature - 86°C; Time - 6 hrs.

| [$(C_2H_5)_4NCl$] | $[\eta]^{THF}_{25°C}$ |
|---|---|
| CONTROL | ---- |
| 0.00179 mole | 0.15 |
| 0.0179 mole | 0.15 |

Even a fraction mole percent of PTC effectively provides a certain concentration of the phenate ion-pair in the organic phase for rapid consumption. This will permit an efficient recycling of PTC for the continuous interfacial process, provided that sufficient base is present in the aqueous layer to regenerate the quaternary ammonium hydroxide from the halide which is formed in the course of reaction. The role of PTC here seems to be truly catalytic as theoretically envisaged for a typical PTC reaction.

Stoichiometric reactant balance seems to be necessary but the effect here is less pronounced than that usually observed in several other systems. (Table IV).

As long as the polymer remains soluble or at least well dispersed, solvent composition was not observed to affect the intrinsic viscosity, [η], of the resulting polymer (Table V).

Figure 4 shows the course of reaction in terms of the intrinsic viscosity of the polymer. These results indicate that the polymerization reaction is essentially complete in 6 hrs.

TABLE IV Interfacial Synthesis of Polyhydroxyether: Effect of Stoichiometry

$[C_6H_5CH_2N(C_2H_5)_3(Cl)]$ - 0.02 mole/mole Bisphenol-A;
[KOH] - 2.55 mole/mole Bisphenol-A; [Bisphenol-A] - 0.02 mole;
Solvent - $H_2O$ (40)/Dioxane (20);
Temperature - 86°C; Time - 6 hrs.

| Epichlorohydrin | $[\eta]^{THF}_{25°C}$ | % Yield |
|---|---|---|
| 100% (Theory) | 0.34 | 70.0% |
| 99% (Theory) | 0.24 | 70.4% |
| 101% (Theory) | 0.22 | 68.66% |

TABLE V Interfacial Synthesis of Polyhydroxyether: Effect of Solvent Composition

[Bisphenol-A] = [EPI] - 0.04 mole; [NaOH] - 0.084 mole; $[C_6H_5CH_2N(C_2H_5)_3(Cl)]$ - 0.0008 mole; temp - 85-87°C; Time - 6 hrs.

| $H_2O$ (ml.) | Dioxane (ml.) | $[\eta]^{THF}_{25°C}$ | % Yield |
|---|---|---|---|
| 60 | 60 | 0.24 | 70.0 |
| 80 | 80 | 0.24 | 70.4 |
| 80 | 60 | 0.24 | 72.0 |
| 80 | 40 | 0.25 | 72.3 |
| 40 | 40 | 0.25 | 71.8 |

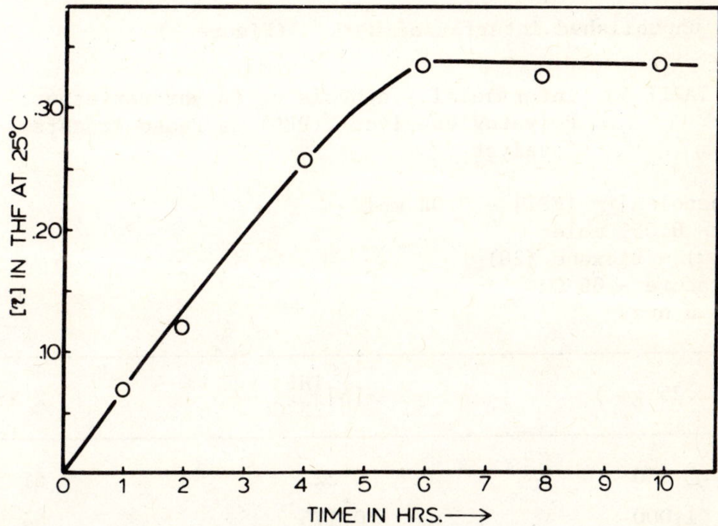


FIG. 4. Effect of Reaction Time on the Course of Interfacial Polymerization of Polyhydroxyether. [Bisphenol-A] = [EPI] - 0.02 mole, $[C_6H_5CH_2N(C_2H_5)_3(Cl)]$ - 0.0004 mole, $H_2O$ (40)/Dioxane (20), Temperature - 86°C, [KOH] - 0.051 mole.

However, much more investigation of the parameters influencing molecular weight is required.

Polyethylene glycol (MW 300-14,000) was found to be an effective PTC for polyhydroxyether synthesis in the biphasic system. (Table VI). The successful use of linear polyethers are particularly of great interest due to their low toxicity and relatively low cost. Essentially, there is only one other report on the successful use of PEG as PTC in interfacial polymer synthesis [40]. The molecular weight of PEG has no significant effect both on the percentage yield and the intrinsic viscosity of the polyhydroxyether. It seems that in the presence of partially aqueous-alkali soluble organic phase, the intrinsic reactivity of these PTC's is similar for all practical purposes.

Unimodel distribution of the Gel-Permeation Chromatography curve of PTC interfacially prepared PHE is in line with our other unpublished interfacial work (Figure 5).

TABLE VI Interfacial Synthesis of Polyhydroxyether: Polyethyleneglycol (PEG) as Phase Transfer Catalyst

[Bisphenol-A] = [EPI] - 0.02 mole;
[KOH] - 0.051 mole;
$H_2O$ (40) - Dioxane (20);
Temperature - 86°C;
Time - 6 hrs.

| PEG (2.25 gm.) | $[\eta]^{THF}_{25°C}$ | % Yield |
|---|---|---|
| MW - 300 | 0.22 | 61 |
| MW - 1,000 | 0.175 | 64 |
| MW - 1,540 | 0.23 | 56 |
| MW - 4,000 | 0.19 | 64 |
| MW - 14,000 | 0.22 | 64 |

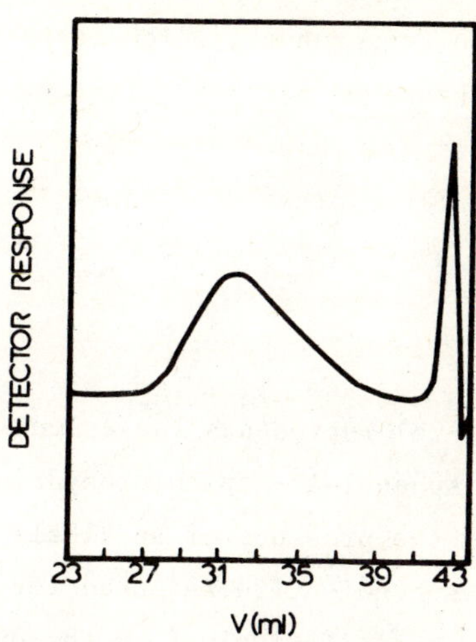


FIG. 5. G.P.C. Curve of an Interfacially Prepared Polyhydroxyether

It will be appropriate here to comment on our related work on quaternary ammonium hydroxide catalyzed synthesis of PHE in an homogeneous medium (dioxane) from bisphenol-A and diglycidylether of bisphenol-A and the commercially available Phenoxy Resin (Union Carbide) of medium molecular weight (Table VII). PHE of comparable intrinsic viscosity can be prepared by PTC synthesis in a biphasic system within a shorter time and at a compara-

TABLE VII Synthesis of Polyhydroxyether: Interfacial and Solution Dioxane Methods

| Catalyst | Phase | Temp. | Time | $[\eta]^{THF}_{25°C}$ |
|---|---|---|---|---|
| $C_6H_5CH_2N(C_2H_5)_3OH$ | Single | ≈ 100°C | 72 hrs. | 0.42 |
| $C_6H_5CH_2N(C_2H_5)_3(Cl)$ | Biphasic | ≈ 86°C | 6 hrs. | 0.34 |
| NaOH* | -------- | ------- | ------- | 0.43 |

* Phenoxy Resin (U.C.C.)

tively lower reaction temperature, which certainly should reduce the possiblity of branched polymer formation. Furthermore, in-depth studies pertaining this system may provide us also a convenient method for PTC synthesis of linear high-molecular weight polyhydroxyether.

Mechanistic Aspects

Equations (15-23) represents the essential reaction sequences for PTC bisphenol-A - epichlorohydrin reaction in a biphasic system. In the presence of an alkali metal hydroxide and a suitable PTC, bisphenol-A (represented for convenience as a model phenol) will quantitatively form the quaternary ammonium ion pair (Eq. 15).

$$C_6H_5\text{-OH (Aq.)} + OH^- + Q^+ \rightleftharpoons C_6H_5\text{-}O^-Q^+\text{ (Aq.)} + HOH \qquad ...(15)$$

Due to its better solubility in the organic phase, the latter will then migrate preferentially into the organic phase (eq. 16).

$$C_6H_5\text{-}O^-A^+\text{(Aq.)} \rightleftharpoons C_6H_5\text{-}O^-Q^+\text{(0)} \qquad ...(16)$$

and react with epichlorohydrin to form the intermediates chlorohydrin (eq. 17) and monoepoxide (eq. 18).

$$C_6H_5\text{-}O^-Q^+\text{(0)} + \overset{O}{CH_2\text{-}CH}\text{-}CH_2\text{-Cl(0)} \longrightarrow C_6H_5\text{-O-}CH_2\text{-}\overset{O^-Q^+}{CH}\text{-}CH_2\text{-Cl(0)} \qquad ...(17)$$

$$C_6H_5\text{-O-}CH_2\text{-}\overset{O^-Q^+}{CH}\text{-}CH_2\text{-Cl(0)} \longrightarrow C_6H_5\text{-O-}CH_2\text{-}\overset{O}{CH\text{-}CH_2}\text{(0)} + Q^+\text{Cl(0)} \qquad ...(18)$$

the regenerated quaternary salt will be recycled back to the aqueous phase (eq. 19).

$$Q^+Cl^-\text{(0)} \rightleftharpoons Q^+Cl^-\text{(Aq.)} \qquad ...(19)$$

The monoepoxide can further react with quaternary ammonium phenoxide ion-pair to provide the necessary chain extension (eq. 20, 21).

$$\text{C}_6\text{H}_5\text{-O-CH}_2\text{-}\underset{\backslash O /}{\text{CH-CH}_2}(0) + \text{C}_6\text{H}_5\text{-O}^-\text{Q}^+_{(0)} \longrightarrow \text{C}_6\text{H}_5\text{-O-CH}_2\text{-}\underset{\text{O}^-\text{Q}^+}{\text{CH}}\text{-CH}_2\text{-O-C}_6\text{H}_5 \quad \ldots(20)$$

$$\text{C}_6\text{H}_5\text{-O-CH}_2\text{-}\underset{\text{O}^-\text{Q}^+}{\text{CH}}\text{-CH}_2\text{-O-C}_6\text{H}_5 \ (0) \longrightarrow \text{Q}^+(0) + \text{C}_6\text{H}_5\text{-O-CH}_2\text{-}\underset{\text{OH}}{\text{CH}}\text{-CH}_2\text{-O-C}_6\text{H}_5 \ (0) \quad \ldots(21)$$

The quaternary cation will then partition itself between the aqueous and organic phase (eq. 22), the relative position depending upon the characteristics of the particular system in question.

$$\text{Q}^+(0) \rightleftharpoons \text{Q}^+(\text{Aq.}) \quad \ldots(22)$$

The relatively fast alkylation step (eq. 17) under these mild conditions will ensure that the chain-branching reaction (eq. 23) is at an absolute minimum.

$$\sim\sim\text{C}_6\text{H}_4\text{-O-CH}_2\text{-}\underset{\text{O}}{\text{CH}}\text{-CH}_2\text{-O-C}_6\text{H}_4\sim\sim(0) + \underset{\backslash O /}{\text{CH}_2\text{-CH}}\text{-CH}_2\text{-Cl}(0) \longrightarrow \sim\sim\text{C}_6\text{H}_4\text{-O-CH}_2\text{-}\underset{\text{O-CH}_2\text{-}\underset{\text{O}^-}{\text{CH}}\text{-CH}_2\text{-Cl}}{\text{CH}}\text{-CH}_2\text{-O-C}_6\text{H}_4\sim\sim \quad \ldots(23)$$

Catalysis by micellar [41] and reverse micellar systems [41, 42] are known not to be decisive for the course of reaction in the presence of concentrated alkali. Besides, in the present case quaternary compounds bearing a symmetrical structure and also having small substitutents e.g., small quaternary salts like $(C_2H_5)_4N^+Br^-$, are poor surfactants, but still behaved as good phase transfer catalysts. Similar reactions carried out in the presence of a good surfactant such a sodium lauryl sulfate demonstrated no significant effect on the intrinsic viscosity of the polymer. Considering these factors, no attempt has been made at the moment to consider the micelle-catalyzed reaction mechanism. However, it will be interesting to examine the role of both the large quaternary salts e.g. $(C_{12}H_{25})_4N^+X^-$, which are poor surfactants but are known to be good phase-transfer catalysts along with good surfactants such as $C_{16}H_{33}N^+(CH_3)_3X^-$, which are not always good phase-transfer agents. Furthermore, attempts are being made to utilize a water-immicible solvent by itself or, in combination with a small percentage of water-miscible solvent like THF or dioxane. The approach will test the question of whether use of a separate phase which is a good solvent for polyhydroxyether will permit higher molecular weights to be achieved.

## REFERENCES


[1] C. M. Starks, J. Amer. Chem. Soc., 93, 195 (1971).

[2] C. M. Starks and R. M. Owens, J. Amer. Chem. Soc., 95, 3613 (1973).

[3] C. L. Liotta and H. P. Harris, J. Amer. Chem. Soc., 96, 2250 (1974).

[4] A. W. Herriott and D. J. Picker, J. Amer. Chem. Soc., 97, 2345 (1975).

[5] D. J. Sam and H. E. Simmons, J. Amer. Chem. Soc., 94, 4024 (1972); 96, 2252 (1974).

[6] S. Akobori and M. Ohtomi, Bull. Chem. Soc., (JAP), 48, 2991 (1975).

[7] E. V. Dehmlov, Angew. Chem. Int. Edit. (Eng.), 16, 493 (1977).

[8] W. P. Weber and G. W. Gokel "Phase Transfer Catalysis in Organic Synthesis", Springer-Verlag, New York (1977).

[9] J. Dockx, Synthesis, 441 (1973).

[10] D. Landini and F. Montanari, J. Chem. Soc. Chem. Commun., 879 (1974).

[11] F. Millich and C. E. Carraher, Jr., Ed., "Interfacial Synthesis" Vol. I and II, Marcel Dekker Inc., New York (1977).

[12] P. W. Morgan, "Interfacial Polycondensation", Wiley, New York (1965).

[13] J. A. Moore, Ed., "Macromolecular Syntheses", Collective Vol. I, Wiley (1978).

[14] P. W. Morgan, Macromolecules, 5, 536 (1970).

[15] H. H. Freedman and R. A. Dubois, Tetrahedron Letters, 30, 1379 (1974).

[16] A. McKillop, J. C. Fiaud and R. P. Hug, Tetrahedron, 30, 1379 (1974).

[17] M. Makosza and A. Zonczyk, Organic Synth., 55, 91 (1976).

[18] J. Uglestad, T. Ellingsen and A. Beige, Acta Chem. Scand., 20, 1593 (1966).

[19] W. F. Hale, Encyclopedia of Polymer Sci. and Technology., 10, 111 (1969).

[20] H. Lee, D. Stoffey and K. Neville "New Linear Polymers" McGraw-Hill Book Company, New York, (1967).

[21] E. L. Eliel in Steric Effects in Organic Chemistry, M. S. Newman, Ed., Wiley, New York, 1956, pp 106-114.

[22] S. Winstein and R. B. Henderson, in Heterocyclic Compounds, Vol. I, R. C. Elderfield, Ed., Wiley, New York, 1950, pp. 22-46.

[23] H. C. Chitwood and B. J. Freure, J. Amer. Chem. Soc., 68, 680 (1946).

[24] N. B. Chapman, N. B. Isaacs and R. E. Parker, J. Chem. Soc., 1925 (1959).

[25] P. D. Bartlett and S. D. Ross, J. Amer. Chem. Soc., 70, 926 (1948).

[26] G. L. Brode and J. Wynstra, J. Polym. Sci. Part A-1, 4, 1045 (1966).

[27] L. Shechter and J. Wynstra, Ind. Eng. Chem., 48(1), 86 (1956).

[28] Nikles, Epoxide Resin Composition, U. S. Pat., 3,006,892 (1961).

[29] Linear Polyether Resins and Processes for Their Production (To Shell Internationale Research), Brit. Pat., 980,509 (1965).

[30] A. K. Banthia and J. E. McGrath, Polymer Preprints, 20(2), 629 (1979).

[31] F. B. Alvey, J. Appl. Polym. Sci., 13, 1473 (1969).

[32] P. N. Son and C. D. Weber, J. Appl. Polym. Sci., 17, 2415 (1973).

[33] C. Starks, Chem Tech, 110 (1980).

[34] R. V. Pagilagon and W. E. McEwen, Chem. Commun., 652 (1966).

[35] M. Zanger, C. A. Vanderwerf and W. C. McEwen, J. Amer. Chem. Soc., 81, 3805 (1959).

[36] W. E. McEwen, K. F. Kumli, A. Blade-Font, M. Zanger and C. A. Vanderwerf, J. Amer. Chem. Soc., 86, 2378 (1964).

[37] W. E. McEwen, A. Axelrad, M. Zanger and C. A. Vanderwerf, J. Amer. Chem. Soc., 87, 3948 (1965).

[38] B. E. Conway, R. E. Verall, and J. E. Desnoyers, Trans. Faraday Soc., 62, 2738 (1966) and references contained therein.

[39] L. Pauling "The Nature of the Chemical Bond", Appendix 3.1, Cornell Univ. Press, Ithaca, New York (1940).

[40] Y. Imai, M. Ueda and M. I., J. Polym. Sci., Polym. Lett. Ed., 17, 85 (1979).

[41] J. H. Fendler, E. J. Fendler, "Catalysis in micellar and Macromolecular systems, Academic Press, New York (1975).

[42] J. H. Fendler, Acc. Chem. Res., 9, 153 (1976).

# Interfacial Synthesis Part II: Phase-Transfer Catalyzed Synthesis of Polycarbonate/Polysiloxane Block Copolymers

J. S. Riffle, R. G. Freelin, A. K. Banthia, and J. E. McGrath

Chemistry Dept. and Polymer Materials and Interfaces Laboratory
Virginia Polytechnic Institute and State University, Blacksburg, Virginia 24061

ABSTRACT

A phase transfer catalyzed (PTC) method for the preparation of bisphenol-A polycarbonate/polydimethylsiloxane multi-block or segmented copolymers of the structure shown below has been investigated.

$$\sim\!\!\sim\left[-\overset{O}{\overset{\|}{C}}-O-\langle\bigcirc\rangle-\underset{CH_3}{\overset{CH_3}{\overset{|}{\underset{|}{C}}}}-\langle\bigcirc\rangle-O-\right]_M-\overset{O}{\overset{\|}{C}}-(CH_2)_3-\left[-\underset{CH_3}{\overset{CH_3}{\overset{|}{\underset{|}{Si}}}}-O-\right]_N\sim\!\!\sim$$

The synthesis route involves the reaction of a preformed carboxypropyl-terminated polydimethylsiloxane oligomer of various molecular weights with bisphenol-A, base, phosgene, and the PTC. The final copolymer contains a direct ≡Si-C- link between the block structures as compared to the relatively hydrolytically unstable ≡Si-O-C bond discussed previously by the authors and others. A phosgenation "pre-step" to the interfacial reaction was found to be desirable. This procedure involves bubbling phosgene through an anhydrous concentrated solution of the carboxypropyl-terminated polydimethylsiloxane in methylene chloride. Subsequently, a two-phase $CH_2Cl_2/H_2O$ reaction medium with the phase transfer catalyst is employed in the interfacial step. The phosgenation pre-step is believed to enhance the conversion of carboxyl to acid chloride, which then promotes the efficient incorporation of the polysiloxane into the copolymer. Moreover, anhydride linkages along the chain are reduced to undetectable levels. The copolymers could be of interest as thermoplastic elastomers, elastoplastics, or biomaterials.

## INTRODUCTION

Thermoplastic elastomers based on polycarbonate-polysiloxane block copolymers display some intrinsic properties which may be of interest as biomaterials [1]. The generalized structure of these copolymers is shown below.

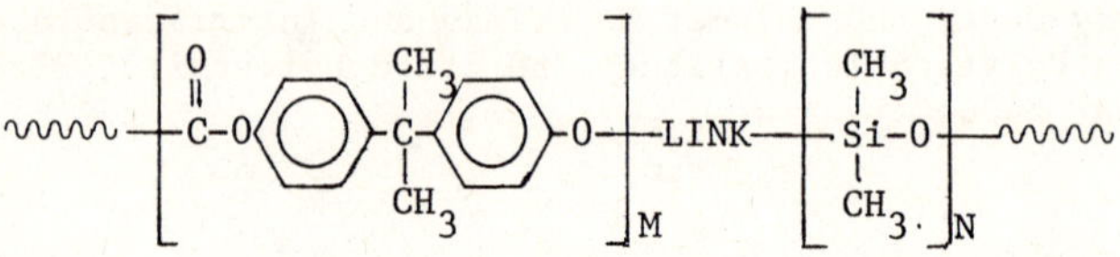

$$\sim\sim\left[\overset{\overset{O}{\|}}{C}-O-\langle\bigcirc\rangle-\underset{CH_3}{\overset{CH_3}{\overset{|}{\underset{|}{C}}}}-\langle\bigcirc\rangle-O\right]_M-\text{LINK}-\left[\underset{CH_3}{\overset{CH_3}{\overset{|}{\underset{|}{Si}}}}-O\right]_N\sim\sim$$

There are at least two basic requirements for materials which one might consider for biological end-uses. First, the polymers should be "compatible"(e.g. not coagulate) with blood, and secondly, the materials must also be very stable in the physiological environment. The latter requirement specifically requires that the copolymers possess good hydrolytic stability.

Earlier publications from this laboratory[2-4] and elsewhere[1,5-7] demonstrate that the subject block copolymers and related structures easily undergo a microphase separation above segment molecular weight of a few thousand. This was recognized as being a function of the inherent immiscibility of bisphenol-A polycarbonate and polydimethylsiloxane block structures. Such phenomena even affect the melt rheology of these systems [7-10]. Quantification of this immiscibility has been probed by inverse gas chromatography [11,12]. Moreover, in polystyrene-polydimethylsiloxane block copolymer systems [13-15], in our preliminary work [2,16] and earlier work by Gaines and LeGrand [17], it was demonstrated that

the polydimethylsiloxane component migrates to the surface of films of these materials in preference to the other block.
This phenomenon was attributed[13-16] to the fact that the polydimethylsiloxane component has a considerably lower surface free energy relative to polystyrene or polycarbonate. The pure polydimethylsiloxane homopolymer surface is known to be quite "compatible" with blood [18]. The fact that surfaces of the films of these copolymers could be tailored to be composed predominantly of siloxane [2,3] suggested that the surface and, hence, the materials, could be tailored to be biologically "compatible"in the sense defined earlier.

In view of the requirement for good hydrolytic stability, block copolymers with two different types of bonds between the block structures have been investigated.

1. ≡Si-O-C- links
2. Direct ≡Si-C- links

The relative hydrolytic stabilities of the ≡Si-O-C bond versus the ≡Si-C- bond in some siloxane copolymers has been extensively studied and it is generally agreed that the ≡Si-C bond is much better, even though the ≡Si-O-C bond[18] is the link of choice when thermal stability is required. However,Noshay et al.[1,19] demonstrated that perfectly alternating block copolymers composed of poly(aryl ethers) or poly(aryl esters) and siloxanes were remarkably resistant to hydrolysis. This was thought[19] to be due to the hydrophobic siloxane nature and the relatively few links between blocks. Consequently, there is some reason to doubt whether the presence of ≡Si-O-C links between blocks in the strictly alternating $(A-B)_N$ block copolymer architecture would detract from hydrolytic stability. In order to further assess these questions, materials were prepared in this study which consisted of both perfectly alternating block copolymers containing the ≡Si-O-C and also, randomly coupled block copolymers containing the direct ≡Si-C bond between the blocks.

Perfectly alternating block copolymers were prepared essentially by Noshay and Matzner's procedure[1] in refluxing chlorobenzene solution by reacting preformed dimethylamine-terminated polydimethylsiloxane and hydroxyl-terminated polycarbonate oligomers. Two series of copolymers produced via this method are briefly discussed herein. Both copolymer composition at constant block length and block lengths at given compositions were investigated.

The preparation of copolymers bearing the direct $\equiv$Si-C- bond between the blocks has proven to be quite challenging. The general synthetic route utilized a preformed carboxypropyl-terminated polydimethylsiloxane and in situ generated polycarbonate segments as schematically shown in equation 1.

Equation 1

$$\underset{\mathbf{1}}{HO-C_6H_4-C(CH_3)_2-C_6H_4-OH} + \underset{\mathbf{2}}{HOOC-(CH_2)_3-\left[Si(CH_3)_2-O\right]_N-Si(CH_3)_2-(CH_2)_3-COOH}$$

$$\xrightarrow[\text{phase transfer catalyst}]{H_2O/CH_2Cl_2,\ COCl_2,\ \text{base}}$$

$$\underset{\mathbf{3}}{\sim O-C_6H_4-C(CH_3)_2-C_6H_4-O-\left[\overset{O}{\overset{\|}{C}}-O-C_6H_4-C(CH_3)_2-C_6H_4-O\right]_M-\overset{O}{\overset{\|}{C}}-(CH_2)_3-\left[Si(CH_3)_2-O\right]_N-Si(CH_3)_2-(CH_2)_3-\overset{O}{\overset{\|}{C}}\sim}$$

Randomly Coupled
Block copolymer, $(\equiv Si-C)$ linked

Note that the reaction of two moles bisphenol-A with one mole of phosgene yields a carbonate $(-O-\overset{O}{\overset{\|}{C}}-O-)$ link. On the other hand, the corresponding reaction of one mole bisphenol-A, one mole of acid—terminated polysiloxane oligomer, and one mole of phosgene is shown producing an ester ($-\overset{O}{\overset{\|}{C}}-O-$) bond between the terminal carboxypropylsiloxane and the bisphenol. Both solution and interfacial (or rather, phase transfer catalyzed) methods have been investigated. It was reasoned that the interfacial route would insure a cleaner reaction product which in turn would be very important for biomedical applications. Solution processes might require pyridine which would be difficult to remove quantitatively after copolymer formation. Residual traces could degrade either segment. For example, the copolymer product from the solution reaction might result in rearrangements of the siloxane portions with time. The main thrust of this paper is the development of procedures and the results obtained using the interfacial method for preparation of the randomly coupled block copolymers. The perfectly alternating copolymers are discussed in detail elsewhere [4].

The synthesis of polycarbonates derived from bisphenol-A in a two-phase medium via quaternary ammonium halide catalysis was one of the earliest applications of phase transfer catalysis [20-26]. Typically, bisphenol-A and the quaternary ammonium halide catalyst are dissolved in an aqueous caustic solution followed by addition of the quaternary ammonium halide.

The aqueous phase is rapidly mixed with a (usually) water immiscible solvent such as dichloromethane. Phosgene is then introduced in stoichiometric quantities. The phase transfer catalysis cycle illustrated for the specific case of the polycarbonate homopolymer is reviewed in equation 2. The synthesis is well known to be both rapid, facile and insensitive to minor variations in reaction conditions. However, the addition of a preformed, acid-terminated polydimethylsiloxane oligomer to the system produced numerous complications and necessitates a much more defined set of reaction parameters.

The phase transfer process envisioned as being operative in the formation of these polycarbonate/polysiloxane random block

Equation 2

$$\mathrm{HO{-}C_6H_4{-}C(CH_3)_2{-}C_6H_4{-}OH} + \mathrm{KOH} \longrightarrow \mathrm{K^{+}\,{}^{-}O{-}C_6H_4{-}C(CH_3)_2{-}C_6H_4{-}O^{-}\,{}^{+}K}$$

$$\Big\Updownarrow \; \overset{+\;-}{(C_2H_5)_4NCl} \quad \text{(Aqueous Phase)}$$

$$\mathrm{\overset{+-}{(C_2H_5)_4NO}{-}C_6H_4{-}C(CH_3)_2{-}C_6H_4{-}\overset{-+}{O\;N(C_2H_5)_4}} + \mathrm{KCl}$$

$$\Big\Updownarrow$$

$$\mathrm{\overset{+-}{(C_2H_5)_4NO}{-}C_6H_4{-}C(CH_3)_2{-}C_6H_4{-}\overset{-+}{O\;N(C_2H_5)_4}} \quad \text{(Organic Phase)}$$

$$\Big\downarrow \; COCl_2$$

$$\overset{+\;-}{(C_2H_5)_4NCl} + \left[ \mathrm{\overset{O}{\overset{\|}{C}}{-}O{-}C_6H_4{-}C(CH_3)_2{-}C_6H_4{-}O} \right]_N$$

copolymers is an extension of the established sequence occurring in the production of the bisphenol-A polycarbonate homopolymers. It first involves dissolution of the bisphenol-A in an aqueous caustic solution. This is followed by reaction of a portion of the alkali metal salt of bisphenol-A with a phase transfer catalyst (P.T.C.) (equation 3). The latter usually possesses a relatively large organic cation.

The alkyl quaternary ammonium salt of bisphenol-A will then partition itself between the aqueous and an organic phase. As it

Equation 3

$$K^{+-}O-C_6H_4-C(CH_3)_2-C_6H_4-O^{-+}K + (C_2H_5)_4\overset{+}{N}\overset{-}{Cl}$$

$$\rightleftharpoons$$

$$(C_2H_5)_4\overset{+-}{N}O-C_6H_4-C(CH_3)_2-C_6H_4-\overset{-+}{O}\ N(C_2H_5)_4 + KCl$$

moves into the organic phase, the highly reactive salt is brought into contact with both phosgene and the acid -terminated polydimethylsiloxane oligomers and may react to form the copolymer (equation 4). The by-product of equation 4, which is the original phase transfer catalyst, is then transferred back to the aqueous phase which completes the catalytic cycle.

The preparation of bisphenol-A polycarbonate homopolymer via such a mechanism can be complicated by the fact that polymerization within the organic phase proceeds via a two-step reaction (equation 5).

For the attainment of high molecular weight polymers, the rate of formation of chloroformate endgroups must be balanced against the rate of their further reaction with bisphenol-A or with the growing polycarbonate chain. Excessive build-up of aryl chloroformate endgroups which occurs if $k_1 >> k_2$, can be controlled via adjustment of a number of reaction parameters.

The reaction of the carboxylic acid-terminated siloxane to form an ester bond in the reaction scheme adds at least one (and possibly more) series of steps to the overall reaction. Although the mechanistic pathways have not been clearly established, two plausible routes to ester formation are envisioned (equations 6, and 7a and 7b).

Mixed carbonic-carboxylic anhydrides of the type illustrated in equations 6, 7a and 7b (4) are known to decompose under anhy-

Equation 4

$$(H_2O) \quad (C_2H_5)_4\overset{+-}{N}O-C_6H_4-C(CH_3)_2-C_6H_4-\overset{-+}{O}\,N(C_2H_5)_4$$

$$\rightleftharpoons$$

$$(CH_2Cl_2) \quad (C_2H_5)_4\overset{+-}{N}\,O-C_6H_4-C(CH_3)_2-C_6H_4-\overset{-+}{O}\,N(C_2H_5)_4$$

$$+$$

$$HOOC-(CH_2)_3-\left[Si(CH_3)_2-O\right]_N-Si(CH_3)_2-(CH_2)_3-COOH + COCl_2$$

$$\downarrow$$

$$HO-\left[\left[\overset{O}{\overset{\|}{C}}-O-C_6H_4-C(CH_3)_2-C_6H_4-O\right]_M-\overset{O}{\overset{\|}{C}}-(CH_2)_3-\left[Si(CH_3)_2-O\right]_N-Si(CH_3)_2-(CH_2)_3-COOH\right]_y$$

$$+$$

$$(C_2H_5)_4\overset{+\,-}{N}Cl$$

Equation 5

Step 1:

$$(C_2H_5)_4\overset{+-}{N}\,O-C_6H_4-C(CH_3)_2-C_6H_4-\overset{-+}{O}\,N(C_2H_5)_4 + Cl-\overset{O}{\overset{\|}{C}}-Cl \xrightarrow{k_1}$$

$$(C_2H_5)_4\overset{+-}{N}\,O-C_6H_4-C(CH_3)_2-C_6H_4-O-\overset{O}{\overset{\|}{C}}-Cl + (C_2H_5)_4\overset{+\,-}{N}Cl$$

Step 2:

$$(C_2H_5)_4\overset{+-}{N}\,O-C_6H_4-C(CH_3)_2-C_6H_4-O-\overset{O}{\overset{\|}{C}}-Cl + (C_2H_5)_4\overset{+-}{N}\,O-C_6H_4-C(CH_3)_2-C_6H_4-\overset{-+}{O}\,N(C_2H_5)_4$$

$$\xrightarrow{k_2} (C_2H_5)_4\overset{+-}{N}\,O-C_6H_4-C(CH_3)_2-C_6H_4-O-\overset{O}{\overset{\|}{C}}-O-C_6H_4-C(CH_3)_2-C_6H_4-\overset{-+}{O}\,N(C_2H_5)_4$$

Equation 6

$(Et)_4\overset{+}{N}\overset{-}{O}$–(bisphenol)–$\overset{-}{O}\ \overset{+}{N}(Et)_4$ (1) + Cl–C(=O)–Cl $\xrightarrow{\text{fast}}$ $(Et)_4\overset{+}{N}\overset{-}{O}$–(bisphenol)–O–C(=O)–Cl

$(Et)_4\overset{+}{N}\overset{-}{O}$–(bisphenol)–O–C(=O)–Cl + HO–C(=O)–$(PSX)_N$–C(=O)–OH (2) → $(Et)_4\overset{+}{N}\overset{-}{O}$–(bisphenol)–O–C(=O)–O–C(=O)–$(PSX)_N$–C(=O)–OH (3)

$(Et)_4\overset{+}{N}\overset{-}{O}$–(bisphenol)–O–C(=O)–O–C(=O)–$(PSX)_N$–C(=O)–OH (4)

Path 1 ($H_2O$ or $^{-}OH$): → HO–C(=O)–$(PSX)_N$–C(=O)–OH + $(Et)_4\overset{+}{N}\overset{-}{O}$–(bisphenol)–O–C(=O)–OH $\xrightarrow{-CO_2}$ $(Et)_4\overset{+}{N}\overset{-}{O}$–(bisphenol)–OH

Path 2 (Disproportionation): → [HO–C(=O)–$(PSX)_N$–C(=O)–]$_2$O + [$(Et)_4\overset{+}{N}\overset{-}{O}$–(bisphenol)–O–]$_2$C=O + $CO_2$

Path 3 ($-CO_2$): → $(Et)_4\overset{+}{N}\overset{-}{O}$–(bisphenol)–O–C(=O)–$(PSX)_N$–C(=O)–OH

Equation 7a

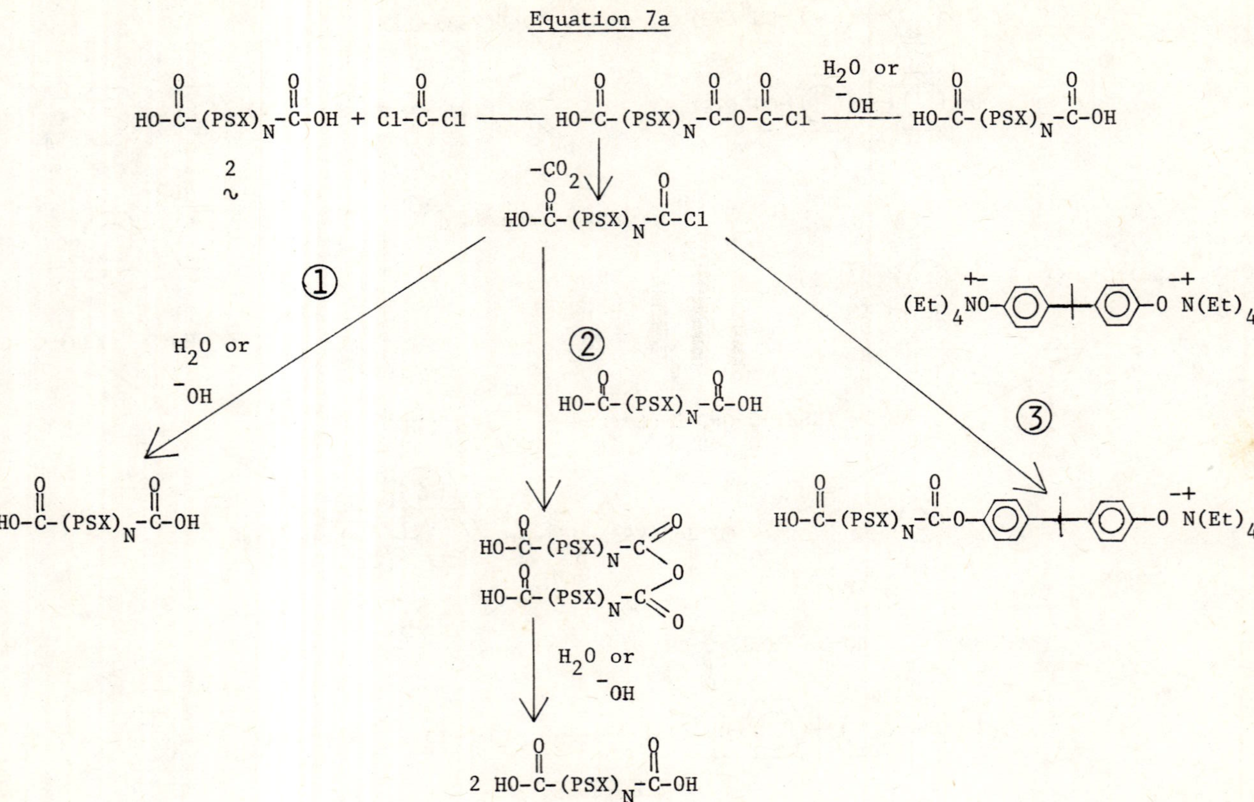
HO-C(=O)-(PSX)N-C(=O)-OH + Cl-C(=O)-Cl
2
HO-C(=O)-(PSX)N-C(=O)-O-C(=O)-Cl
H2O or -OH
HO-C(=O)-(PSX)N-C(=O)-OH
-CO2
HO-C(=O)-(PSX)N-C(=O)-Cl
1
H2O or -OH
HO-C(=O)-(PSX)N-C(=O)-OH
2
HO-C(=O)-(PSX)N-C(=O)-OH
HO-C(=O)-(PSX)N-C(=O)-O-C(=O)-(PSX)N-C(=O)-OH
H2O or -OH
2 HO-C(=O)-(PSX)N-C(=O)-OH
3
(Et)4N+ -O-⟨⟩-C-⟨⟩-O- +N(Et)4
HO-C(=O)-(PSX)N-C(=O)-O-⟨⟩-C-⟨⟩-O- +N(Et)4

Equation 7b

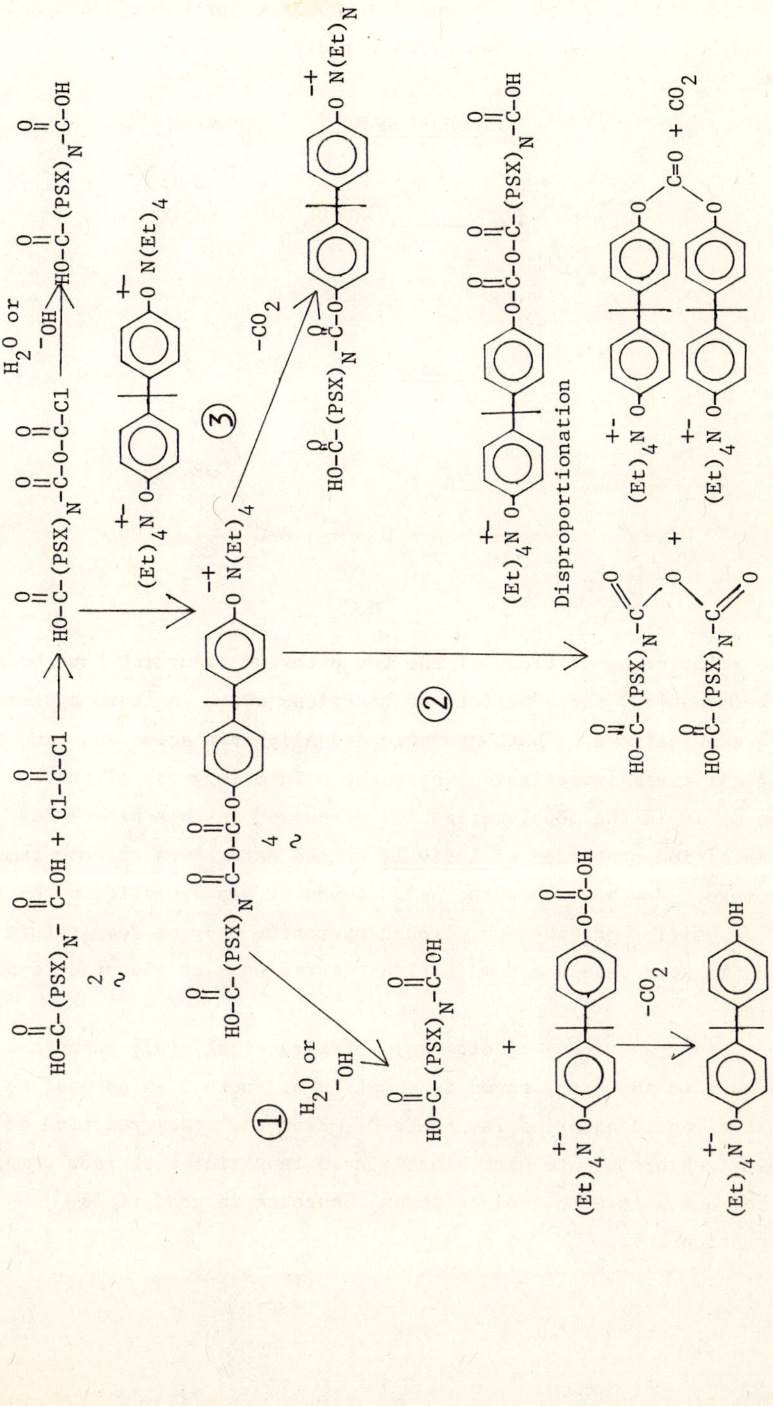

HO-C(=O)-(PSX)$_N$-C(=O)-OH (**2**) + Cl-C(=O)-Cl → HO-C(=O)-(PSX)$_N$-C(=O)-O-C(=O)-Cl → ($H_2O$ or $^-OH$) → HO-C(=O)-(PSX)$_N$-C(=O)-OH

$(Et)_4\overset{+-}{N}$ O-⟨bisphenol⟩-$\overset{+-}{O}$ $N(Et)_4$

HO-C(=O)-(PSX)$_N$-C(=O)-O-C(=O)-O-⟨bisphenol⟩-$\overset{-+}{O}$ $N(Et)_4$ (**4**)

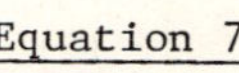

$H_2O$ or $^-OH$

HO-C(=O)-(PSX)$_N$-C(=O)-OH

+

$(Et)_4\overset{+-}{N}$ O-⟨bisphenol⟩-O-C(=O)-OH

$-CO_2$

$(Et)_4\overset{+-}{N}$ O-⟨bisphenol⟩-OH

②

$(Et)_4\overset{+-}{N}$ O-⟨bisphenol⟩-O-C(=O)-O-C(=O)-(PSX)$_N$-C(=O)-OH

Disproportionation

[HO-C(=O)-(PSX)$_N$-C(=O)]$_2$O + [$(Et)_4\overset{+-}{N}$ O-⟨bisphenol⟩-O]$_2$C=O + $CO_2$

③

$-CO_2$

HO-C(=O)-(PSX)$_N$-C(=O)-O-⟨bisphenol⟩-$\overset{-+}{O}$ $N(Et)_N$

drous conditions via pathways analogous to 2 and 3 in both equations 6 and 7b [27-30]. Proposed mechanisms for these decompositions[30] are given in equations 8 and 9.

Equation 8

$$R_1\text{-C}(=\text{O})\text{-O-C}(=\text{O})\text{-O-C}\!\!\lesssim \longrightarrow R_1\text{-C}(=\text{O})\text{-O-C}\!\!\lesssim + CO_2$$

Equation 9

$$2\; R_1\text{-}\overset{\text{O}}{\overset{\|}{\text{C}}}\text{-O-}\overset{\text{O}}{\overset{\|}{\text{C}}}\text{-OR}_2 \longrightarrow (R_1\text{-C}(=\text{O}))_2\text{O} + \text{O=C}(\text{O-R}_2)_2 + CO_2$$

The relative proportions of the two pathways occurring have been determined[30] for a variety of reactions which included systematic combinations of both aromatic and aliphatic acids and alcohols. For all cases investigated where the acid moiety was aliphatic (as it is in the copolymerization discussed in this paper) decomposition proceeded exclusively to the ester even at room temperature. However, when the acid component was aromatic, both decomposition pathways were found operative at room temperature and, in some cases, decomposition occured only at elevated temperatures.

Under anhydrous conditions, Goldberg et al. [31] supported the pathway to the ester shown in equation 6, route 3 as opposed to that given in equation 7a, route 3. From the model reaction of phenyl chloroformate with benzoic acid in pyridine at room temperature he was able to isolate phenyl benzoate in good yield (equation 10).

Equation 10

$$\text{C}_6\text{H}_5\text{-O-}\overset{\text{O}}{\overset{\|}{\text{C}}}\text{-Cl} + \text{HO-}\overset{\text{O}}{\overset{\|}{\text{C}}}\text{-C}_6\text{H}_5 \xrightarrow{\text{pyridine}} \text{C}_6\text{H}_5\text{-O-}\overset{\text{O}}{\overset{\|}{\text{C}}}\text{-C}_6\text{H}_5 + CO_2$$

Moreover, when phosgene was bubbled into a solution of phenol and benzoic acid in pyridine he obtained the same product. Goldberg et al. dismissed the possibility of the acid chloride intermediate from being plausible for two reasons: 1) acid chlorides are less reactive than chloroformates $\left(\sim\overset{\text{O}}{\overset{\|}{\text{C}}}\text{-O-}\overset{\text{O}}{\overset{\|}{\text{C}}}\text{-Cl}\right)$, and 2) no evidence of carboxylic anhydrides was found in the infrared spectra of the products. While Goldberg's evidence does provide support for the feasibility of the pathway shown in equation 6, route 3, it would appear to not preclude the possibility of alternate routes. For example, some reactions not considered could be occurring simultaneously (ie. equation 7b, route 3). Nevertheless, one must agree that good yields of the desired model compound were isolated. Of course, the analytical techniques employed were probably not sensitive enough to detect minor in chain anhydride imperfections. Quite recently, Prevorsek et al. [32] have reinvestigated the reaction of bisphenol-A, terepthalic acid, and phosgene under anhydrous conditions in pyridine/dichloromethane co-solvent. In contrast to Goldberg's work, they did observe carboxylic anhydrides in the infrared spectra of their products. Compared to the current work herein, both of these papers dealt with aromatic acid reactions which may indeed produce relatively more stable anhydride intermediates. It is clear that additional studies would be desirable.

## EXPERIMENTAL

$\alpha,\omega$ - carboxypropyl-terminated polydimethylsiloxane oligomers (2) of controlled number average molecular weights were produced in our laboratory via a siloxane redistribution reaction. Details of the synthetic procedure and analytical aspects will be provided separately [33]. However, a typical synthesis is briefly described below.

Preparation of Carboxypropyl-Terminated Polydimethylsiloxane Oligomers:

Carboxypropyltetramethyldisiloxane was obtained from Silar Laboratories, Scotia, N. Y. The octamethylcyclotetrasiloxane (D-4) was obtained from either Union Carbide, Silar Labs, or Petrarch, (Levittown, PA). Trifluoroacetic acid was used as the redistribution catalyst and was reagent grade from Fisher Chemical.

The reaction apparatus consisted of a three-necked, round-bottomed flask equipped with a mechanical stirrer, argon inlet, condenser with a drying tube and heated with a silicone oil bath. The required proportions of the bis(1,3-carboxypropyl)-tetramethyldisiloxane and D-4 were charged to the reaction vessel and heated to 60°C. In the reaction, the cyclic monomers (D-4) are opened and inserted into the 1,3-carboxypropyltetramethyldisiloxane. Therefore, the necessary ratio of D-4 to linear dimer is determined by the final <Mn> desired. Typically, trifluoroacetic acid (∿ 0.013 moles per 0.024 moles starting material) was added at 60°C for 24 hours. Rapid agitation was maintained throughout the reaction. After the equilibration, the reaction mixture was washed three times with water to remove the residual trifluoroacetic acid catalyst. Remaining traces of water were azeotroped with tetrahydrofuran under vacuum.

Oligomer structure was characterized by both high resolution proton NMR (Varian EM-390 spectrometer) and $^{13}$C NMR (JEOL-PFT-100 spectrometer) to insure that the endgroup structure present in the dimer was retained in the final oligomers. Traces of residual cyclics can be assessed by liquid chromatography, GLC, or TGA. The progress of the redistribution of the carboxypropyl-terminatd siloxane was followed by gel permeation chromatography (Waters 6000-A pump, Waters R-400 D.R.I. detector, micro-styragel columns - 500 Å, $10^3$ Å, $10^4$ Å, $10^5$ Å, toluene solvent). Molecular weights, <Mn>, of these oligomers were assessed via potentiometric titration in isopropanol with an isopropanolic potassium hydroxide titrant [33].

The development of the synthetic procedures for the randomly placed block copolymers required that a number of bisphenol-A polycarbonate homopolymers be produced under different phosgene flow rates and catalyst concentrations. The apparatus for these reactions is identical to that used for the copolymerization (Figure 1). Typically, 7.5 gm of bisphenol-A, a stoichiometric quantity of NaOH (2 moles per mole bisphenol-A), and the appropriate amount of the tetraethylammonium chloride catalyst were dissolved in 75 ml water. An equal volume of dichloromethane was then added. Approximately 18 N. aqueous sodium hydroxide was

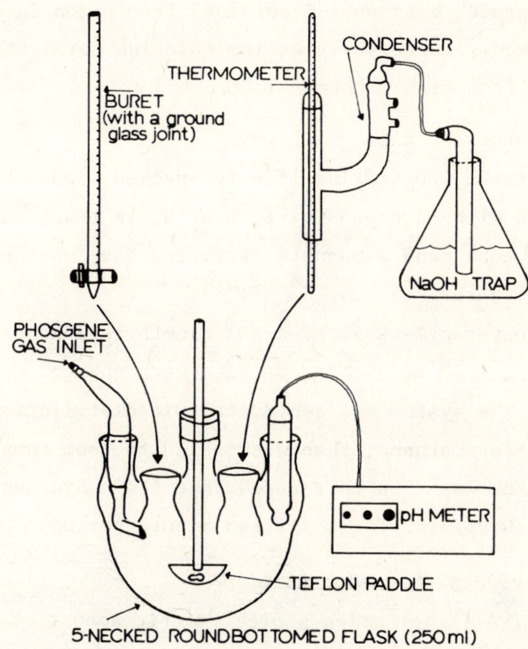


FIG. 1. Copolymerization apparatus

added throughout the 30 minutes of phosgene addition. This allowed the pH to be maintained between 11-12.

Randomly placed block copolymerization was studied using principally a 2000 <Mn> carboxypropyl-terminated polydimethylsiloxane oligomer. Results demonstrate that an anhydrous prestep (hereafter called a "pre-phosgenation" step) to the interfacial reaction is desirable. A typical 2-stage copolymer formulation using optimized experimental parameters is detailed below:

Preparation of Randomly Placed Polycarbonate/Polydimethylsiloxane Block Copolymers:

The preformed polydimethylsiloxane oligomer was prepared and characterized according to the methods previously described. High purity "UCAR grade" bisphenol-A obtained from Union Carbide was used as received. Tetraethylammonium chloride catalyst used was reagent grade from Fisher Chemical.

Step 1: "Pre-Phosgenation"

The apparatus, consisting of a two-necked, round-bottomed flask equipped with an argon inlet, Dean Stark trap, condenser with a drying tube, and a magnetic stirring bar, was heated with a silicone oil bath.

The carboxypropyl-terminated polydimethylsiloxane oligomer (7.5 grams) and ∿ 40 ml dichloromethane were charged to the reaction vessel. The system was dehydrated via distillation of ∿ 20 ml of the dichloromethane, then was cooled to room temperature. Gaseous phosgene was then introduced into the anhydrous solution at a rate of 30 cc/min. for a fifteen minute period.

Step 2: Interfacial Copolymerization

Bisphenol-A (7.5g) and a stoichiometric amount of potassium hydroxide (2 moles KOH per mole bisphenol-A) were charged to the reaction vessel and dissolved in 75 ml water. Tetraethylammonium chloride catalyst (1.1g, 20 mole percent based on moles of bisphenol-A) was added to the mixture. The siloxane/dichloromethane solution from the "pre-phosgenation" step, followed by 55 ml additional dichloromethane were added. Phosgene was again introduced to the mixture at a rate of 40-58 cc/min. for one hour. Throughout the hour of phosgene addition, pH was monitored using a combination pH electrode (glass body, silver/silver chloride reference) and constrained to a range of 11-12 by the continuous addition of 12 N. aqueous potassium hydroxide solution. For the concentrations employed, the reaction required approximately 2 ml KOH solution every 3 minutes. Following the hour of phosgene addition with the pH maintained between 11-12, phosgene flow was continued at the same rate without any further addition of KOH for an additional ten minutes. This generally desirably produced a neutral or slightly acidic pH. The organic phase was washed several times with water, coagulated in isopropanol, isolated by filtration and vacuum dried.

Copolymers and bisphenol-A polycarbonate homopolymers were analyzed for intrinsic viscosity at 25°C in tetrahydrofuran. Values given in the tables are reported in deciliters per gram.

Copolymer compositional analyses were performed using high resolution proton NMR spectroscopy (Varian EM-390 spectrometer) via comparison of the areas of the aromatic and silicon methyl peaks (see Fig. 2). The precision of the compositional analysis was estimated to be ± 1.0%.

Copolymer fine structure was also elucidated by comparison of both $^{13}C$ NMR (JEOL-FX60Q instrument) and IR (Perkin-Elmer model 283 instrument) spectra of the copolymers with that of model compounds and homopolymers. The diphenyl ester of 1,3-carboxypropyltetramethyldisiloxane (5) was prepared as a model compound in a manner closely paralleling that of Mulvaney and Marvel. [34]

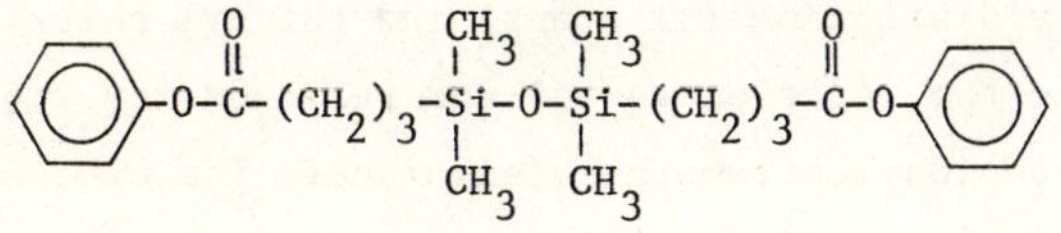

$$C_6H_5\text{-O-}\overset{O}{\overset{\|}{C}}\text{-}(CH_2)_3\text{-}\underset{CH_3}{\overset{CH_3}{Si}}\text{-O-}\underset{CH_3}{\overset{CH_3}{Si}}\text{-}(CH_2)_3\text{-}\overset{O}{\overset{\|}{C}}\text{-O-}C_6H_5$$

5

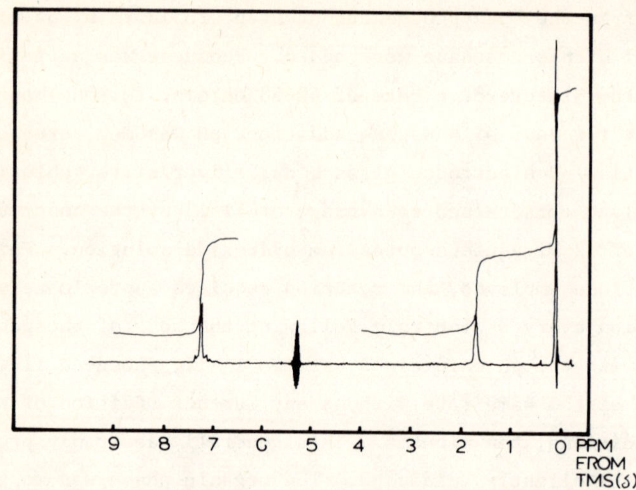


FIG. 2. Proton NMR spectrum of a randomly coupled polycarbonate/polydimethylsiloxane copolymer

Assignments: ∿ 0.17 P.P.M. - Silicon Bonded Methyls
∿ 1.75 P.P.M. - Bisphenol-A Methyls
∿ 7.1 - 7.4 P.P.M. - Aromatic Protons
∿ 5.30 P.P.M. Methylene Chloride
(Lock Signal and Reference)

Preparation of $\text{Ph-O-}\overset{\text{O}}{\overset{\|}{\text{C}}}\text{-}(\text{CH}_2)_3\text{-}\left(\text{Si}(\text{CH}_3)_2\text{-O}\right)_{n=1}\text{-Si}(\text{CH}_3)_2\text{-}(\text{CH}_2)_3\text{-}\overset{\text{O}}{\overset{\|}{\text{C}}}\text{-O-Ph}$ :

Thionyl chloride (Fisher reagent grade) and phenol (Fisher reagent ACS grade - 0.5% maximum water) were used as received.

The apparatus consisted of a two-necked, round-bottom flask equipped with an argon inlet, condenser with a drying tube, magnetic stirring bar and a silicone oil heating bath. The 1,3-carboxypropyltetramethyldisiloxane (11 grams) and thionyl chloride (50 ml) were charged to the reaction vessel and refluxed for 2.5 hours to produce 1,3-propionyltetramethyldisiloxane. The excess thionyl chloride was stripped off under vacuum, and then a 20 mole percent excess of phenol was added. The mixture was heated at 85°C for three hours to produce the diphenyl ester. Following the reaction, the product mixture was dissolved in 60 ml of diethyl ether, washed with water, then with 100 ml 10% aqueous NaOH, and finally with water (two more times). Traces of water were removed via a tetrahydrofuran/water azeotrope under vacuum. Vacuum distillation of the product produced a small forecut, followed by a major fraction of the diester (B.P. 175°C, 0.2 mm).

## RESULTS AND DISCUSSION

The investigation described in this paper is perhaps the first simultaneous reaction of a carboxylic acid, a phenol, and phosgene to be studied in an aqueous PTC type medium. The approach taken for development of the synthesis of these block copolymers was to first investigate the formation of the homopolycarbonate. Copolymerizations including the siloxane oligomers were then conducted under analogous conditions. This method allowed a systematic study of the major differences between homopolycarbonates and copolymer formation. Consequently, results and interpretations pertinent to the homopolymer study will first be discussed, followed by aspects related to copolymerization.

A series of homo-polycarbonates were produced where the amount of phase transfer catalyst was systematically varied from 4 –40 mole percent tetraethylammonium chloride based on number of moles of bisphenol-A (cf. Table 1, #1-3). These values are seemingly high, but were based on some earlier work by Morgan, et al.[35,36]. A maximum in intrinsic viscosity was observed at the middle catalyst concentration (20%) initially studied. This suggested that a 4% catalyst concentration might be too low to promote migration of bisphenol-A to the organic phase.
Hence, a slower reaction ensues which results in a lower molecular weight. Alternatively, the 40% catalyst concentration probably causes a relatively rapid transfer of the bisphenol to the dichloromethane phase. Such a rate of phase transfer might allow

TABLE 1. Interfacial Polycarbonates: Investigation of the Effect of Catalyst Concentration and Phosgene Flow Rate

| Phosgene Flow Rate (cc/min.) | pH | Reaction Time | $(Et)_4\overset{+}{N}\overset{-}{Cl}$ [a] | 25°C [η] (dℓ/g) THF |
|---|---|---|---|---|
| 1. 40-58 | 11-12 | 1/2 hr. | 4% | 0.34 |
| 2. 40-58 | 11-12 | 1/2 hr. | 20% | 1.20 |
| 3. 40-58 | 11-12 | 1/2 hr. | 40% | 0.22 |
| 4. 120-140 | 11-12 | 1/2 hr. | 4% | 1.50 |
| 5. 120-140 | 11-12 | 1/2 hr. | 40% | 0.23 |

[a] Expressed in mole % of $(Et)_4\overset{+}{N}\overset{-}{Cl}$ per mole bisphenol.

formation of nearly complete chloroformate endgroup functionality. The result would be that the correct stoichiometry needed to produce high molecular weight polymers would never by achieved.

It was reasoned that if chloroformate endgroups could be built up through the use of a large concentration of catalyst, then the presence of a large excess of phosgene should also enhance the effect. Results of reactions run with a high phosgene gas flow rate ( see Table 1) support this hypothesis. Indeed, high molecular weights are achieved at lower catalyst concentration as phosgene flow rate is increased.

Copolymerizations were normally performed under basically the identical conditions which were used for the homopolycarbonates.

The polydimethylsiloxane oligomer was added to the reaction to yield a theoretical copolymer composition of approximately 50% polysiloxane and 50% polycarbonate by weight. The reactions also were continued for longer periods of time than the 30 minutes used for the homopolymers.

Block copolymers were first prepared utilizing reaction parameters which had been shown to produce high molecular weight homopolymers (cf. No. 2 and No. 4, Table 1). After one hour of reaction, a moderate portion of the theoretical amount of siloxane (90% and 85%, respectively - see Table 2) had been incorporated into both copolymers, but only relatively low molecular weights were obtained. These results suggested a difficulty in the incorporation of the acid-terminated siloxane oligomer.

If the mechanism shown in equation 6 was operative, then formation of the ester would be enhanced by first producing a high concentration of aryl chloroformate endgroups. By contrast, the routes shown in equation 7 would be promoted by a large excess of phosgene in the organic phase. It was conceived that copolymerization might be assisted if either or both of the abovementioned conditions could be achieved. Recall that the conditions of reaction 3, Table 1 were postulated to have produced a nearly exclusive aryl chloroformate endgroup. It appears reasonable that if the above is true, then parameters used for reaction 5, Table 1 (same as for reaction 3, Table 1 except with a much faster phosgene flow rate) should produce both of the desired situations.

Results obtained in the copolymerization under the conditions of reaction 3, Table 1 were encouraging (see Table 2, reaction 3). After three hours of reaction, a relatively high intrinsic viscosity of 0.79 dℓ/g was obtained. Moreover, the stoichiometric amount of siloxane was essentially incorporated. Copolymerization under the analogous conditions at a higher phosgene flow rate was much less successful. Unfortunately, the use of a large excess of phosgene results in excessive phosgene hydrolysis and hence renders the technique impractical. For example, neutralization of the HCl by-product with base produces very large quantities of salt.

A second attempt was made to define conditions where the mechanism shown in equation 7 might be favored. In this case, an anhydrous "pre-phosgenation" step was used. This consisted of bubbling an excess of phosgene through a dichloromethane solution

Table 2. Interfacial Polycarbonate/Polydimethylsiloxane Copolymers: Investigation of the Effect of Catalyst Concentration and of Phosgene Flow Rate on Intrinsic Viscosity and Polysiloxane Incorporation.

| Phosgene Flow[c] Rate (cc/min.) | [a] $(Et)_4\overset{+}{N}\overset{-}{Cl}$ Conc. | $[\eta]^{25°C}_{THF}$ 1 hr. | 2 hrs. | 3 hrs. | [b] % of Theoretical PSX Incorporated into the Copolymer 1 hr. | 2 hrs. | 3 hrs. |
|---|---|---|---|---|---|---|---|
| 1. 40-58 | 20% | 0.56 | 0.50 | 0.41 | 90% | 93% | 92% |
| 2. 120-140 | 4% | 0.44 | --- | --- | 85% | | |
| 3. 40-58 | 40% | 0.42 | 0.53 | 0.79 | 98% | 92% | 90% |
| 4. 120-140 | 40% | 0.48 | --- | --- | 87% | | |

(a) Figures are expressed in mole percent based on bisphenol.

(b) 100% would result in 50% of polysiloxane by weight in the copolymer.

(c) pH was held in the range of 11-12.

of the siloxane oligomer prior to the interfacial reaction. It was anticipated and then confirmed that one or more of the possible intermediates, $HO-\overset{O}{\overset{\|}{C}}-(PSX)_N-\overset{O}{\overset{\|}{C}}-O-\overset{O}{\overset{\|}{C}}-Cl$ and/or $HO-\overset{O}{\overset{\|}{C}}-(PSX)_N-\overset{O}{\overset{\|}{C}}-Cl$, would be formed via this approach. The carboxypropyl-terminated disiloxane or rather, an anionic form of it, is soluble in the aqueous medium used for the interfacial reaction. Experiments were conducted using the precursor siloxane dimer for the copolymerization. In the absence of the anhydrous pre-step, almost no disiloxane could be incorporated into the copolymers (see reactions 1 and 2, Table 3). By contrast, when the pre-step was used, the theoretical and actual copolymer compositions were extremely close. Differences are believed to reflect analytical errors and possibly some non-functional siloxane (cf. reactions 1 and 2, Table 3, with reaction 3, Table 3). It is hypothesized that when the dimer is used in the interfacial reaction (no pre-step), it does not react because it is easily extracted into the basic aqueous phase. If this is indeed the case, then the fact that it does react when the pre-step is utilized suggests than the "dimer" was not in an extractable form when the aqueous phase was added, and must have been in the form of one of the intermediates proposed earlier. Reactions performed with the siloxane oligomer using the "pre-phosgenation"step showed definite improvements in both molecular weights and copolymer compositions (ideal copolymer composition is 100% of the theoretical amount of siloxane incorporated into the copolymer) in all cases investigated (see Table 3).

To insure that the ester bond was indeed forming in the interfacial copolymerizations, both $^{13}C$ NMR and infrared spectra were run on the copolymers and compared with the analogous spectra of the following,

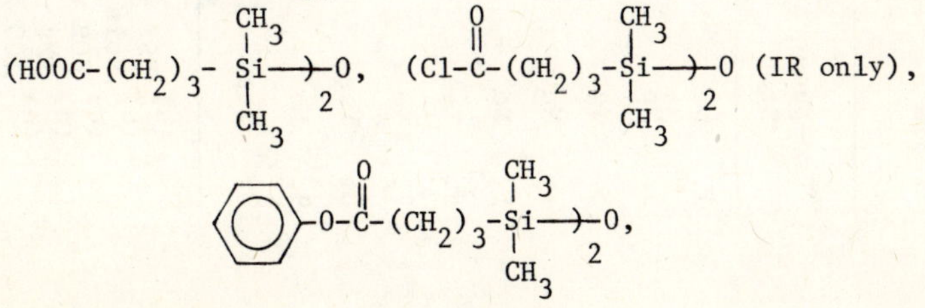

$$(HOOC-(CH_2)_3-\underset{CH_3}{\overset{CH_3}{\underset{|}{\overset{|}{Si}}}}\!\!-\!\!)_2O,\quad (Cl-\overset{O}{\overset{\|}{C}}-(CH_2)_3-\underset{CH_3}{\overset{CH_3}{\underset{|}{\overset{|}{Si}}}}\!\!-\!\!)_2O \text{ (IR only)},$$

$$(C_6H_5-O-\overset{O}{\overset{\|}{C}}-(CH_2)_3-\underset{CH_3}{\overset{CH_3}{\underset{|}{\overset{|}{Si}}}}\!\!-\!\!)_2O,$$

Table 3. Interfacial Polycarbonate/Polydimethylsiloxane Copolymers: Investigation of the Effect of Using a "Pre-phosgenation" Step.

| | "Pre-phosgenation Step" | (a) PSX | Phosgene Flow Rate (cc/min.) | pH | (b) $(Et)_4\overset{+}{N}\bar{Cl}$ Conc. | $[\eta]^{25°C}_{THF}$ | (e) % of Theoretical PSX Incorporated into the Polymer |
|---|---|---|---|---|---|---|---|
| 1. | No | Dimer | 120-140 | 9-10 | 40% | 0.68 | 15% |
| 2. | No | Dimer | 40-48 | 11-12 | 40% | ---- | 33% |
| 3. | Yes | Dimer | 40-58 | 11-12 | 40% | ---- | 95% |
| 4. | No | (d) Oligomer | 40-58 | 9-10 | 40% | 0.29 | 76% |
| 5. | No | Oligomer | 40-58 | 9-10 | 40% | 0.28 | 92% |
| 6. | Yes | Oligomer | 40-58 | 9-10 | 40% | 0.50 | 94% |
| 7. | No | Oligomer | 40-58 | 11-12 | 20% | 0.56 | 90% |
| 8. | Yes | Oligomer | 40-58 | 11-12 | 20% | 0.67 | 93% |
| 9. | (c) No | Oligomer | 40-58 | 11-12 | 20% | 0.28 | 85% |
| 10. | (c) Yes | Oligomer | 40-58 | 11-12 | 20% | 0.67 | 93% |

(a) PSX = polydimthylsiloxane
(b) Mole percent based on moles bisphenol-A
(c) KOH was used as the base, NaOH was used in all other cases
(d) <Mn> of all oligomers ≃ 2000g/mole
(e) 100 % would result in 50% of polysiloxane by weight in the copolymer

and of a polycarbonate homopolymer. The carbonyl region of the infrared spectra of the model compounds shows a progression of an absorption at $\sim$ 1713 $cm^{-1}$ for the carboxylic acid group to $\sim$ 1803 $cm^{-1}$ for the acid chloride (this compound was not isolated, its IR spectrum also contains acid starting material) to $\sim$ 1767 $cm^{-1}$ for the phenyl ester (see Fig. 3). Unfortunately, the polycarbonate carbonyl absorption occurs at $\sim$ 1771 $cm^{-1}$ (see Fig. 4), so close to the ester absorption that resolution of the two peaks in the copolymer (see Fig. 4) is difficult. The progression of the absorptions shown in the model compounds together with the fact that the boiling point of the diester very closely matched the previous literature value (175°C at 0.2 mm as compared to 190°C at 0.3 mm [17]) did however confirm the ester structure below,

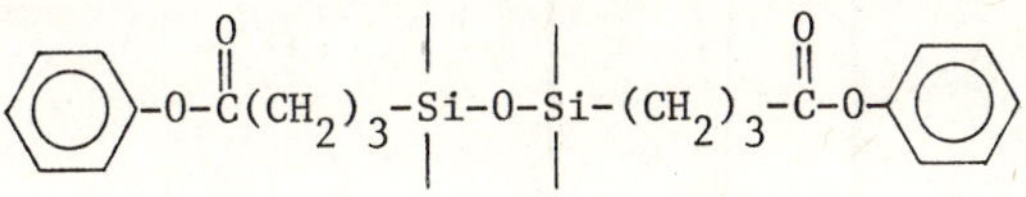

$$C_6H_5\text{-O-}\overset{\overset{O}{\|}}{C}(CH_2)_3\text{-}\underset{|}{\overset{|}{Si}}\text{-O-}\underset{|}{\overset{|}{Si}}\text{-}(CH_2)_3\text{-}\overset{\overset{O}{\|}}{C}\text{-O-}C_6H_5$$

and, hence, provided a model ester for $^{13}C$ NMR analysis.

The diphenyl ester, a copolymer structurally analogous to the interfacial copolymer discussed herein but produced in an anhydrous medium (dichloromethane/pyridine) and a copolymer produced in the interfacial reaction all exhibit $^{13}C$ NMR (see Fig. 5) ester carbonyl resonance at 171.8 ppm from TMS. The carbonyl resonances due to the polycarbonate and to the acid-terminated polysiloxane (see Fig. 5) are all well resolved from the ester region.

Upon expansion of the abscissa scale of the $\sim$ 140-180 ppm (from TMS) region of the copolymer, several small unidentified peaks in the $\sim$ 147-149 ppm (from TMS) region (see Fig. 6) became apparent. This is the region in which the aromatic carbon next to the carbonate group and the aromatic carbon adjacent to the isopropylidene group in the bisphenol-A polycarbonate homopolymer

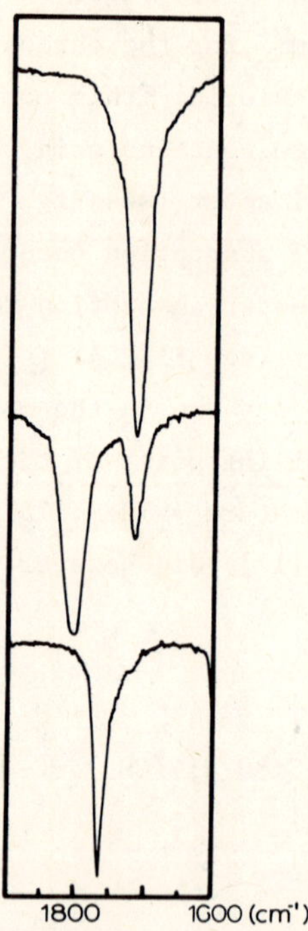


FIG. 3. Model compound infrared spectra showing the progression of the carbonyl absorbance band as the diphenyl ester of 1,3-carboxypropyltetramethyldisiloxane is produced

Assignments:

Top, $\left(HO-\overset{O}{\overset{\|}{C}}-(CH_2)_3-\underset{CH_3}{\overset{CH_3}{\overset{|}{\underset{|}{Si}}}}-\right)_2 O$, $\sim 1713\ cm^{-1}$

Middle, $\left(Cl-\overset{O}{\overset{\|}{C}}-(CH_2)_3-\underset{CH_3}{\overset{CH_3}{\overset{|}{\underset{|}{Si}}}}-\right)_2 O$, $\sim 1803\ cm^{-1}$

Bottom, $\left(C_6H_5-O-\overset{O}{\overset{\|}{C}}-(CH_2)_3-\underset{CH_3}{\overset{CH_3}{\overset{|}{\underset{|}{Si}}}}-\right)_2 O$, $\sim 1767\ cm^{-1}$

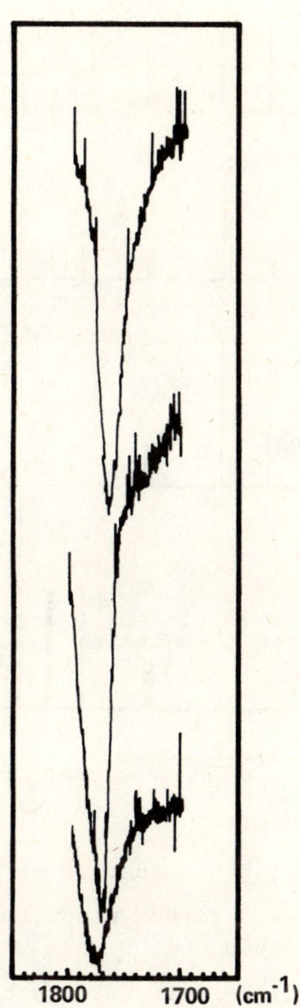


FIG. 4. Comparison of the carbonyl region of the infrared spectra of:

Top: The diphenyl ester of 1,3-carboxypropyltetramethyldisiloxane, ∿ 1767 cm$^{-1}$

Middle: Homo-polycarbonate, ∿ 1771 cm$^{-1}$

Bottom: Polycarbonate/polysiloxane copolymer containing both the carbonate and the ester bond

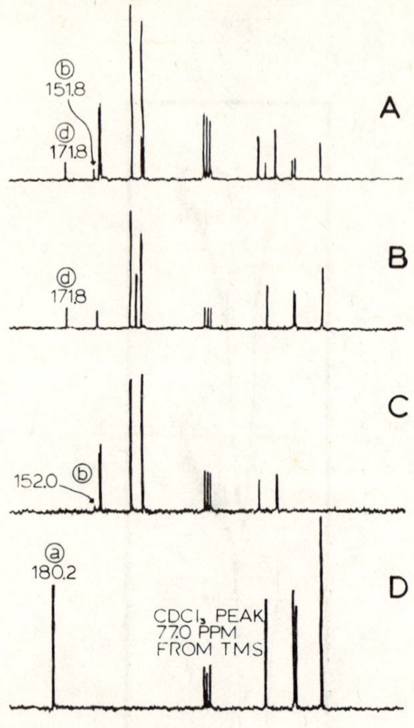


FIG. 5. Comparison of $^{13}$C NMR spectra: (top to bottom) Carbonyl carbon assignments:

A: Copolymer spectrum, ⓑ = carbonate carbon, ⓓ = ester carbon

B: Diphenyl ester of 1,3-carboxypropyltetramethyldisiloxane, ⓓ = ester carbon

C: Homopolycarbonate, ⓑ = carbonate carbon

D: 1,3-carboxypropyltetramethyldisiloxane, ⓐ = carboxylic acid carbon

resonate. It seems probable that the unidentified peaks are due to random copolymer sequence distributions but the proximity of this area in the spectrum to the carbonate carbonyl region means that the possibility of residual mixed carbonic-carboxylic anhydrides (not visible in the infrared spectra) cannot yet be ruled out. Carboxylic anhydrides normally appear ∿ 10 ppm upfield

500 Hz.

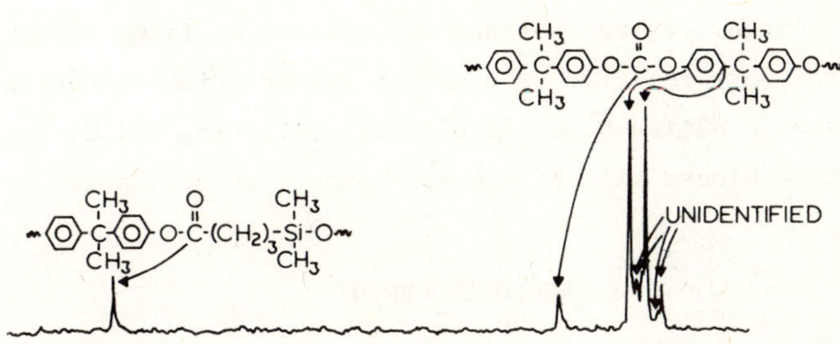


FIG. 6. Expansion of the carbonyl region of the $^{13}C$ NMR spectrum of polycarbonate/polysiloxane copolymer containing both carbonate and ester links

from the resonance of their parent acid [37]. The fact that the unidentified peaks all appear ∿ 30 ppm from the acid peak suggests that, although carboxylic anhydrides may form via several pathways (see equations 6, route 2, 7a, route 2, and 7b, route 2), they apparently cannot survive the conditions of the reaction.

In conclusion, if copolymerization is indeed benefited by an excess of aryl chloroformate groups as our results imply, then this suggests that the reaction pathway proceeding through the mixed carbonic-carboxylic anhydride (4) to the ester (equation 6, route 3) is operative to some extent. Likewise, the fact that higher molecular weight copolymers (and with improved incorporation of the siloxane component) are formed when the "pre-phosgenation" step is added, implies that one or more of the pathways to the ester shown in equation 7 (equation 7a, route 3, and/or 7b, route 3) are also viable. The potential extent to which the interfacial copolymerization can be guided through a particular mechanism is not yet clearly understood. Copolymers with reasonably high molecular weights ($[\eta] = 0.6\text{-}0.7$), good incorporation of the siloxane, and with good film properties can however, be produced via utilization of the "pre-phosgenation" step and the reaction conditions suggested in the experimental sections herein. Infrared and $^{13}C$ NMR spectra confirm the copolymer structure with the presence of the desired ester link between the block types. The hydrolytic stability and surface composition of films of these block copolymers relative to those of the solution-synthesized perfectly alternating block copolymers containing the Si-O-C link between blocks will be the subject of future papers.

ACKNOWLEDGEMENT

The authors would like to thank the National Science Foundation, Polymers Program, for partial support of this research.

REFERENCES

[1] A. Noshay and J. E. McGrath, Block Copolymers: Overview and Critical Survey, Academic Press, New York, (1977).

[2] J. E. McGrath, D. W. Dwight, J. S. Riffle, T. F. Davidson, D. C. Webster, and R. Viswanathan, Polym. Prepr., 20(2), 528 (1979).

[3] J. E. McGrath, T. C. Ward, D. W. Dwight, J. S. Riffle, T. F. Davidson, and A. J. Wnuk, Preprints, Makro Mainz, 1, 405 (1979).

[4] J. E. McGrath, J. S. Riffle, D. W. Dwight, and D. C. Webster, Macromolecules, in press, (1981).

[5] R. P. Kambour, in Block Copolymers, (S. L. Aggarwal, Ed., Plenum Press, New York-London, (1970).

[6] P. C. Juliano, G. E. Technical Information Series, Report No. 74CRD172, August, (1974).

[7] M. Matzner, A. Noshay, and J. E. McGrath, Polym. Prepr., 14(1), 68 (1973).

[8] M. Matzner, A. Noshay, and J. E. McGrath, Trans. Soc. Rheol., 21(2), 273 (1977).

[9] S. H. Tang, and E. A. Meinecke, Paper #39 Presented at the Rubber Division Meeting of the A.C.S., Las Vegas, Nevada, May 20-23, 1980.

[10] E. A. Meinecke, S. H. Tang, J. S. Riffle, and J. E. McGrath, J. Appl. Polym. Sci., in press, (1980).

[11] T. C. Ward, D. P. Sheehy, and J. E. McGrath, Polym. Prepr., 21(2), 70 (1980).

[12] T. C. Ward, D. P. Sheehy, J. S. Riffle, and J. E. McGrath, Symposium in Honor of Michel Shen, (A. Eisenberg, Ed.,), in press, (1980).

[13] D. T. Clark, and J. Peeling, J. Polym. Sci., Polym. Chem., Ed., 14 543 (1976).

[14] D. T. Clark, A. Dilks, J. Peeling, and H. R. Thomas, Faraday Disc., Chem. Soc., 60, 183 (1976).

[15] M. J. Owen, and T. C. Kendrick, Macromolecules, 3, 458 (1970).

[16] D. W. Dwight, B. Beck, J. S. Riffle, and J. E. McGrath, Polym. Prepr., 20(1), 702 (1979).

[17] D. G. LeGrand, and G. L. Gaines, Jr., Polym. Prepr., 11, 442 (1970).

[18] W. Noll, Chemistry and Technology of Silicones, Academic Press, New York,(1968).

[19] A. Noshay, and M. Matzner, Makromol. Chem., 37, 215 (1974).

[20] H. Schnell, Angew. Chem., 68, 633 (1954).

[21] H. Schnell, L. Bottenbruch, and H. Krimm, (To Farbenfabriken Bayer, A. G.) Ger. Pat. 971, 790 (1959), Belgian Pat. 532, 543.

[22] H. Schnell, Chemistry and Physics of Polycarbonates, Wiley-Interscience, (1964).

[23] P. W. Morgan, S.P.E. (Society of Plastics Engineers) J., 15, 548 (1959).

[24] W. P. Weber, and G. W. Gokel, Phase Transfer Catalysis in Organic Synthesis, Springer-Verlag, New York, (1977).

[25] O. V. Smirnova, E. Khorvat, and I. P. Losev, Vysokomol. Soedin., Ser. A, 12, 424 (1970).

[26] H. Vernaleken, in Interfacial Synthesis, Vol. II, Polymer Application and Technology, (F. Millich and C. E. Carraher, Jr., ed., Marcel-Dekker, Inc., New York and Base, (1977).

[27] D. S. Tarbell and N. A. Leister, J. Org. Chem., 23, 1149 (1958).

[28] D. S. Tarbell, and E. J. Longosz, J. Org. Chem., 23, 2044 (1958).

[29] T. B. Windholz, J. Org. Chem., 23, 2044 (1958).

[30] T. B. Windholz, J. Org. Chem., 25, 1703 (1960).

[31] E. P. Goldberg, S. F. Strause, and H. E. Munro, Polym. Prepr., 5, 233 (1964).

[32] D. C. Prevorsek, B. T. Debona, and Y. Kesten, J. Polym. Sci., Polym. Chem. Ed., 18, 75 (1980).

[33] J. E. McGrath, J. S. Riffle, A. K. Banthia, and W. Steckle, to be published, J. S. Riffle, Ph.D Thesis, VPI & SU, Blacksburg, Va., June, 1981.

[34] J. E. Mulvaney, and C. S. Marvel, J. Polym. Sci., 50, 541 (1961).

[35] P. W. Morgan, Macromolecules, 3,(5), 536 (1970).

[36] P. W. Morgan, Condensation Polymers, Wiley-Interscience, New York, (1965).

[37] G. C. Levy, and G. L. Nelson, Carbon-13 Nuclear Magnetic Resonance for Organic Chemists, Wiley-Interscience, New York, (1972).

# Polycondensation of Long-Chain Esters of $\alpha$-Amino Acids in Monolayers at Air/Water Interface and in Multilayers on Solid Surface

KIYOSHIGE FUKUDA, YOSHIO SHIBASAKI, and HIROO NAKAHARA

Department of Chemistry
Saitama University
Urawa, 338, Japan

## ABSTRACT

Polycondensations of long-chain esters of glycine and alanine in the monolayers spread at air/water interface and in the multilayers deposited on solid surface have been investigated by surface viscosity and IR spectra. These compounds form stable monolayers on neutral and alkaline subphases. The polycondensation can occur rather quickly in the monolayers at room temperature without any catalyst or initiator. The polymerization rate is markedly influenced by the area per molecule and the maximum polymerizability is obtained at the area somewhat larger than the condensed region. Furthermore, polymerizability in the multilayer prepared by Blodgett's technique is superior to those in the bulk powder at the same temperature and in the molten state at higher temperature. These results lead to the conclusion that the polycondensation is accelerated by the regular arrangement of the monomer molecules at the interface, where the functional groups are concentrated and situated most effectively for the reaction. Two probable mechanisms for the polycondensation of the long-chain esters of $\alpha$-amino acids in the Y-type multilayer have been proposed.

## INTRODUCTION

Polymerization reactions at the interface are of particular interest with respect to the effects of molecular orientation on the reaction rates and on the configuration or conformation of the resultant macromolecules [1-6].

This paper deals with polycondensation of long-chain esters of α-amino acids in the monolayers spread on water surface and in the multilayers deposited on solid surface. From the quantitative analysis of the reaction by infrared spectra, it has been found that the polycondensation is markedly accelerated by regular arrangement of the monomer molecules at the interface.

## EXPERIMENTAL

### Materials

Materials used in this study are given in Table 1.

These compounds were prepared by alcoholysis of the corresponding methyl esters in hydrochlorinated form with hexadecanol or octadecanol, followed by dehydrochlorination with ammonia and purified by repeated recrystallizations from ethanol solutions. The melting points of the racemic compounds are somewhat higher than those of the optical isomers.

### Procedures

Monolayers of these compounds were spread from benzene solutions on aqueous subphases. Surface pressure was measured by a film balance of hanging plate type; its sensitivity was within

Table 1. Long-Chain Esters of α-Amino Acids

| Compounds | | Abbrev. | mp(°C) |
|---|---|---|---|
| Glycine hexadecyl ester | $NH_2CH_2COOC_{16}H_{33}$ | Gly-$C_{16}$ | 51.3 - 51.8 |
| Alanine hexadecyl ester | $NH_2CH(CH_3)COOC_{16}H_{33}$ | d-Ala-$C_{16}$ | 32.0 - 32.5 |
| | | dl-Ala-$C_{16}$ | 36.0 - 36.5 |
| Alanine octadecyl ester | $NH_2CH(CH_3)COOC_{18}H_{37}$ | l-Ala-$C_{18}$ | 37.3 - 38.4 |
| | | dl-Ala-$C_{18}$ | 39.1 - 39.4 |

0.025 dyn $cm^{-1}$. Polymerization process in the monolayers was followed by the increase of surface viscosity and also by the analysis of infrared spectra of the scraped films. Surface viscosity was measured by the damping of rotatory oscillation of a platinum ring suspended on the monolayer covered water surface.

Multilayers were prepared by Langmuir-Blodgett's technique on calcium fluoride plate which had been precoated with ferric stearate monolayer. Under appropriate conditions, it was possible to deposit more than 300 layers of glycine hexadecyl ester in the alternating Y-type film within 15 minutes, during this period no appreciable polymerization has yet proceeded. Deposition conditions were as follows: the substrate water contains $10^{-3}$M calcium chloride, the pH is adjusted in the range of 9 to 10 with ammonium hydroxide, the piston pressure is 30 dyn $cm^{-1}$ and temperature is below 15°C. With alanine esters, succesive deposition of the multilayers is somewhat difficult. It was necessary to dry up water included between adjacent two layers before next dipping, unless a higher salt concentration, a higher pH value of about 11 and a lower temperature below 10°C were used. Polycondensation in the multilayers was followed by infrared spectra.

## RESULTS AND DISCUSSION

### Monolayer Properties

Surface pressure-area isotherms of the long-chain esters of α-amino acids depend markedly on pH values of the aqueous subphase, as shown in Fig.1. These compounds can form stable monolayers on neutral and alkaline subphases, but on acidic subphase the films tend to expand and to become unstable because of increasing solubility due to ionization of the amino group. The monolayer of glycine ester is a condensed type, whereas those of alanine derivatives are somewhat expanded and the transition from the expanded to the condensed films occurs at the molecular area of about 30 $Å^2$ or more. Some differences are observed between the racemic com-

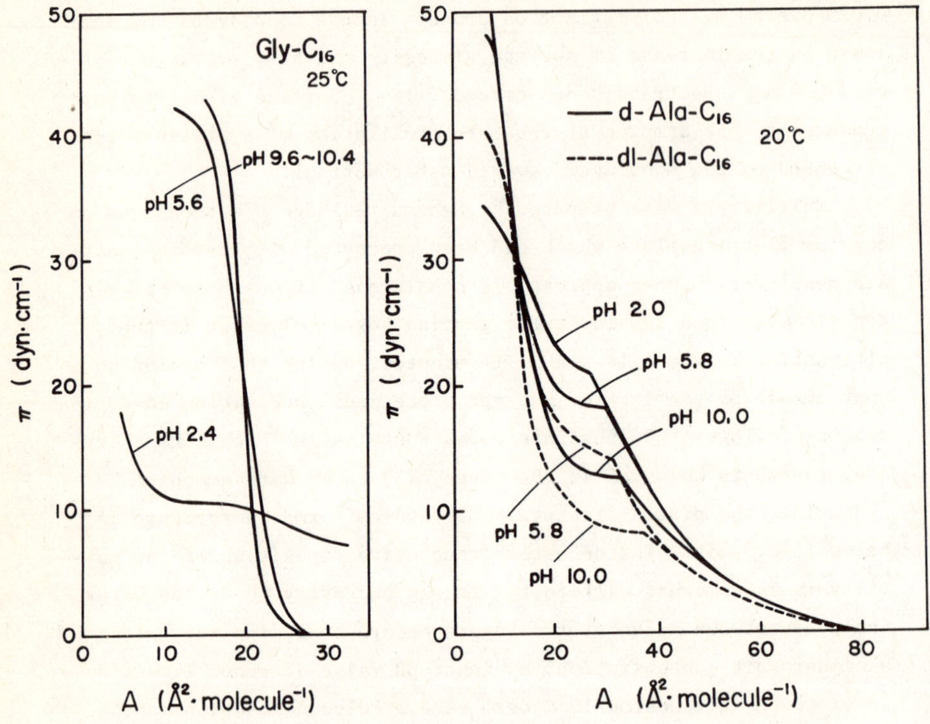


Fig.1. Surface pressure-area isotherms for hexadecyl esters of glycine (A) and alanine (B).

pound and the optical isomer; the racemic compound seems to be packed more closely.

Polycondensation in Multilayers

In Fig. 2, surface viscosity $\eta_s$ is plotted as a function of time for hexadecyl esters of glycine and alanine spread on distilled water at various area per molecule. A remarkable increase of surface viscosity indicates that the polycondensation can occur rather quickly in the monolayers at the air/water interface without any catalyst or initiator. When the area per molecule was changed, the most significant increase of surface viscosity was

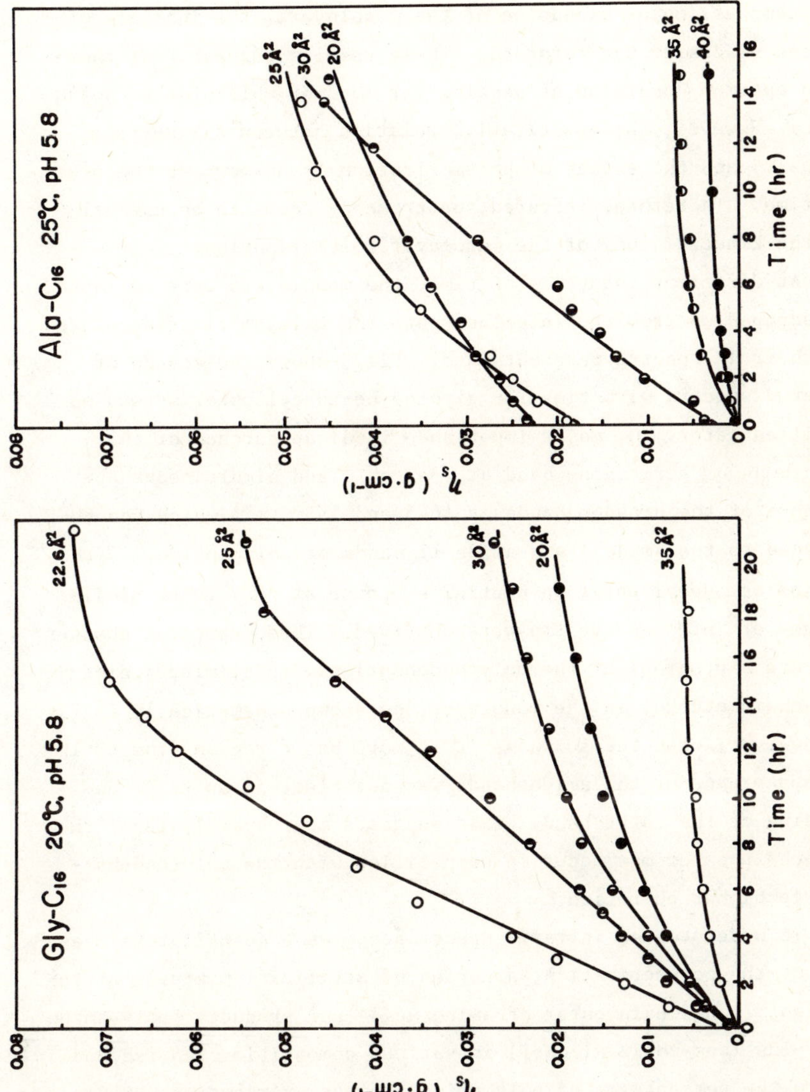


Fig.2. Change of surface viscosity ($\eta_s$) with time for hexadecyl esters of glycine (A) and alanine (B) spread on distilled water at various areas per molecule.

observed at the area of 22.6 $Å^2$ for glycine ester (A). On the other hand, with alanine ester (B) the rate of viscosity change was highest in the region of 25 to 30 $Å^2$ per molecule. On further compression or expansion of the monolayers, the increase of surface viscosity was retarded. These results suggest that there is an optimum condition of packing for monomer molecules to polymerize. However, the quantitative relation between the surface viscosity and the extent of polymerization is unknown at the present time. Therefore, infrared spectroscopy seems to be usefull for the kinetic study of the monolayer polymerization.

At different intervals of time, the monolayers were collapsed and scraped up from the interface onto the calcium fluoride plates and their IR spectra were obtained. Fig.3 shows the change of infrared spectra with time for glycine hexadecyl ester spread on distilled water. Of major importance is disappearance of the ester carbonyl stretching band at 1743 $cm^{-1}$ and simultaneous appearance of the broader bands at 1645 and 1550 $cm^{-1}$ which can be assigned to the amide I and amide II bands of polypeptide. With alanine octadecyl ester on neutral subphase at pH 7 to 8, similar changes of infrared spectra were observed. These spectral changes indicate a progress of the polycondensation with elimination of the long-chain alcohol in the monolayer, as shown schematically in Fig. 4. However, when the subphase is in more basic region from pH 10 up, appearance of the amide bands was not clear in spite of decreasing of the ester band. This suggests that hydrolysis of the long-chain ester may occur in competition with the polycondensation in higher pH region.

In order to use infrared spectroscopy as a quantitative measure of the polymerization, a series of artificial mixtures of the reactant (long-chain ester of amino acid) and products (poly-amino acis plus long-chain alcohol) in various compositions corresponding to the percentages of polymerization was examined by KBr disc method. The results showed that the integrated intensity of the ester band decreases linearly with increasing extent of polymeri-

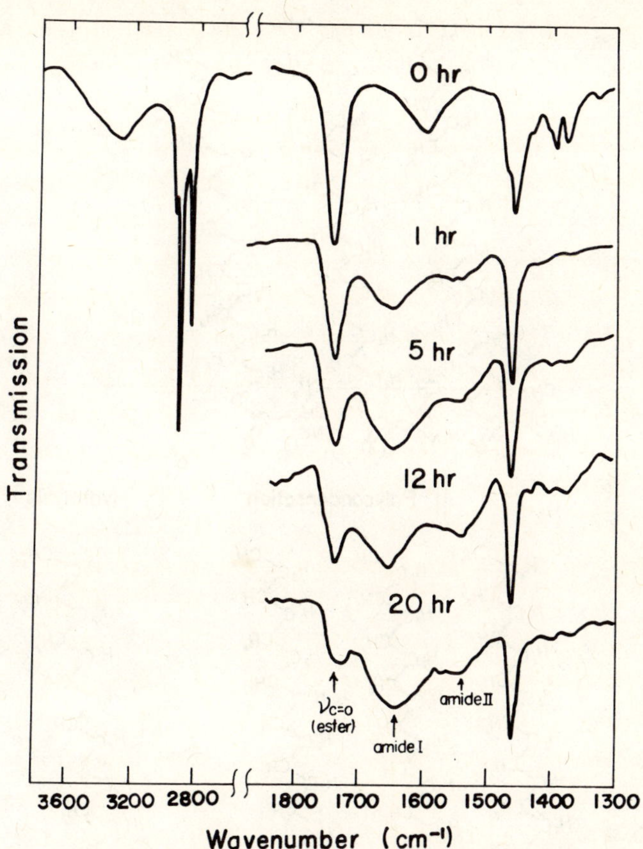


Fig.3. IR spectra for monolayer of glycine hexadecyl ester spread on distilled water and scraped up at different time intervals.

zation while that of the amide I band increases. With amide I band, however, the measurement of band intensity is less accurate because of an overlap with $NH_2$ deformation band of the monomer. On the basis of these data, the percentage of polymerization can be estimated by the quantity $100(A_o - A_t)/A_o$, where $A_o$ and $A_t$ are the integrated intensities of the ester band at start and time t,

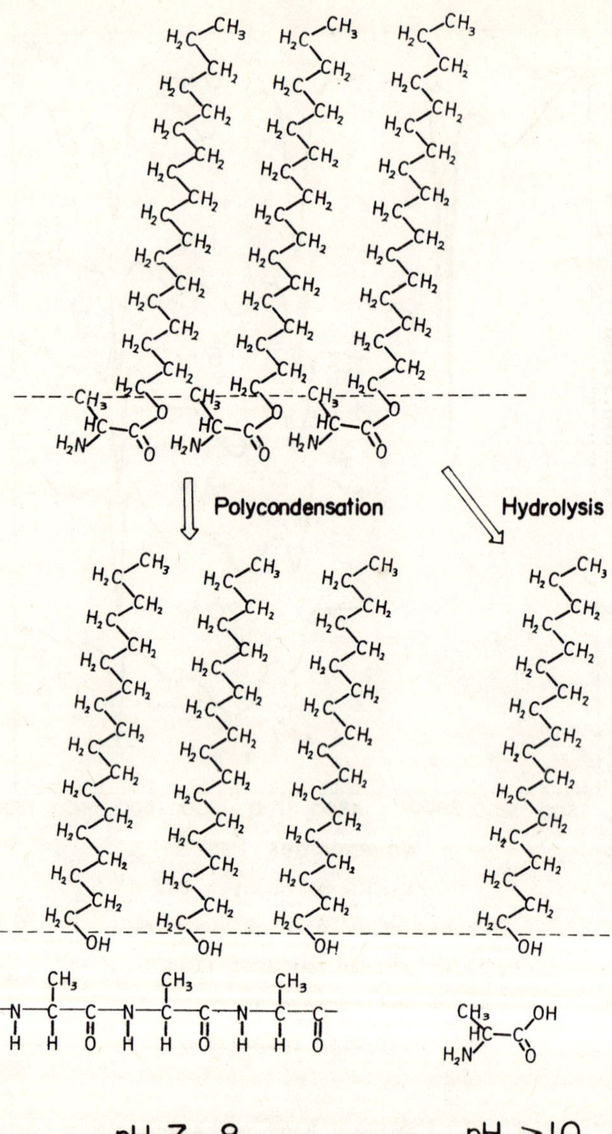


Fig.4. Schematic illustration for two probable reactions in the monolayer of long-chain ester of α-amino acid.

respectively. Thus, polymerization curves under various conditions can be compared.

In Fig.5(A), the percent conversion are plotted against the reaction time for glycine hexadecyl ester spread on distilled water at various areas per molecules. It is found that the rate of polymerization in the monolayer is markedly influenced by the molecular packing. In Fig.5(B), the conversions at 2 and 20 hours after spreading are plotted as functions of the area per molecule. The maximum polymerizability was obtained at the area of about 22 Å$^2$ for glycine hexadecyl ester. On the other hand, in the case of dl-alanine octadecyl ester the maximum polymerizability was observed in the region of 25 to 30 Å$^2$ per molecule. These results are in well agreement with those obtained from surface viscosity measurements. Thus, it has been confirmed that there is an optimum condition for the polymerization with respect to the packing and orientation of the monomer molecules at the interface, which corresponds to just before the condensed region, in other words, to the nearly close packing but with some freedom. Particular orientation and suitably close packing together with some rotational freedom of the monomer molecules in the monolayer seem to be very favorable for the polymerization reactions.

## Polycondensation in Multilayers

Fig.6 shows the change of IR spectra for the multilayers of glycine hexadecyl ester kept at 40°C for various time periods. In the top trace obtained immediately after the deposition, the ester band at 1743 cm$^{-1}$ is very strong, indicating no appreciable polymerization at this stage. As time advances the amide bands appear and increase at the expense of the ester band. These spectral changes represent a progress of the polycondensation in the multilayers. The bottom trace (full line) shows the spectrum obtained after dipping of the polymerized film into n-hexane for two minutes, in which the characteristic bands due to hydrocarbon chain are markedly diminished; this indicates the dissolution of unreacted

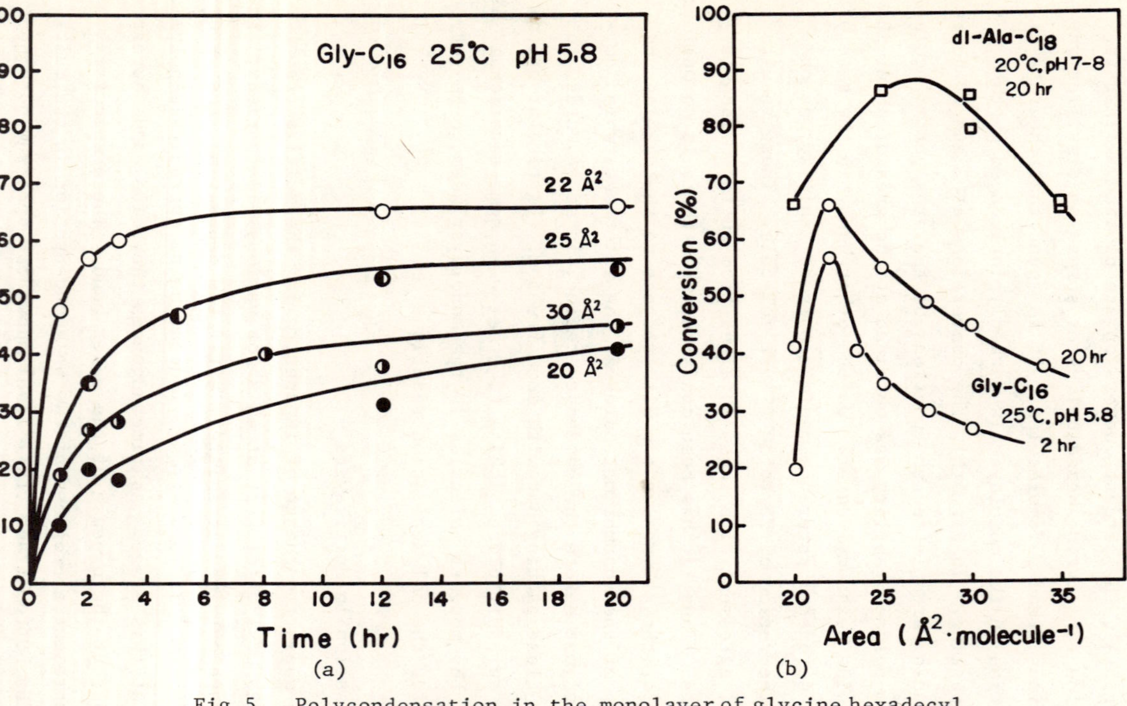


Fig.5. Polycondensation in the monolayer of glycine hexadecyl ester spread on distilled water at various areas per molecule (A). Effect of molecular area on the polycondensation in the monolayers of long-chain esters of glycine and alanine (B).

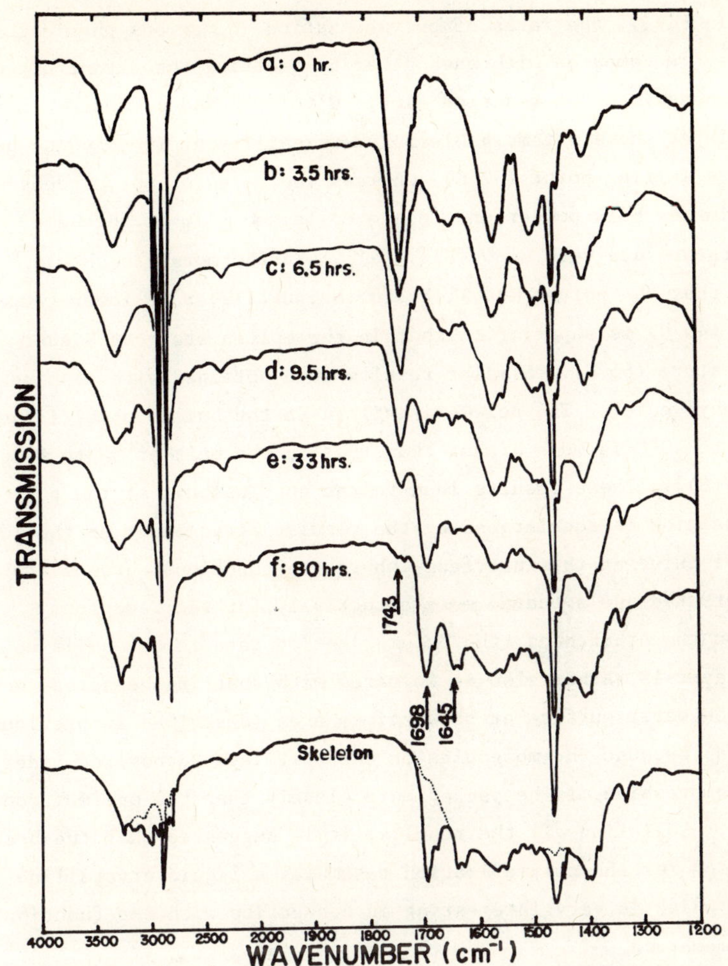


Fig.6. Change of IR spectra with time for the multilayer of glycine hexadecyl ester kept at 40°C.

monomer and resultant hexadecanol. Furthermore, in the spectrum obtained after additional dipping into ethanol for two minutes (dotted line) the band at 1698 $cm^{-1}$, as well as $-CH_2-$ bands, almost disappears, whereas the amide I and II bands remain invariably. From these facts, the band at 1698 $cm^{-1}$ may be assigned to the oligomers of glycine.

In Fig.7, the rates of polymerization in various physical states are compared with each other by plotting the extent of conversion against the reaction time. Glycine hexadecyl ester in the multilayer shows a remarkable polymerizability at 30 to 40°C, below its melting point (52°C), whereas the rates of polycondensation in the bulk powder and in the collapsed film at the same temperature are very slow [Fig.7(a)]. Furthermore, it should be noted that the polymerizability in the multilayer at lower temperature (40°C) is superior to that in the molten state at higher temperature (65°C). Similar results were obtained with alanine octadecyl ester. The polymerizability in the built-up multilayer at 30 to 40°C is much higher than that in the melt at 50 to 80°C [Fig.7(B)]. These results lead to the conclusion that the polycondensation is accelerated by the regular arrangement of the monomer molecules at the interface, where the functional groups are concentrated and situated most effectively for the reaction.

On the other hand, the polymerization rate in the built-up multilayer is rather slow as compared with that in the spread monolayer on water surface at the optimum area (described in previous section) because the molecules in the multilayer deposited under high compression may be packed more closely than the optimum condition. Collating all the results, it is suggested that the best condition for the polymerization resembles a liquid crystalline state, which is very interesting in connection with the function of biomembrane.

Next, the reaction kinetics has been examined. As shown in Fig.8, when logarithms of the ratio of the initial and the remaining concentrations of the monomer [$\log(A_0/A_t)$] are plotted against the time, approximately straight lines are obtained up to about 70 % conversion for both the polycondensations of alanine octadecyl ester in the monolayer (A) and of glycine hexadecyl ester in the multilayer (B). Therefore, the polycondensation in the interfacial films is considered to proceed by the first-order reaction mechanism at least in the initial stage. The rate constant k at

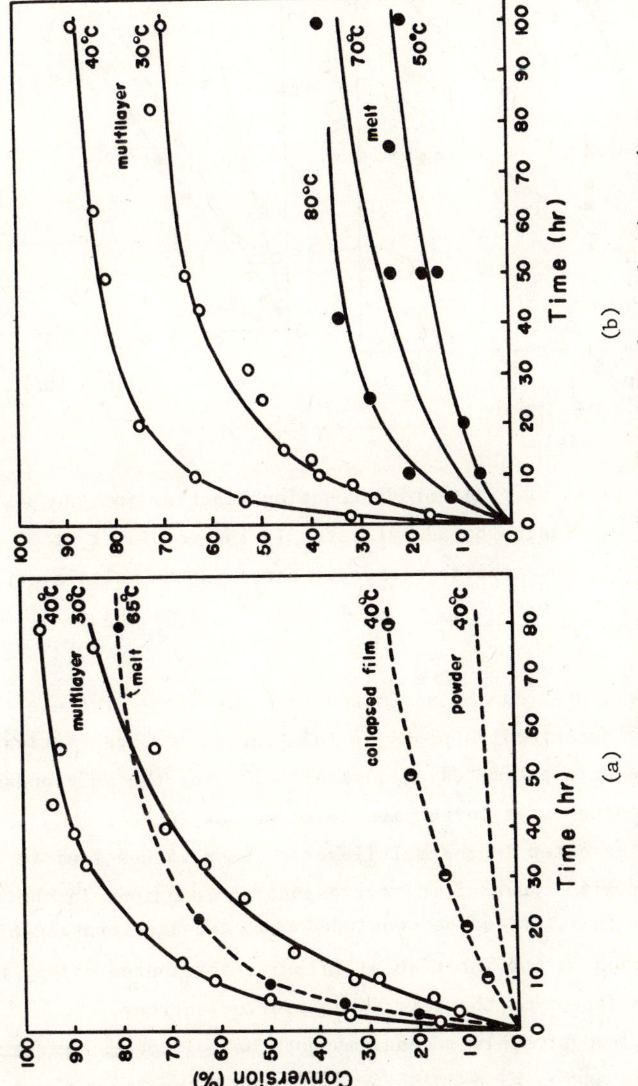


Fig.7. Comparison of the rates of polycondensation in various physical states for glycine hexadecyl ester (A) and alanine octadecyl ester (B).

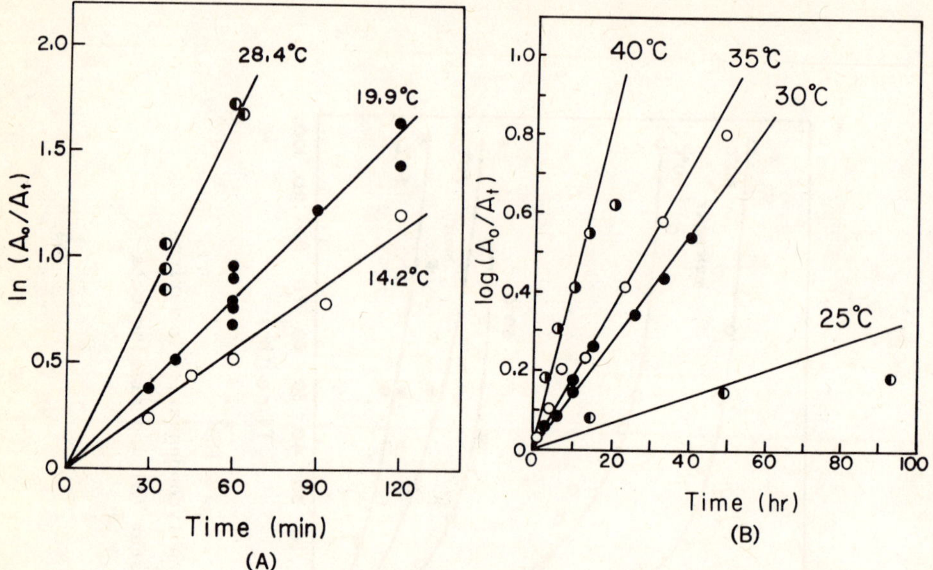


Fig.8. Plots of first-order reaction kinetics for the polycondensation of alanine octadecyl ester in the monolayer at 30 Å$^2$ per molecule (A) and glycine hexadecyl ester in the multilayer (B).

various temperatures can be estimated from the slope of these lines. From the Arrhenius plots of ln k versus 1/T, the activation energies were estimated at 14 kcal mol$^{-1}$ for the polycondensation of alanine ester in the monolayer and at 20 kcal mol$^{-1}$ for that of glycine ester in the multilayer. These values seem to be of same order with those of the corresponding reactions in the bulk system. Thus, it can be concluded that the acceleration of polycondensation in the interfacial films is attributed mainly to the frequency factor or the term of activation entropy.

Finally, two probable mechanisms for the polycondensation of the long-chain esters of α-amino acids in the Y-type multilayer have been proposed, as shown schematically in Fig.9. If the reaction propagates in a manner of sewing up alternately the functional

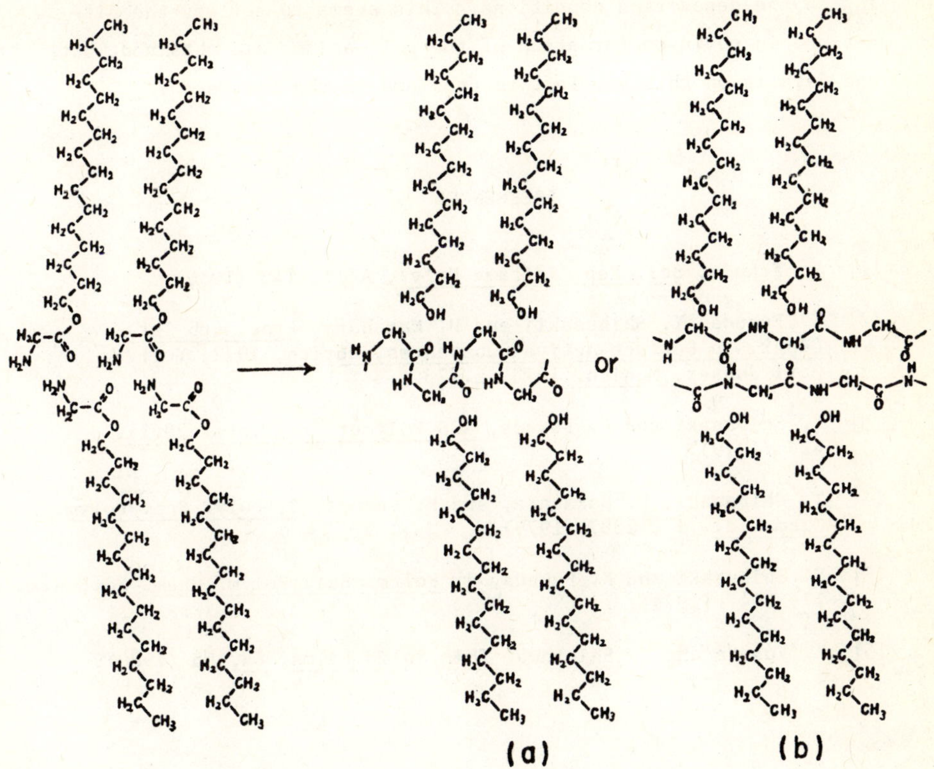


Fig.9. Schematic illustration of two probable mechanisms for the polycondensation in Y-type multilayer of long-chain ester of α-amino acid.

groups aligned head to head in the two adjacent layers, polypeptide chains of some helical structure are expected to be obtained (a). On the other hand, when the polycondensation can proceed two-dimensionally in each single layer of the built-up multilayer, as well as in the monolayer on water surface, poly-amino acids of extended zig-zag chain (β-form) may be formed (b). In fact, the amide I band in IR spectra for the multilayers after polymerization was observed at two different wave numbers, 1635 and 1645 $cm^{-1}$, depending on the deposition conditions. This seems to reflect the difference in chain conformation of the poly-amino acid obtained, but the details on this are left to problems in the future.

## REFERENCES


[1] K. Fukuda, Sci. Rep. Saitama Univ., AIII, 143 (1959).

[2] K. Fukuda, Y. Shibasaki, and H. Nakahara, Proc. 6th Int. Congr. on Surface-Active Substances, Zürich, 1972, Vol.2, Carl Hanser, Munich, 1973, p.273.

[3] Y. Shibasaki and K. Fukuda, ACS Polymer Preprints, 20(1), 427 (1979).

[4] Y. Shibasaki, H. Nakahara, and K. Fukuda, J. Poly. Sci. Polym. Chem. Ed., 17, 2387 (1979).

[5] Y. Shibasaki and K. Fukuda, J. Polym. Sci. Polym. Chem. Ed., 17, 2947 (1979).

[6] K. Fukuda and T. Shiozawa, Thin Solid Films, 68, 55 (1980).

# Polymerization of Substituted Butadienes at the Gas-Water Interface

H. Ringsdorf, H. Schupp, Institute of Organic Chemistry, University of Mainz, 6500 Mainz, West-Germany

ABSTRACT

In order to investigate the polymerization behaviour of long-chain amphiphilic compounds in dependence of their orientation, 1,4-disubstituted butadienes of the type:

$$CH_3\text{-}(CH_2)_{12}\text{-}CH{=}CH\text{-}CH{=}CH\text{-}X \qquad X = \text{-}COOH,\ \text{-}CHO,\ \text{-}CH_2OH$$

have been synthesized and investigated in a monolayer at the gas-water interface. The orientation of the compounds has been studied by means of a Langmuir film balance. The alcohol exhibits an orientation similar to that of saturated compounds. The orientation of the aldehyde and the acid differ from the expected behaviour because of the enhanced hydrophilicity of the headgroup due to the conjugation of the carbonyl group to the diene unit. Attempts to polymerize the substances radically have been made in melt, in solution, in the crystalline state and in monolayers. A polymer was only obtained in the case of monolayer polymerization. This fact can be explained with the necessity of a defined orientation of the monomers before the polymerization. The polyreaction was followed by the contraction of the film. The kinetics of the polymerization show a different reactivity for the alcohol and the carbonyl compounds. The resulting polymer is assumed to have a poly-1,4-trans backbone.

INTRODUCTION

Up to now there have been numerous studies to influence a polyreaction by orienting amphiphilic molecules in a monolayer (1-4), which offers the unique possibility to orient and vary the packing of molecules without a suppression of the mobility (5). Whereas the former attempts were concerning with systems, the polymerizable unit of which is located in the hydrophilic headgroup (e.g. Octadecylmethacrylate (2,3), Vinylstearate (1) etc.), substances with a polymerizable group within the aliphatic chain were investigated more recently. The most intensive studies have been made with the Diacetylene carbonic acids (4). The solid-state polymerization of Diacetylenes has long been studied (5), and the orientation of the molecules within the crystal has then been substituted by a similar orientation in the monolayer or multilayer, respectively.

Therefore the question arose to prove the model considerations for polymerization behaviour by a system which is unable to polymerize neither in the crystalline state nor in solution. As model compounds, 1,4-disubstituted butadienes of the type:

$$CH_3\text{-}(CH_2)_{12}\text{-}CH\text{=}CH\text{-}CH\text{=}CH\text{-}X \qquad X = \text{-}COOH,\ \text{-}CHO,\ \text{-}CH_2OH$$

<u>1a-c</u>

have been synthesized and investigated in a monolayer.

## EXPERIMENTAL

<u>Octadecadienal (1b)</u> was prepared according to the method of E.L. Pippen and M. Nonaka (9):
1-Methoxybut-1-en-3-yne, 41.9 g (0.51 mole), dissolved in 280 ml of tetrahydrofuran was added dropwise with

stirring to a Grignard reagent prepared from 11.7 g (0.48 mole) of magnesium and 52.48 g (0.48 mole) of ethyl bromide in 300 ml of tetrahydrofuran while the mixture was maintained at about 40°C. After an additional hour of stirring at room temperature, the reaction flask was cooled and there was added over a period of 30 min a solution of 84.95 g (0.4 mole) of tetradecanal dissolved in 150 ml of tetrahydrofuran. After 2 hours of stirring at room temperature, the mixture was heated to reflux. 30 min later, the flask was cooled and the mixture was treated with 18.4 ml of ethanol. Twenty minutes later solid sodium aluminium hydride, 20 g (0.5 mole), was added in small portions over a period of 30 min. The mixture was stirred for 2 hours and permitted to stand overnight. It was then treated successively with ethyl acetate (20 ml), water (90 ml), and 4 N sulfuric acid (450 ml). The organic layer was separated and the aqueous layer was extracted with ether. The combined organic layers were washed with 10% sodium carbonate solution, 3 times with water and dried over sodium sulfate. The solvent was distilled under reduced pressure and the resulting solid was recrystallized several times from hexane. Yield: 48.5 g (46%) of a white solid with melting point of 33 °C.

| Elemental analysis: | calc: | C: 81.75% | H: 12.20% |
|---|---|---|---|
| | found: | C: 81.54% | H: 12.34% |

Octadecadienoic acid (1a): A solution of 8.16 g (0.05 mole) of silver nitrate in 100 ml of water was treated with a solution of 2.2 g (0.08 mole) sodium hydroxide in 20 ml of water. The precipitated silver oxide was washed with water and suspended in a solution of 10.4 g (0.26 mole) of sodium hydroxide in 150 ml of water. To this suspension 13.2 g (0.05 mole) of Octadecadienal in

40 ml ethanol was added over a period of 20 min at a temperature of 50-60 °C. The mixture was stirred for 3 hours and permitted to stand overnight. Then the mixture was acidified with hydrochloric acid and extracted several times with ether. From the ether extract the salt of the acid was removed by extraction with concentrated sodium carbonate solution; the water extract was reacidified and extracted with ether. The ether layer was dired over sodium sulfate and the solvent was allowed to evaporate. The resulting crystals were recrystallized from hexane. Yield: 4 g (29%), MP: 71 °C.

Elemental analysis: calc: C: 77.09% H: 11.50%
found: C: 77.24% H: 11.54%

Octadecadienol (1c): A solution of 10 g (0.04 mole) of Octadecadienal in 50 ml of ether was added dropwise to a suspension of 0.86 g (0.02 mole) of lithium aluminium hydride in 50 ml of ether. After 3 hours of stirring, the mixture was cooled and the remained hydride was destroyed by addition of ethyl acetate. The mixture was washed with water, the ether layer was separated and dried over sodium sulfate. After evaporation of the solvent, a white solid was obtained. Recrystallization from hexane yielded the pure product in 67% yield (7.2 g). MP: 58 °C.

Elemental analysis: calc: C: 81.13% H: 12.86%
found: C: 81.14% H: 13.04%

Monolayers were spread from chloroform solutions all having a concentration of approximately 1 mg/ml. The films were spread on a Lauda film balance, surface pressure and area were automatically recorded. Polymerization was carried out via UV-irradiation (254 nm) with an energy of 5 $mW/cm^2$ at the water surface. The enclosed film balance was purged with nitrogen 1 hour to and during polymerization. The water substrate was ion exchanged and then double distilled in an all-glass-apparatus.

The polymerization attempts in solution were carried out under nitrogen in sealed ampoules with 1 mole % azobisisobutyronitrile in benzene for 24 hours at 60 °C.

Irradiation experiments were made with an 200 W UV-lamp in a quartz ampoule for 8 hours. No change could be detected by TLC and IR-spectroscopy.

## RESULTS AND DISCUSSION

The orientation of the molecules at the gas-water interface was studied by recording the pressure-area diagrams:

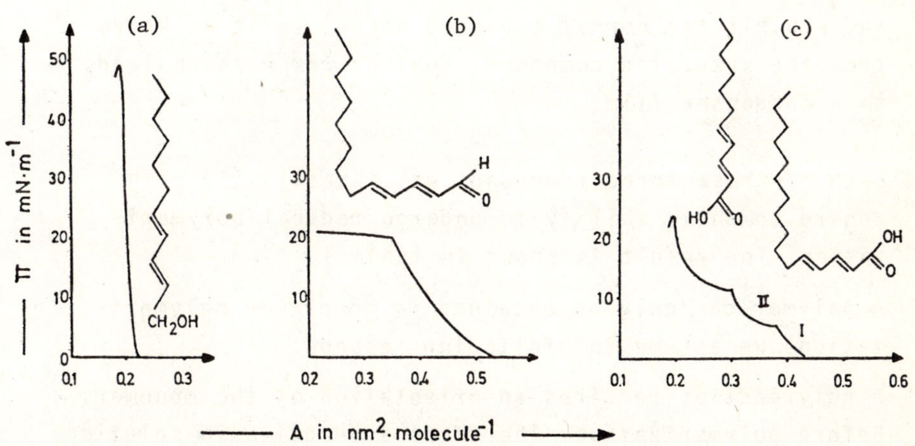


FIGURE 1a-c: Surface pressure-area diagrams at 20 °C of Octacecadienol (a), Octadecadienal (b) and Octadecadienoic acid (c)

The alcohol shows the same pressure-area diagram which was known from saturated compounds: The strong chain-chain interaction leads to a condensed film with an occupied area of about 0.20 $nm^2$/molecule, which is near to the closest packing of aliphatic chains, thus indicating a high degree of orientation and a dense packing.

The aldehyde forms an expanded film. The occupied area at the collapse point (0.35 $nm^2$/molecule) leads to the assumption that the whole diene group acts as a part of the headgroup and lies flat on the water surface. This orientation is caused by the conjugation of the diene unit to the carbonyl group, which polarizes the whole headgroup. At further compression of the film collapse occurs without forming a condensed phase.

In the case of the acid three phases can be seen depending on the temperature. The occupied area at the first transition point shows again, that the whole diene group lies flat on the water surface. When compressing the film the molecules are forced into an upright position and exhibit the normal expanded phase, as it is known from the saturated compound. Further compression leads to a condensed film.

Each of these three compounds was investigated with regard to their ability to undergo radical polymerization. The result is shown in table 1.

A polymer can only be obtained in monolayer polymerization. We assume the following reasons:

A polyreaction requires an orientation of the monomers before polymerization. Therefore a reaction in solution or melt is impossible. A polyreaction in the crystalline state, however, does not take place, because the diene units are not in the required close distance from each

TABLE 1
Polymerization Attempts of Disubstituted Butadienes

| STATE | INITIATION | POLYM. |
|---|---|---|
| solution | AIBN | - |
| solution | UV | - |
| melt | UV | - |
| crystal | UV | - |
| monolayer | UV | + |

other and the mobility of the molecules is too much suppressed.

The polymerization can be followed by measuring the contraction of the film versus the polymerization time. In the next figures the pressure-area diagrams of the monomeric and polymeric compounds and the area change versus time are shown.

Although in the case of the alcohol the change of the area per molecule is not very big, it can easily be measured. The reaction is finished after about 20 min. The resulting area is 0.23 $nm^2$/monomer unit, which corresponds to a close packing of the molecules. The reaction takes place without a change in the orientation of the molecules during the polyreaction.

The aldehyde has been polymerized at 15 °C at a constant pressure of 7 mN/m and 15 mN/m. In each case the same polymer was obtained. Two results of this reaction are remarkable:

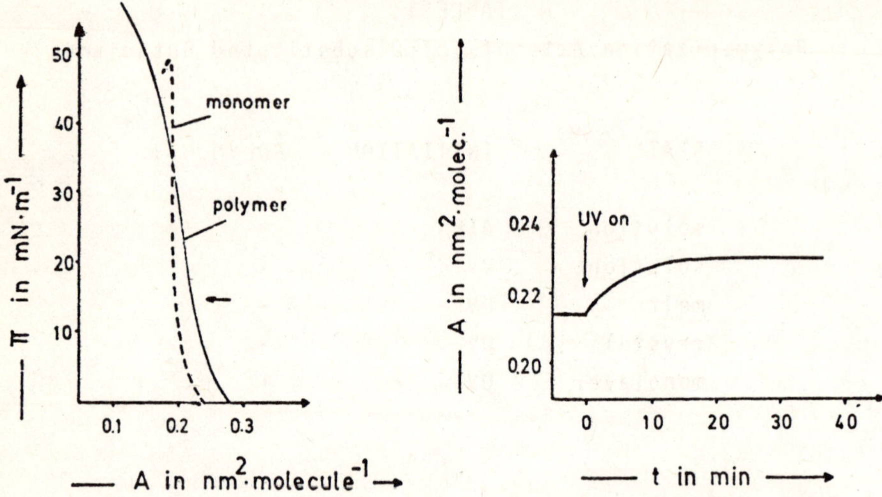


FIGURE 2: Pressure-area diagram and film contraction of Octadecadienol

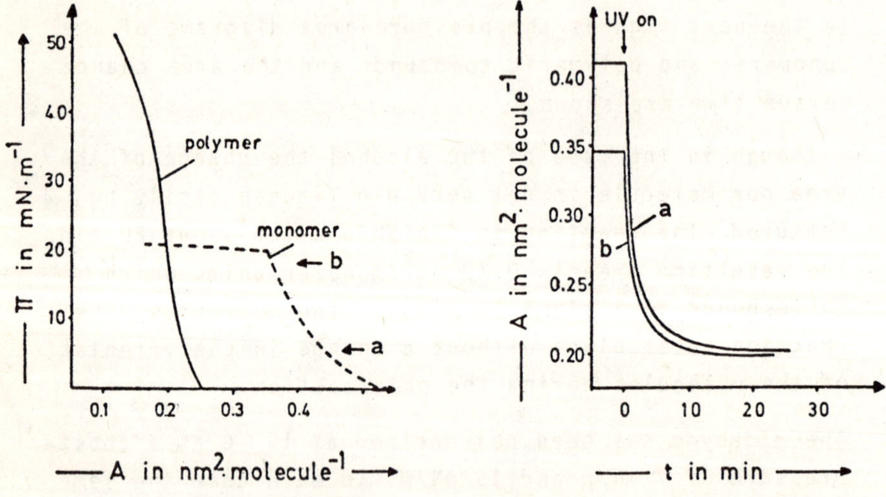


FIGURE 3: Pressure-area diagram and film contraction of Octadecadienal

1. The unexpected high polymerization rate; the contraction starts within a few seconds after switching on the UV-lamp and is finished after about 5 min.

2. The contraction curve of the films shows, that the resulting polymer has a very dense packing in spite of the comparably great occupied area of the monomer.

This can be explained by following assumptions:
The monomeric molecules in the monolayer and especially the headgroup which lies flat on the water, are oriented. This allows the very fast reaction. The polymerization rate is higher than in the case of the alcohol. This is caused, as mentioned before, by the conjugation of the carbonyl group to the diene units, which activates the diene moiety. During the reaction this conjugation is lost; only the aldehyde group can act as a headgroup, the occupied area of which is considerably lower; the molecules are therefore forced to stand up during the polyreaction, and the resultin polymer is closely packed.

The investigation of the acid was more complicated than in the case of the alcohol and the aldehyde, because the acid exhibits a condensed and two expanded phases. Therefore the polymerization has been carried out at different temperatures and in all three phases.
Polymerization in the condensed phase leads to a polymer with an occupied area of about 0.23 $nm^2$/monomer unit. The same polymer is obtained by polymerization in the normal expanded phase, for which we proposed an erected position of the monomer molecules. Polymerization in the second expanded phase leads to a polymer with a less dense packing of the molecules, which confirms our assumption of an orientation with the diene group lying

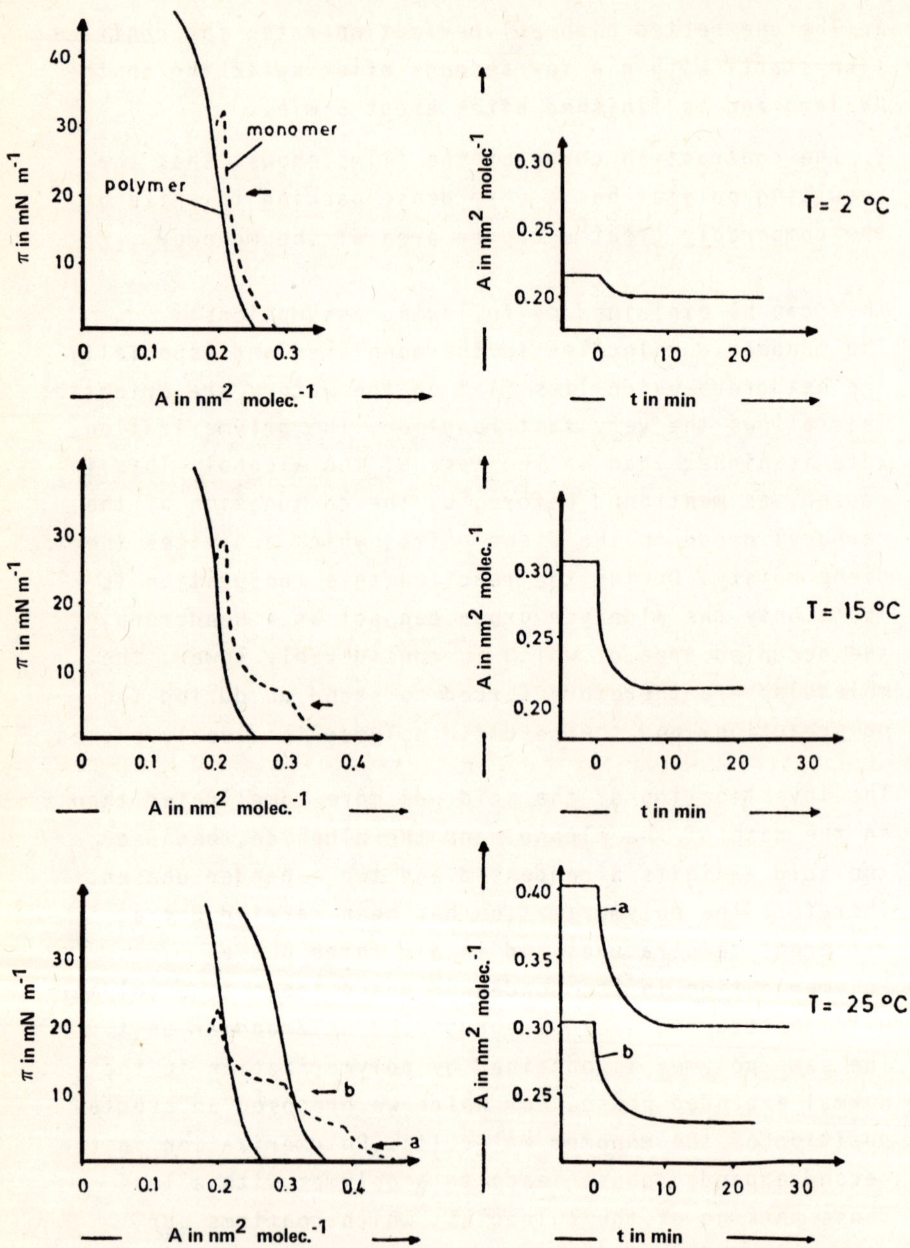


FIGURE 4: Pressure-area diagrams and film contraction of Octadecadienoic acid

flat on the water surface. In each case the polyreaction starts within seconds after switching on the UV-lamp and is finished after about 5 min.

The structure of the resulting polymer is not yet fully established. IR-measurements together with theoretical considerations leads to the assumption, that the polymer has a 1,4-trans backbone:

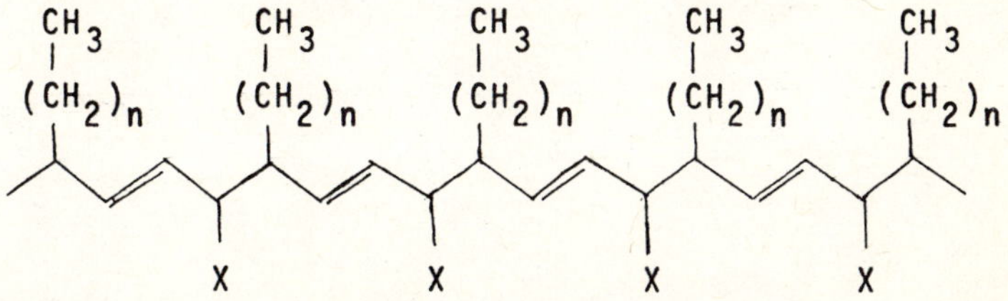

$CH_3$ $CH_3$ $CH_3$ $CH_3$ $CH_3$

$(CH_2)_n$ $(CH_2)_n$ $(CH_2)_n$ $(CH_2)_n$ $(CH_2)_n$

X X X X

The possible use of the monolayer polymers as synthetical membranes is already discussed elsewhere (4). In contrast to the Diacetylenes which form a rigid film due to the conjugated backbone, the polydienes are much more flexible and a better approach to biological membrane properties should be possible.

## REFERENCES

1. S.A. Letts, T. Fort Jr., J.B. Lando, J. Coll. Interface Sci., 56, 64 (1976)

2. R. Ackermann, O. Inacker, H. Ringsdorf, Kolloid-Z. u.Z. Polym. 249, 1118 (1971)

3. M. Hatada, M. Nishii, J. Polym. Sci. 15, 927 (1977)

4. D. Day, H. Ringsdorf, J. Polym. Sci. Polym. Lett. 16, 205 (1978)

5. G. Wegner, Makromol. Chem. 154, 35 (1972)

6. G.L. Gaines, Insoluble Monolayers at Liquid-Gas Interfaces, Interscience Publishers New York, 1966, p.184

7. G.C. Nutting, W.D. Harkins, J. Am. Chem. Soc. 61, 2040 (1939)

8. G.L. Gaines, Insoluble Monolayers at Liquid-Gas Interfaces, Interscience Publishers New York, 1966, p.162

9. E.L. Pippen, M. Nonaka, J. Org. Chem. 23, 1580 (1958)

# Novel Preparations of Specialty Polyamides by Interfacial and Solution Methods

W. Deits

GTE Laboratories, Inc.
40 Sylvan Road
Waltham, Massachusetts 02154

S. Grossman

The Upjohn Company
410 Sackett Point Road
North Haven, CT 06473

O. Vogl

Polymer Science and Engineering Department
University of Massachusetts
Amherst, Massachusetts 01003

ABSTRACT

A series of polyoxamide homopolymers and regular copolyoxamides was prepared by liquid-liquid interfacial, gas-liquid interfacial, and solution polycondensation techniques. Thus, poly(hexamethylene oxamide) {Nylon 62}, poly(ethylene oxamide) {Nylon 22}, and poly(p-phenylene oxamide) were prepared in moderate molecular weight by reacting gaseous oxalyl chloride with aqueous solutions of 1,6-hexanediamine, N,N'-bis(2-aminoethyl) oxamide, and p-phenylenediamine, respectively, using a gas-liquid interfacial technique developed for the continuous production of polyoxamides. A number of regular copolyoxamides were prepared by condensing various aliphatic diamine-diamides with both aliphatic and aromatic acid chlorides using both liquid-liquid interfacial and solution polymerization procedures. Both the yield and molecular weight of the polymers produced were found to be dependent on factors such as polymerization technique, monomer solubility, reaction temperature, reactant concentration, type of solvent, and type of acid acceptor.

## INTRODUCTION

Polyamides have been prepared by a number of different techniques including the dehydration of ammonium salts, the condensation of amino acids, the ring opening polymerization of lactams, and the condensation of specific acid derivatives, particularly acid chlorides with diamines.[1] The high reactivity of acid chlorides has enabled these types of polymerization reactions to be carried out at low temperatures with a minimum of side reactions. Consequently, the use of the well known Schotten-Baumann reaction for the preparation of high molecular weight polyamides has become very important, particularly when sequential regularities in the polymer chain must be maintained. The reaction of acid chlorides and other reactive carboxylic acid derivatives with amines has been utilized in a variety of different techniques, most notably interfacial polymerization and solution polymerization in solvents of high dielectric constants.

Liquid-Liquid Interfacial Polycondensation: In an interfacial polycondensation, the reaction takes place at or near the interface between a pair of liquids with limited miscibilities, generally water and an organic solvent. The aqueous phase usually contains the diamine and an added acid acceptor while the organic phase generally contains the diacid chloride. Polymer formation has been shown to occur in the organic phase in that monofunctional reactants dissolved in the aqueous phase do not always decrease the molecular weight in the proportion to which they are used.[2] The primary function of the aqueous phase is to provide a solvent medium for the diamine and acid acceptor and to remove the by-product acid from the polymerization zone. The controlled introduction of the water soluble diamine into an excess of diacid chloride in the organic phase is carried out by the interface.

In order to maintain a rapid rate of reaction, an efficient transfer and neutralization of by-product hydrogen chloride is essential, which is generally done by adding an appropriate acid acceptor. In the absence of an acid acceptor, however, a portion of the reactant diamine reacts with the by-product acid to form the unreactive hydrochloride salt which consequently decreases both the yield and molecular weight of the resultant polymer due to the insufficient amount of diamine subsequently available for reaction. The amount and type of acid acceptor, therefore, has a great influence on both the yield and molecular weight of the polymer produced.

The major side reaction in interfacial polymerization reactions is the hydrolysis of the diacid chloride reactant. This is particularly prevalent with the more reactive lower aliphatic diacid chlorides.[3] It has been shown, for example, that terephthaloyl chloride undergoes only 1% hydrolysis in 10 minutes at pH 10 while 50% of adipoyl chloride hydrolyzes in 3 minutes, even at a pH value of 4.5.[4] Hydrolysis rates of aliphatic diacid chlorides were also shown to decrease with increasing chain length and increase with rising temperature.[5] Oligomers end-capped with acid chlorides are also susceptible to hydrolysis, which has frequently been found to occur when the polymer itself is substantially hydrophilic and/or where polymerization occurs close to the aqueous portion of the interface.

Because of the nature of interfacial polymerization reactions, the presence of an equivalence of reactants is less critical than in other methods, however, it has been shown that a 12% excess of diacid chloride in the preparation of Nylon 610 decreases the inherent viscosity by 15% while the same excess of diamine decreases the viscosity by 25%.[6] The method of addition of the reactants as well as the stirring have also been shown to have a profound effect on the polymerization reaction as both yield and solution viscosity are effected.[7] Although interfacial polymerization reactions can tolerate some impurities in the reaction media, some monofunctional species such as the ethanol stabilizer in chloroform, and/or certain slow reacting difunctional materials will decrease the molecular weight of the polymer formed.[8]

Gas-Liquid Interfacial Polycondensation: In a gas-liquid interfacial polycondensation, an acid chloride diluted with an inert gas is bubbled through a hot aqueous solution of diamine. Although the subsequent reaction is interfacial in nature, it differs in several respects from typical liquid-liquid interfacial reactions.[9] In the first place, the reaction cannot take place in the organic phase as this phase is absent. Secondly, one of the phases is a gas which in turn alters many of the characteristics of the reaction itself. In normal liquid-liquid interfacial techniques, for example, the use of highly reactive acid chlorides such as oxalyl chloride leads to a significant amount of hydrolysis and a subsequent low molecular weight product,[2] whereas the limited solubility of gaseous oxalyl chloride at the elevated temperatures employed in the gas-liquid interfacial technique reduces hydrolysis and favors polymerization.[9-12] Consequently, polyamides that are prepared from highly reactive and volatile acid chlorides can be obtained in high yield and with high molecular weights.

In addition, gas-liquid interfacial polymerization techniques have a number of advantages over conventional methods of producing condensation polymers. A variety of otherwise difficult to utilize monomers may be employed, particularly the lower dicarboxylic acids, such as oxalyl chloride, phosgene, thiophosgene, and perfluoroadipic acid dichloride, as well as the relatively inflexible aromatic diamines such as p-phenylenediamine.[9] The method is rapid, and the need for organic solvents, high temperatures, and vacuum equipment is eliminated. The procedure can also be easily adapted for the production of polymer in a continuous manner.

Solution Polycondensation: In solution polycondensations, the polymerization reaction is carried out in a single liquid phase. The liquid may consist of a single solvent or a mixture or solvents inert to the reactants employed. Solution processes include not only systems where both reactants and products are soluble, but also systems where one or both of the reactants are undissolved in the early stages of the reaction, but dissolve as the reaction proceeds as well as systems where the polymer product precipitates during the course of the reaction.[2] In contrast to interfacial polymerization techniques, polymerization in solution avoids the possibility of acid chloride hydrolysis. The major disadvantage of solution techniques, however, is the necessity of utilizing systems with low levels of impurities.

The solvent used in solution polymerization reactions must be not only unreactive towards both the diamine and diacid chloride, but it must also be a good solvent for both reactants and polymer. It has, in fact, been shown that the molecular weight of a polyamide can be controlled through the use of good or poor solvents for the polymer.[13] As in interfacial procedures, an acid acceptor is usually necessary for most solution polymerization reactions involving diamines and diacid chlorides, although they are sometimes unnecessary when amide solvents such as DMAc are used. The major side reaction in solution polymerization is the destruction of the diacid chloride either by impurities or by the acid acceptor. Consequently, a high purity of reactants must be maintained as most reactive impurities will decrease the molecular weight of the polymer produced. A majority of side reactions can be avoided, however, by carrying out the reaction at low temperatures. An equivalence of reactants is also important in solution polymerization reactions as this method is more sensitive to reactant balance than are interfacial methods.

Despite these well known advances in polycondensation procedures, certain novel polyamides with unusually rigid backbone structures have remained difficult to prepare except under the most exacting conditions. These include the aromatic polyamides such as poly(p-phenylene terephthalamide),[14-18] poly(p-benzamide),[19-21] and poly(m-phenylene isophthalamide),[2,14] as well as the aliphatic polyamides based on the oxamide linkage. Neither polyoxamide homopolymers or copolymers have ever reached commercial importance in spite of their potential as high temperature nylons[22] or as materials for efficient desalination membranes.[23-29] This is due, in part, to difficulties involved in their production and fabrication. Standard methods of preparing polyamides have been found to be of limited use for the preparation of polyoxamide homopolymers due to a number of factors among which are their high melting point and limited solubility.

Polyoxamide homopolymers generally cannot be prepared directly from oxalic acid salts of aliphatic diamines, for example, as the oxalate salts tend to decompose to CO and $CO_2$ at the high temperatures necessary for melt polycondensations. The ring opening polymerization of oxalic acid lactams, on the other hand, have been shown to give polyamides, although under melt polymerization conditions some loss of CO and $CO_2$ again takes place, thus limiting the molecular weights obtainable.[30-32] Furthermore, the relative insolubility of most polyoxamides prevents the formation of high molecular weight polymer when using conventional solution techniques, whereas the use of typical liquid-liquid interfacial polycondensation procedures is limited due to the rapid hydrolysis of the oxalyl chloride starting material. Polymerization at a gas-liquid interface, on the other hand, is a polymerization technique initially developed by Sokolov[10-12] for the preparation of polymers based on acid chlorides that are both reactive and volatile. On the other hand, certain solution and liquid-liquid interfacial techniques can be used for the preparation of many regular copolyoxamides.[23]

## EXPERIMENTAL

### Poly(hexamethylene oxamide) {Nylon 62}

Using a procedure similar to that described earlier,[33] a solution of hexamethylene diamine (465g, 4.00 moles) and sodium carbonate (848g, 8.00 moles) in 20 liters of water was added slowly to a reactor which was equipped so as to allow continuous addition of fresh diamine solution while maintaining a constant volume of solution. The apparatus also allowed for the continuous removal of polymer. The aqueous diamine solution was preheated to 95$^{\circ}$C by passing it through a heating coil prior to introducing it into the reactor. The polymerization was carried out by dispersing a gaseous mixture of oxalyl chloride and dry nitrogen in the hot diamine solution with a 70 μm to 100 μm fritted glass filter. Polymer formation was almost immediate, and a white foam consisting of small microballoons of poly(hexamethylene oxamide) could be seen floating on top of the diamine solution within a few minutes. The reaction was carried out for several hours while about 100g of oxalyl chloride was passed through the solution. The resultant moist polymer foam was washed repeatedly with water in a one-gallon Waring blender. The white fibrous material was then washed twice with methanol and dried at room temperature for 48 hr at 20 mm. The polymer yield was 28g (31% based on oxalyl chloride) and the inherent viscosity (0.5% in trifluoroacetic acid) was 0.65 dL/g.

Poly(ethylene oxamide) {Nylon 22}

Nylon 22 with an inherent viscosity (0.5% in trifluoroacetic acid) of about 0.4 dL/g was prepared from N,N'-bis(2-aminoethyl) oxamide[7] in a manner similar to that used for the preparation of Nylon 62.

Poly(p-phenylene oxamide)

p-Phenylene diamine was recrystallized and sublimed immediately prior to use. Poly(p-phenylene oxamide) was prepared in a manner similar to that described above. The reaction was carried out in the absence of oxygen in order to obtain a nearly colorless polymer of reasonable molecular weight. Polymers with an inherent viscosity (0.5% in $H_2SO_4$) of 0.8 to 1.1 dL/g were obtained.

Polymerization of N,N'-bis(6-aminohexyl)oxamide (m-6-2-6) with Sebacyl Chloride: (p-6-2-6-10)

(1) Interfacial Polymerization.

N,N'-Bis(6-aminohexyl)oxamide (0.86g, 0.003 mole) and sodium carbonate (0.64g, 0.006 mole) was dissolved in 300 ml of distilled water. The organic solution was prepared by mixing sebacyl chloride (0.72g, 0.003 mole) with tetrachloroethylene (300 ml) in a 500 ml Erlenmeyer flask. To a one-quart Waring blender containing the aqueous diamine solution, was poured the organic solution while the blender was stirring at high speed. The blending was continued for 4 min., the mixture was filtered through a 600 ml sintered glass filter funnel and the solid polymer was washed with acetone (300 ml), water (400 ml), and acetone (100 ml). The residue was extracted in a Soxhlet extractor for 3 hrs. with tetrahydrofuran and dried for 20 hrs. at $80^{o}C/1$ mm. The polymer weighed 0.86g (63% yield), had a m.p. of $257^{o}C$ (by DSC) and decomposed at $360^{o}C$ (by TGA). The inherent viscosity (0.5% $H_2SO_4$ solution) of this polymer was 1.10 dL/g. Anal. Calcd. for $(C_{24}N_{44}O_4N_4)_n$: N, 12.4%; Found: N, 12.6%.

(2) Solution Polymerization.

To a 500 ml three-neck round bottom flask equipped with a mechanical stirrer, a condenser closed with a drying tube and a dropping funnel, was added m-6-2-6 (0.90g, 0.003 mole), and triethylamine (0.30g, 0.003 mole) in chloroform (40 ml). The flask was flushed with nitrogen and a slow flow of nitrogen was maintained during the reaction. The mixture was cooled to $10^{o}C$ and sebacyl chloride (0.72g, 0.003 mole) in chloroform (10 ml) was added from a dropping funnel. Stirring was continued for 3 hrs., 300 ml of n-hexane was added and the suspension was filtered, washed consecutively with acetone (200 ml), water (400 ml) and acetone (200 ml), and dried at $60^{o}C/1$ mm for 20 hrs. The polymer weighed 0.85g (62% yield) and had an inherest viscosity (0.5% $H_2SO_4$ solution) of 0.45 dL/g.

Polymerization of N,N'-Bis(2-aminoethyl)oxamide (m-2-2-2) with Isophthaloyl Chloride: (p-2-2-2-I)

(1) Interfacial Polymerization.

N,N'-Bis(2-aminoethyl)oxamide (32.1g, 0.18 mole) and sodium carbonate (39.0g, 0.37 mole) in distilled water (1600 ml) were allowed to react with isophthaloyl chloride (37.4g, 0.18 mole) in chloroform (1000 ml). The reaction time was 3 min., and a polymer yield of 55.0g (99%) was obtained. The polymer had a m.p. of 367°C (by DSC), and decomposed above 370°C (by TGA). The polymer was soluble in sulfuric acid, trifluoroacetic acid and warm DMAc containing 3% LiCl. The inherent viscosity (0.5% $H_2SO_4$ solution) of this polymer was 1.32 dL/g. Anal. Calcd. for $(C_{14}H_{16}N_4O_4)_n$: N, 18.4%; Found: N, 18.4%.

(2) Solution Polymerization.

To a 2000-ml four-neck round-bottom flask was added N,N'-bis (2-aminoethyl)oxamide (57.0g, 0.33 mole), anhydrous triethylamine (100g, one equivalent excess), LiCl (48g), and anhydrous DMAc (120 ml). The mixture was warmed to 45°C until everything dissolved and isophthaloyl chloride (67.3g, 0.33 mole) in 300 ml of DMAc was added over a 1-hr. period. The temperature rose spontaneously to 55°C, where it was maintained for 18 hrs. Triethylamine hydrochloride precipitated almost immediately. The warm mixture was poured into 1200 ml of ice water, the polymer precipitated and was filtered and washed with water, a water/acetone mixture and acetone, and dried at 40°C/20 mm for 48 hrs.; yield 82.0g (82%). The polymer decomposed above 355°C (by DSC). The inherent viscosity (0.5% $H_2SO_4$ solution) was 0.90 dL/g.

Polymer of N,N'-Bis(4-aminobutyl)oxamide (m-4-2-4) with Isophthaloyl Chloride: (p-4-2-4-I)

(1) Interfacial Polymerization.

N,N'-Bis(4-aminobutyl)oxamide (2.3g, 0.01 mole) and sodium carbonate (2.1g, 0.2 mole) in 300 ml of distilled water were allowed to react with isophthaloyl chloride (2.0g, 0.01 mole) in 200 ml of chloroform; reaction time 5 min.; polymer yield 3.0g (84%); m.p. 302°C (by DSC). The polymer decomposed above 350°C (by TGA). The inherent viscosity of a 0.5% solution of this polymer was 1.6 dL/g. Anal. Calcd. for $(C_{18}H_{24}N_4O_4)n$: N, 15.5%. Found: N, 15.6%.

(2) Solution Polymerization.

To a 250 ml three-neck round-bottom flask was added DMAc (50 ml) containing 5% LiCl, triethylamine (4.0g, 0.04 mole), and m-4-2-4 (4.5g, 0.02 mole). The mixture was warmed slightly until everything was dissolved. Isophthaloyl chloride (4.1g, 0.02 mole) in 50 ml of DMAc was added slowly at 0°C over a period of 1 hr. After 2 hrs. the mixture was allowed to warm to 25°C, stirred for 10 hrs., and poured into 500 ml of water. The suspension was filtered, and the solid was washed with water (200 ml), then acetone, and dried; polymer yield 5.8g (81%); m.p. 287°C (by DSC). The polymer decomposed above 350°C (by TGA). The inherent viscosity of a 0.5% solution of the polymer in DMAc containing 3% LiCl was 0.9 dL/g.

## RESULTS AND DISCUSSION

Polymerization at a gas-liquid interface has been shown to be an effective method for the preparation of high molecular weight polyoxamide homopolymers.[10-12] The method of Sokolov was modified as described earlier[8] to provide for higher yields and continuous operation.

During a gas-liquid interfacial polymerization, both the molecular weight and yield increase rapidly as the temperature increases beyond 80°C. Polymerization at a gas-liquid interface must, therefore, be carried out at high temperatures in order to be effective. One would normally expect high temperatures to cause a rapid hydrolysis of oxalyl chloride in the basic media, however, the solubility of the gaseous monomer in the aqueous media is minimal at high temperatures, thus, the rate of acid chloride hydrolysis is reduced. As in the original method, the concentration of gaseous oxalyl chloride in nitrogen and the concentration of the diamine in the aqueous phase also influence the polymerization to a great degree, as does the basicity of both the diamine used and the reaction media itself.

Using this technique, several polyoxamides were produced using various diamines with oxalyl chloride (Equation 1).

$$H_2N\text{-}R\text{-}NH_2 + ClC(=O)C(=O)Cl \longrightarrow \left( NH\text{-}R\text{-}NHC(=O)C(=O) \right)_n \qquad (1)$$

$$R = (CH_2)_6$$

$$-C_6H_4-$$

$$-CH_2CH_2NHC(=O)C(=O)NHCH_2CH_2-$$

Nylon 62 prepared from hexamethylenediamine and oxalyl chloride was obtained in about 40% yield with an inherent viscosity of 0.6 to 0.8 dL/g. The highest molecular weight material was obtained when the concentration of the diamine was approximately 0.2 mole/liter. Nylon 62 prepared by gas-liquid interfacial polymerization was obtained in the form of microballoons of relatively tough polymer with an oriented structure.[33] The microballoons were formed of very thin films of the polymer, consequently the bulk density of the material obtained by this technique was quite low.

Nylon 22 cannot be made directly by the condensation of oxalyl chloride with ethylene diamine, as piperidine-2,3-dione is the normal reaction product (Equation 2). We have, however, succeeded in preparing nylon 22 from N,N'-bis(2-aminoethyl)oxamide (m-222) (Equation 3) in yields of 30% to 50% but with rather low inherent viscosity. Cyclo-condensation of m-222 and oxalyl chloride would give a 12-membered ring which apparently does not occur to a significant extent. The polymer obtained remains suspended in the polymerization mixture and, unlike nylon 62, does not form microspheres. The inherent viscosity and yield depend on the diamine and diacid chloride concentration in much the same manner as described for nylon 62.[10] Nylon 22 forms brittle films, however, as would be expected from the less flexible backbone structure.

Poly(p-phenylene oxamide) has been obtained in 20% to 50% yield with inherent viscosities of up to 1.1 dL/g. High molecular weight material can be produced over a much narrower range of diamine concentrations than is possible with nylon 62 or nylon 22, however. The inherent viscosity of the polymer decreases significantly when concentrations of diamine greater or less than 0.02 to 0.05 mole/liter are used. As with nylon 22, poly(p-phenylene oxamide) is formed as a suspended solid in the aqueous reaction mixture. This is presumably due to the brittle nature of these polymers preventing the formation of intact microballoons of polymer film.

$$H_2NCH_2CH_2NH_2 + Cl\overset{O}{\overset{\|}{C}}\overset{O}{\overset{\|}{C}}Cl \longrightarrow \begin{matrix} & NH & \\ CH_2 & & C{=}O \\ | & & | \\ CH_2 & & C{=}O \\ & NH & \end{matrix}$$

$$\not\searrow \quad -\!\!\left(NHCH_2CH_2NH\overset{O}{\overset{\|}{C}}\overset{O}{\overset{\|}{C}}\right)\!\!-_n \qquad (2)$$

nylon 22

$$\begin{matrix} O{=}C\diagup NHCH_2CH_2NH_2 \\ | \\ O{=}C\diagdown NHCH_2CH_2NH_2 \end{matrix} + Cl\overset{O}{\overset{\|}{C}}\overset{O}{\overset{\|}{C}}Cl \not\longrightarrow \begin{matrix} & NHCH_2CH_2NH & \\ O{=}C & & C{=}O \\ | & & | \\ O{=}C & & C{=}O \\ & NHCH_2CH_2NH & \end{matrix}$$

$$\searrow \quad -\!\!\left(NHCH_2CH_2NH\overset{O}{\overset{\|}{C}}\overset{O}{\overset{\|}{C}}\right)\!\!-_n \qquad (3)$$

Both liquid-liquid interfacial and solution polycondensation procedures were used to prepare a variety of regular copolyoxamides[23-29] as shown in Equation 4, where m and p designate the diamine-diamide monomer used and the resultant polymer, respectively, and where A represents the number of carbon atoms in the diamine portion, B represents the number of carbon atoms in the diacid portion, and D represents the diacid chloride comonomer portion.

Among other factors, both solubility and reaction time were important considerations in the selection of the polymerization technique used with the solubility of the diamine reactant the most important factor. Hydrophilic diamine-diamides such as m-2-2-2 to m-6-2-6 were found to be water soluble and as such were suitable for standard liquid-liquid interfacial polycondensations.[24] Diamine-diamides with longer aliphatic groups in the diamine portion could only be used effectively in solution polycondensations, however.

$$\text{Excess } \underset{A}{H_2N{-}\!(CH_2)_n\!{-}NH_2} + \underset{B}{\text{EtOCOCOOEt}} \longrightarrow$$

$$\underset{\substack{A\ -\ B\ -\ A \\ \text{monomer} \\ \text{m-A-B-A}}}{H_2N{-}\!(CH_2)_n\!{-}NHCONH{-}\!(CH_2)_n\!{-}NH_2} + 2\ \text{EtOH} + {-}\!\!\left[NHCOCONH{-}\!(CH_2)_n\right]\!\!{-}$$

$$+ \underset{D}{ClCO{-}R{-}COCl}$$

$${-}\!\!\left[NH{-}\!(CH_2)_n\!{-}NHCOCONH{-}\!(CH_2)_n\!{-}NHCO{-}R{-}CO\right]\!\!{-}$$

p-A-B-A-D Regular Copolyoxamide

(4)

The polymer p-6-2-6-10 was the material that received the greatest attention as its preparation was used as a model for a study of procedures suitable for the synthesis of regular copolyoxamides by both interfacial and solution methods.[25] The diamine monomer used, m-6-2-6, was found to be only moderately soluble in water (0.8 g/dL at 25$^{\circ}$C), hence interfacial polymerizations had to be carried out either in dilute solution at room temperature or in higher concentrations at elevated temperatures. Sebacyl chloride was found to be suitable under these conditions due to its reduced rate of hydrolysis.

Interfacial polycondensations were carried out using tetrachloroethylene, chloroform, or carbon tetrachloride as the organic solvent. Copolyoxamides containing the more hydrophilic segments tended to swell in the solvents used for the polymerizations which made their isolation difficult. Monomer concentrations were generally kept at about 1% as higher concentrations caused the formation of a thick polymer gel which tended to affect the blending efficiency.

The nature of the acid acceptor utilized was shown to have a pronounced effect on the molecular weight of the polymer produced as evidenced by their inherent viscosities.[25] Sodium bicarbonate, for example, was found to be an inefficient acid acceptor, as was sodium hydroxide. Utilizing sodium carbonate as an acid acceptor, on the other hand, gave p-4-2-2-10 with an inherent viscosity of 1.03 dL/g from m-4-2-4 and sebacyl chloride while sodium hydroxide yielded polymer with an inherent viscosity of only 0.54 dL/g. Similarly inherent viscosities of 1.09 and 0.67 dL/g were obtained when sodium carbonate and sodium hydroxide respectively were used with m-6-2-6 and sebacyl chloride for the preparation of p-6-2-6-10. The amount of acid acceptor used was also found to be critical as the inherent viscosity decreased by 50% when the quantity of acid acceptor was reduced by 20%.

The use of standard liquid-liquid interfacial techniques was found to be unsuitable for the preparation of regular copolyoxamides derived from lower diacid chlorides (e.g., malonyl or succinyl chloride) as those intermediates were readily hydrolyzed. In fact, no polymers of useful molecular weights were obtained using liquid-liquid interfacial techniques when diacid chlorides lower than adipoyl chloride were used. This phenomenon was particularly noticeable when the diamine-diamides used were highly water soluble which caused the resultant polymers to be quite hydrophilic.[34]

Polycondensations performed in solution generally utilized chlorinated hydrocarbons as solvents, however, DMAc both with and without LiCl was used for the preparation of aliphatic/aromatic copolyoxamides.[26] In several cases, the molecular weight of the product prepared by solution techniques was found to be limited by premature precipitation, a problem commonly encountered in the synthesis of many homo and copolyoxamides. The acid acceptor most commonly used for solution polycondensations was triethylamine while the reaction temperature was generally held below 40$^{\circ}$C in order to minimize side reactions. In addition, the concentration of polymer in solution polycondensations was generally kept below 2.5% as the starting materials often could not be kept in solution at higher concentrations.

In an effort to gain a better understanding of the factors involved in synthesizing copolyoxamides using solution polymerization procedures, the sequence of addition of the reactants was varied. The addition of the diamine-diamide to the diacid chloride was found to give the highest molecular weight material in the highest yield. By adding m-6-2-6 to sebacyl chloride, for example, p-6-2-2-10 with an inherent viscosity of 0.43 dL/g was produced in 62% yield, while the reverse order of addition afforded polymer with an inherent viscosity of only 0.13 dL/g in a 34% yield.[25]

Both standard liquid-liquid interfacial and solution polycondensation techniques were compared for the preparation of a series of copolyoxamides based on the condensation of a number of diamine-diamides with the same diacid chloride. Chloroform was used as the organic solvent in the interfacial polymerizations, while DMAc with LiCl was used as the solvent in the solution procedures. When m-0-2-0 was condensed with isophthaloyl chloride using interfacial techniques, an inherent viscosity of 0.23 dL/g was obtained while solution techniques gave polymer with an inherent viscosity of 0.44 dL/g. On the other hand, inherent viscosities of 1.3 and 0.90 dL/g respectively were obtained when m-2-2-2 was used. Similarly, interfacial polymerizations using m-4-2-4 as the diamine-diamide gave an inherent viscosity of 1.6 dL/g while a value of 0.88 dL/g was obtained using solution techniques. A number of copolyoxamides were also prepared by interfacial methods utilizing m-2-2-2 as the diamine-diamide. Inherent viscosities of 0.4 to 0.8 dL/g were routinely obtained. Most aromatic diacid chlorides gave copolyoxamides with inherent viscosities between 0.6 and 0.8 dL/g, however. When the diamine-diamide m-2-4-2, the only diamine-diamide without an oxamide linkage, was condensed with isophthaloyl chloride, on the other hand, a polymer with an inherent viscosity of 2.5 dL/g was readily obtained.[26]

From our studies comparing both gas-liquid and liquid-liquid interfacial techniques with solution techniques for the preparation of both aliphatic and aromatic polyoxamides and aliphatic/aliphatic and aliphatic/aromatic copolyoxamides, it was found that solubility and reactivity (i.e., hydrolyzability) were both essential factors for the preparation of high molecular weight materials. The purity and reactivity of the solvents employed was also found to be critical. The choice of solution versus interfacial polymerization techniques for preparing polymers was thus based on those factors. The gas-liquid interfacial technique was found to be far superior for the preparation of polyoxamide homopolymers, while solution techniques, although more difficult, were generally found to be most suitable for the preparation of regular copolyoxamides. Liquid-liquid interfacial techniques were found to be more readily carried out than either gas-liquid interfacial or solution techniques and thus were useful for the fast screening of reaction conditions and polymer properties.

## REFERENCES

1. W. Sweeny and J. Zimmerman, Encyl. Polym. Sci. and Technol. 10, 484 (1969).

2. P.W. Morgan, Condensation Polymers: By Interfacial and Solution Methods, Interscience Publishers, New York, 1965.

3. V.V. Korshak, T.M. Frunze, S.V. Vinogradova, V.V. Kurashev and A.S. Lebedeva, Polymer Sci. (U.S.S.R.), 3, 109 (1962).

4. P.W. Morgan, Ref. 2, p. 87.

5. W.L. Wasley, R.L. Whitfield, L.A. Miller and R.Y. Kodoni, Textile Res. J., 33, 1029 (1963).

6. P.W. Morgan and S.L. Kwolek, J. Polym. Sci., 62, 45 (1962).

7. T.M. Frunze, V.V. Korshak, V.V. Kurashev and P.A. Alienskii, Polymer Sci. (U.S.S.R.), (English Transl.), 3, 109 (1962).

8. P.W. Morgan and S.L. Kwolek, J. Polym, Sci., 62, 53 (1962).

9. L.B. Sokolov, Synthesis of Polymers by Polycondensation, Israel Program for Scientific Translations, Jerusalem, 1968.

10. L.B. Sokolov, L.V. Turetskii and T.V. Kudim, Vysokomol. Soedin. 2, 1744 (1960).

11. L.B. Sokolov, J. Polym. Sci. 58, 1253 (1962).

12. L.B. Sokolov, T.V. Kudim and L.V. Turetskii, Vysokomol. Soedin, 3, 1370 (1960).

13. P.W. Morgan, J. Polym. Sci., C4, 1075 (1964).

14. S.L. Kwolek, P.W. Morgan and W.R. Sorenson, U.S. Patent 3,063,966 (1962).

15. T.I. Bair and P.W. Morgan, U.S. Patent 3,673,143 (1972).

16. T.I. Bair and P.W. Morgan, U.S. Patent 3,817,941 (1974).

17. A.A. Federov, et al., Vysokomolekul. Soedin., Ser. B., 15, 74 (1973).

18. T.I. Bair, P.W. Morgan and F.L. Killian, Polymer Preprints 17(1), 59 (1976).

19. S.L. Kwolek, U.S. Patent 3,600,350 (1971).

20. S.L. Kwolek, U.S. Patent 3,671,542 (1972).

21. S.L. Kwolek, P.W. Morgan, J.R. Schaefgen, Polymer Preprints 17(1), 59 (1976).

22. J.A. Somers, Man-Made Textiles, Feb., 60 (1956).

23. H.J. Chang, Ph.D. Thesis, University of Massachusetts, 1974.

24. H.J. Chang and O. Vogl, J. Polym. Sci., Polym. Chem. Ed., 15, 311 (1977).

25. H.J. Chang and O. Vogl, J. Polym. Sci., Polym. Chem. Ed., 15, 1043 (1977).

26. D. Stevenson, A. Beeber, R. Gaudiana, and O. Vogl, J. Macromol. Sci., Chem., 11(4), 779 (1977).

27. D. Tirrell and O. Vogl, J. Polym. Sci. Polym. Chem. Ed., 15(8), 1889 (1977).

28. O.F. Vogl and D.R. Stevenson, U.S. Patent 3,935,172 (1976).

29. O.F. Vogl and D.R. Stevenson, U.S. Patent 4,002,563 (1977).

30. O. Vogl and A.C. Knight, Macromolecules 1, 315 (1968).

31. P.W. Morgan, Polymer Reviews 10: Condensation Polymers, Interscience, New York, NY (1965).

32. O. Vogl and A.C. Knight, Macromolecules 1, 311 (1968).

33. Y. Chatani, Y. Ueda, H. Tadokoro, W. Deits, and O. Vogl, Macromol. 11, 636 (1978).

34. D. Tirrell, S. Grossman, and O. Vogl, Makromol. Chem. 180, 721 (1979).

# Topochemical Reactions in Langmuir-Blodgett Multilayers

Bernd Tieke, Volker Enkelmann, Harald Kapp,
Günter Lieser, and Gerhard Wegner

Institut für Makromolekulare Chemie
der Universität Freiburg
Hermann-Staudinger-Haus
Stefan-Meier-Straße 31
D-7800 Freiburg i. Br. (F.R.G.)

ABSTRACT

The topochemical polymerization of diacetylenes and the dimerization of cinnamic acid derivatives have been studied in Langmuir-Blodgett-multilayers. The resulting polymer films are of controlled thickness and morphology and exhibit excellent stability. Structure, morphology and phase changes occuring in monomer and polymer films have been investigated using spectroscopic methods, electron and x-ray diffraction. Possible applications of the polymer films are discussed.

INTRODUCTION

Reactions in monomolecular layers or in multilayers built up by the Langmuir-Blodgett technique (1) have recently become of interest because of the orientational effects obtainable in these structures. Polymerizations, in particular, have been demonstrated to be a way to effectively stabilize these normally quite unstable molecular assemblies (2-4).

Of special interest are topochemical, i.e. strictly lattice controlled reactions where minimal structural disorder is introduced by the reaction and the molecular packing is retained (5,6). Polymer films prepared by this technique could be effectively utilized as stable ultrathin coatings or membranes of controlled thickness in a variety of industrial or biomedical applications.

The topochemical reactions studied in built-up multilayers were the solid-state polymerization of diacetylenes and the four-center-type dimerization of cinnamic acid derivatives (5,6). The monomers used in this investigation were unsaturated long chain fatty acids of the general structure

$$C_mH_{2m+1}\text{-}C{\equiv}C\text{-}C{\equiv}C\text{-}(CH_2)_8\text{-}COOH$$

with m = 9 (1), m = 10 (2) and m = 12 (3), which polymerize upon ultraviolet radiation in a 1,4-addition reaction to form a planar, conjugated polymer backbone of the general structure

$$\cdots{=}\underset{R}{C}\text{-}C{\equiv}C\text{-}\overset{R'}{C}{=}\cdots$$

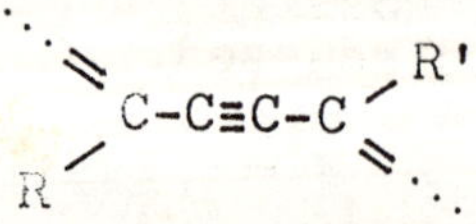

and long chain ether and ester derivatives of the 4-hydroxy-cinnamic acid which dimerize according to

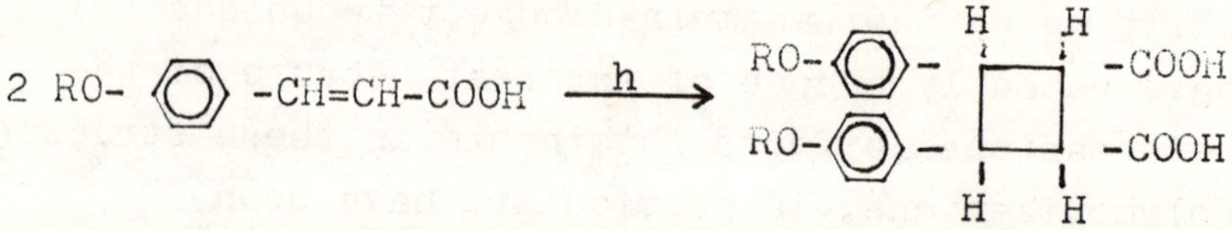

$$2\ RO\text{-}C_6H_4\text{-}CH{=}CH\text{-}COOH \xrightarrow{h} \begin{matrix} RO\text{-}C_6H_4\text{-}CH\text{-}CH\text{-}COOH \\ \vert \qquad \vert \\ RO\text{-}C_6H_4\text{-}CH\text{-}CH\text{-}COOH \end{matrix}$$

with R = $C_{16}H_{33}$ (4), $C_{18}H_{37}$ (5) and $C_{16}H_{33}\text{-}\underset{O}{\overset{\Vert}{C}}\text{-}$ (6).

Both reactions have in common that they are strictly lattice controlled, i.e. both the reactivity and the symmetry of the products are determined by the monomer packing. All reactivity comes about from very specific rotations of the monomer molecules on their lattice sites. Since these reactions take only place in the perfectly packed crystalline state they offer an excellent opportunity to study the state of order in mono- and multilayers.

## EXPERIMENTAL

### Chemicals

Diacetylene monocarbonic acids were prepared as described previously (7). Ether derivatives of the cinnamic acids were prepared according to the method described by Cray and Jones (8).

### Mono-and Multilayer Formation

A commercially available Langmuir trough equipped with a film balance (MGW Lauda) was used. Monolayers were obtained by spreading the acid solutions in $CHCl_3$ (spectroscopic grade, concentration 1 mg/ml) on a subphase of triply distilled water (pH 5,9 - 6,1) or a $10^{-3}$ molar $CdCl_2$-solution. pH-variations were carried out by addition of 0,1 n HCl or NaOH, respectively.

Multilayers were built up by the Langmuir-Blodgett-technique (1). Transfer of the monolayers onto quartz- or germanium plates was performed at a surface pressure of 20 mN/m.

### Polymerization

Diacetylene derivatives were polymerized using a medium-high-pressure mercury lamp (Heraeus Hanau Q

300). Cinnamic acid derivatives were polymerized using a 150 W Xe-lamp.

Multilayer Characterization

The UV-spectra were recorded from multilayers deposited on quartz plates (24 x 12 x 1 mm) using a Perkin-Elmer Hitachi 200 spectrometer.

Structure investigations were performed by SAXS using a Philips powder diffractometer and electron diffraction using a Philips EM 400 electron microscope. Details of sample preparation have been described previously (9,10).

## MONOLAYER PROPERTIES

The monolayer properties of the derivatives 1-6 have been studied by measuring the surface pressure-area-isotherms. Isotherms of 1-3 have been described previously (11,12), those of 4-6 are shown in Fig.1. In general, the stability increases with increasing chain length and is maximal when the subphase contains counterions such as $Cd^{2+}$, which allow a salt formation of the polar head groups.

As it is known that the pH of the subphase has a marked effect on the salt formation in the monolayers (13), it was necessary to study the conditions of the Cd-salt formation more in detail. The results of such studies are shown in Fig. 2 for 3. If the collapse pressure of the monolayer is plotted as a function of the pH in the subphase, the salt formation seems to be essentially complete at a pH of 6,5.

More direct evidence for a salt formation is obtained by means of ATR-IR-spectroscopy (9,10,14). In

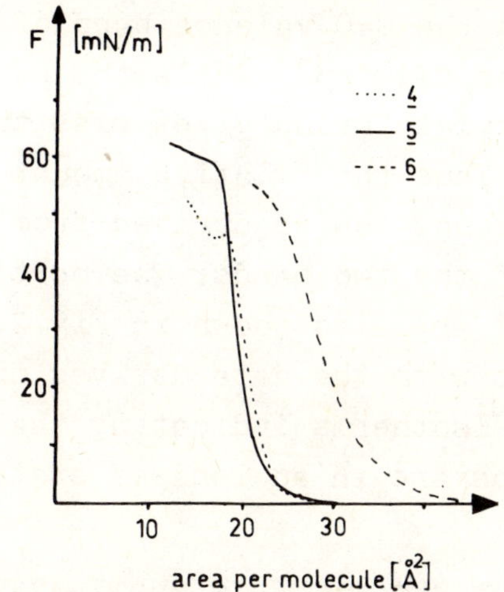


Fig.1. Surface pressure-area-isotherms of 4-6 on a $10^{-3}$ molar $CdCl_2$-solution, T=21°C, pH=6,5.

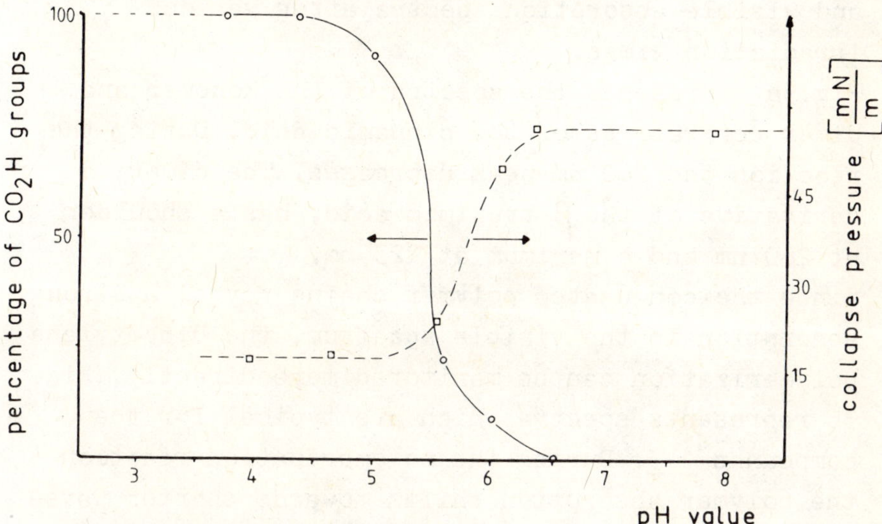


Fig.2. Percentage of free acid 3 in the multilayers (full line) and collapse pressure (broken line) as a function of the pH of the subphase; subphase $10^{-3}$ molar $CdCl_2$-solution, T=20°C.

the IR-spectra the C=O valence band of the free acid group occurs at 1710 $cm^{-1}$ while the presence of an ionized carboxylate group gives rise to a strong band at 1550 $cm^{-1}$. Thus the relative amount of acid to carboxylate groups can be derived from the relative intensities of the two bands. The pertinent results obtained for 3 are also shown in Fig.2. They are in good agreement with the data derived from the surface pressure-area-isotherms indicating that about 95% of all acid groups are in an ionized state at a pH of 6,5.

## REACTIONS IN MULTILAYERS

Both the reactions are initiated by ultraviolet radiation and can be detected by monitoring the uv- and visible absorption spectra after various irradiation times.

Fig.3a represents the spectra of the monomer and dimer derivatives of the cinnamic acid. During the reaction the 260 nm-peak decreases. The dimer, a derivative of the $\beta$-truximic acid, has a shoulder at 280 nm and a maximum at 223 nm.

Since the conjugated polymer chains reveal a strong absorption in the visible spectrum, the diacetylene polymerization can be monitored more directly. Fig. 3b represents spectra which are typical for the compounds 1-3. During the polymerization reaction the polymer absorption shifts towards shorter wavelengths by about 100 nm.

## MORPHOLOGY OF THE MULTILAYERS

Investigations on the morphology were carried out using SAXS and electron diffraction.

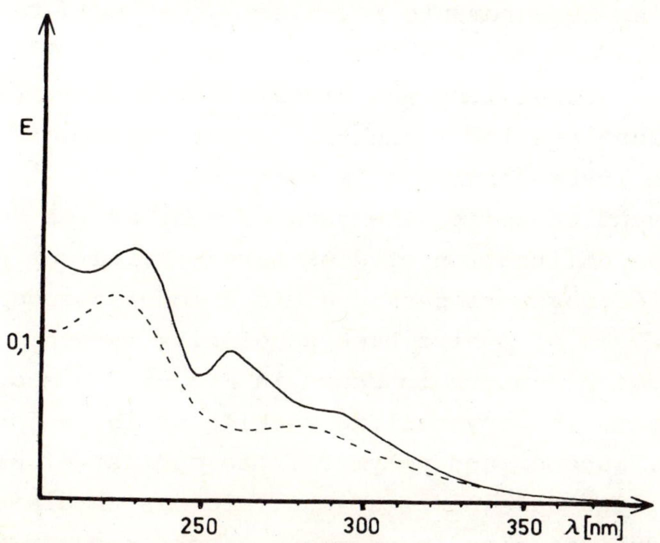


Fig.3a. Transmission spectra of a multilayer of 4 as Cd-salt before (——) and after (----) 30 min of UV-irradiation.

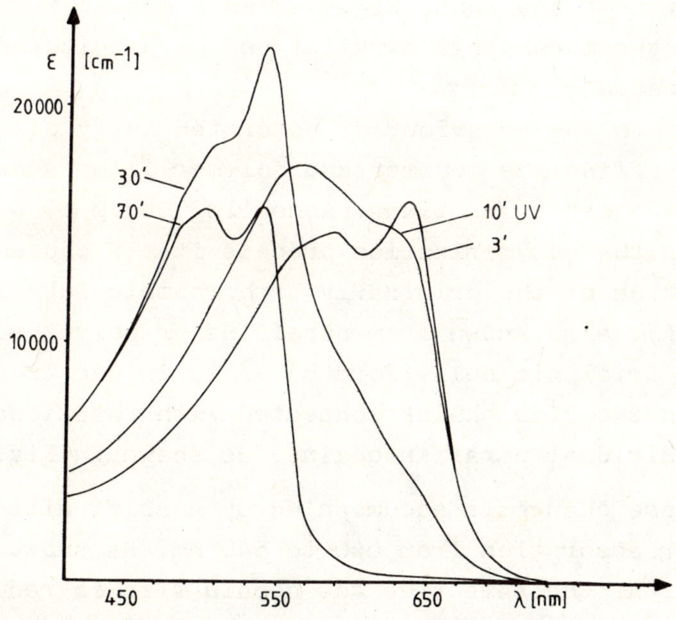


Fig.3b. Transmission spectra of a multilayer of 3 as Cd-salt after different UV-irradiation times.

The x-ray measurements revealed a head-head-tail-tail (Y-) structure of the multilayers (9-11). In both the diacetylene and cinnamic acid derivatives the Cd-Cd-long spacing increases during the reaction, but the layer structure is retained.

To prevent damage to the samples most of the electron diffraction studies were carried out at liquid nitrogen temperature. As a model compound multilayers of 3 have been studied. A typical diffraction pattern is shown in Fig.4. The spots are due to the crystalline packing of the aliphatic groups. Monomer and polymer films consist of perfect-crystalline domains of several microns in diameter. Individual domains in adjacent layers are not in register with each other. The Cd-ions are not located at fixed lattice sites which is probably caused by inclusions of water, due to a partial solvation of the ions. Fig.5 shows a model of the multilayer morphology parallel and perpendicular to the substrate.

Similar to the behaviour of saturated fatty acids and paraffins the monomer and polymer films show various phase transitions. Annealing above 60$^{\circ}$C as well as the polymerization process itself cause a transition of the originally orthorhombic subcell (form A), also known from paraffins or polyethylene, into a triclinic cell (form B). This is due to a slip in the side chains connected with rotations of the individual paraffin chains, as shown in Fig.6.

The phase change is accompanied by a shift of the polymer absorption from 640 to 540 nm, as shown in Fig.3b. At the same time the domain size is reduced

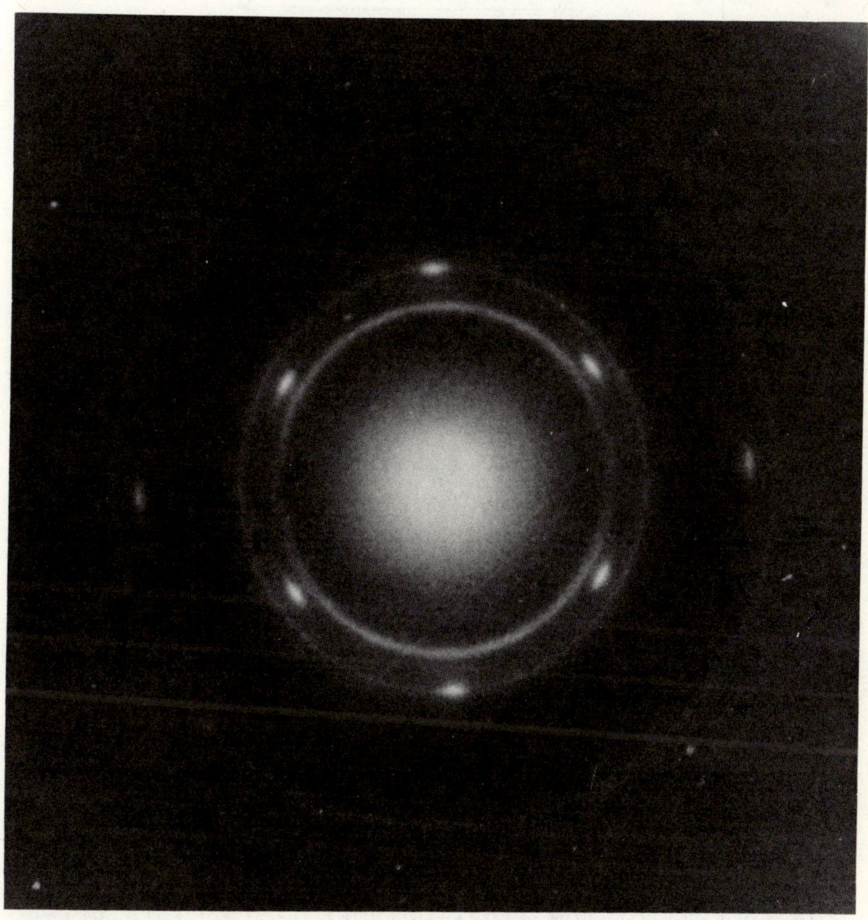

Fig.4. Electron diffraction pattern of a stack of ten monomeric double layers, specimen temperature -150°C.

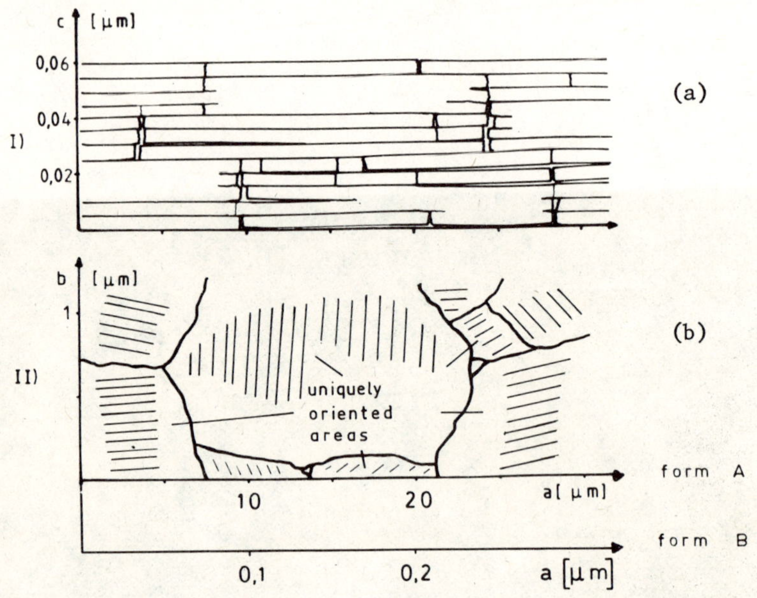


Fig.5. Model of a multilayer: a) parallel and b) perpendicular to the substrate.

and a polycrystalline texture of very small islands, less than 0,1 micron in diameter, is obtained (9,10).

Upon storage as monomers the multilayers tend to reorganize into microcrystals. UV-irradiation of these samples leads to a green-blue polymer with an absorption maximum at 700 nm which also shifts towards shorter wavelengths during the reaction process. The subcell of the final polymer (form C) is triclinic (Fig.6). Fig.7 shows absorption spectra of multilayers of 1-3 which have been uv-irradiated for 5 min. after a storage at 20°C for three days. The

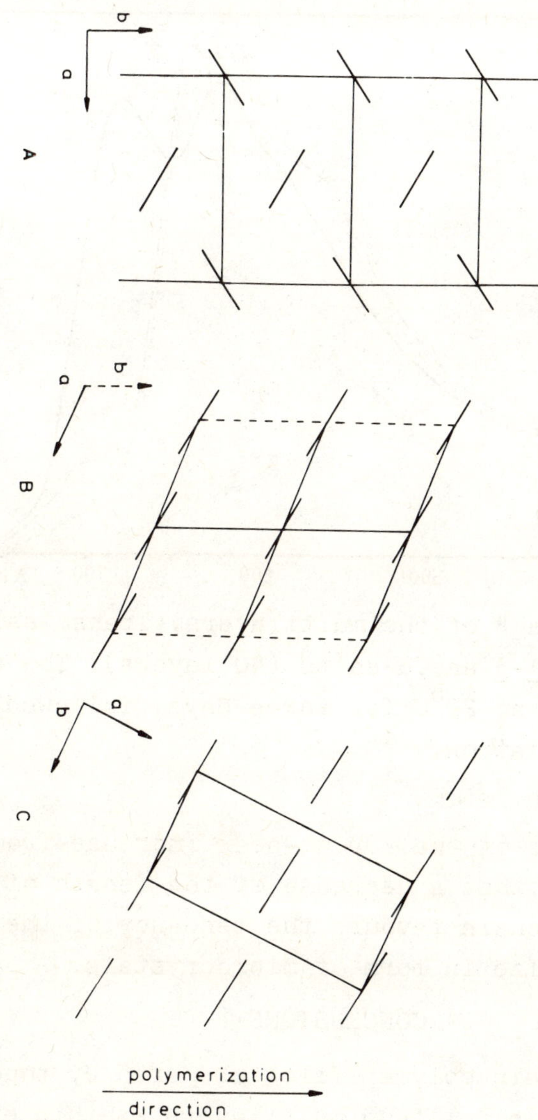


Fig.6. Projection along the paraffin chain axes. The bars represent the zig-zag planes (the setting angle is arbitrary).

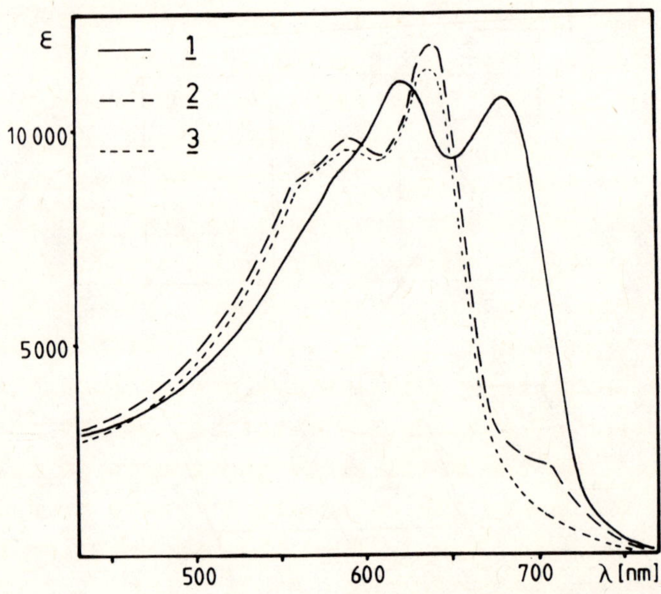


Fig.7. "Aging" of the multilayers. Transmission spectra of 1-3 as Cd-salts (40 layers). The samples were stored at 20°C for three days, followed by 5 min. of UV-irradiation.

intensities of the 700 nm-peak increase from 3 to 1 indicating that a decrease of the length of the aliphatic chain favours the tendency of the compounds to reorganize in form of microcrystals.

## CONCLUSIONS

The ultrathin polymer films prepared by topochemical reactions in built-up multilayers exhibit excellent stability and could show potentially useful applications as coatings or membranes of controlled thickness and morphology. It should also be emphasized that multilayers are ideal systems for the investigation of solid-state reactions by spectroscopic methods. According to the special mechanism of multilayer deposition the number of monomer molecules per unit area and the number of layers are exactly known. The samples available by this technique are so thin that several problems arising in large, three dimensional crystals from inhomogeneous product distribution, effects of crystal optics and high extinction coefficients which complicate a quantitative spectroscopic investigation of solid-state reactions can be safely neglected. In the multilayers conversion to polymer, quantum yield and the detection of phase transitions can be evaluated easily from the transmission spectra (9,10,15).

## ACKNOWLEDGEMENT

The authors gratefully acknowledge financial support for this work by the Deutsche Forschungsgemeinschaft.

## REFERENCES

(1) K.B.Blodgett, I.Langmuir; Phys.Rev. 51,964(1937)

(2) V.Enkelmann, J.B.Lando; J.Pol.Sci.Polym.Chem.Ed. 15,1843(1977)

(3) D.Naegele, J.B.Lando, H.Ringsdorf; Macromolecules 10,1339(1977)

(4) D.Day, J.B.Lando; J.Pol.Sci.Polym.Chem.Ed. 16, 1431(1978)

(5) G.M.J.Schmidt, in Reactivity of the Photoexcited Molecule, Interscience Publ., New York, 1967, p.227 ff

(6) G.Wegner; Pure & Appl. Chem. 49,443(1977)

(7) B.Tieke, G.Wegner, D.Naegele, H.Ringsdorf; Angew.Chem.Int.Ed.Engl. 15,764(1976)
(8) J.W.Cray, P.Jones; J.Chem.Soc. 1954,1467
(9) G.Lieser, B.Tieke, G.Wegner; Thin Solid Films 68,77(1980)
(10) B.Tieke, Thesis, Freiburg 1978
(11) B.Tieke, G.Lieser, G.Wegner; J.Pol.Sci.Polym. Chem.Ed. 17,1631(1979)
(12) D.Day, H.Ringsdorf; J.Pol.Sci.Polym.Lett.Ed. 16,205(1978)
(13) J.A.Spink; J.Colloid Sci. 18,512(1963)
(14) N.J.Harrick; Internal Reflection Spectroscopy, Interscience Publ., New York, 1967
(15) J.P.Fouassier, B.Tieke, G.Wegner; Isr.J.Chem. 18,227(1979)

# Interfacial Polycondensation in Aqueous and Non-Aqueous Systems

Naoya Ogata, Kohei Sanui, Takao Onozaki, and Shoehei Imanishi

Department of Chemistry
Sophia University
7 Kio-Cho, Chiyoda-Ku
Tokyo, Japan 102

ABSTRACT

Interfacial polycondensation in aqueous and non-aqueous systems has been investigated in order to synthesize various polyamides having functional groups. Rigid aromatic polyamides having pyridine moieties (PPy) were synthesized by interfacial polycondensation using an aqueous system and the solution properties of PPy in concentrated sulfuric acid were investigated in terms of solution viscosity and lyotropic behavior. Interfacial polycondensation in a non-aqueous system using two immiscible solvent was found to be useful for the synthesis of aromatic polyesters and copolyesters.

## INTRODUCTION

An interfacial polycondensation reaction takes place at the interface between a solution of a dibasic acid chloride in a water-immiscible organic solvent and a water solution of, for example, a diamine containing an acid acceptor. This reaction is useful to prepare various condensation polymers including polyamides or polyesters under mild conditions.

However, hydrolysis of the acid halide, which is always a possible side reaction, may compete with the acylation reaction to prevent the formation of a high polymer. Moreover, when a monomer has somewhat lower solubility in water, the reaction must be carried out in dilute solution, which leads to a poor yield of the resulting polymer with a low molecular weight. Therefore, an interfacial polycondensation using immiscible binary solvents would appear to be the most useful route for the preparation of condensation polymers from monomers which are less-reactive with acid chloride and less-soluble in water.

This paper deals with the synthesis of various polyamides and polyesters by means of the interfacial polycondensation method in aqueous and non-aqueous systems, especially in terms of the synthesis of aromatic polymers having functional groups. Details were published elsewhere [1,2].

## RESULTS AND DISCUSSION

### Interfacial Polycondensation in Aqueous System

There has been extensive work in the field of lyotropic liquid-crystalline polymers which have a highly rigid structure in concentrated solutions. Such a family of polymers is the fully aromatic polyamides; a typical polymer of this family is poly(p-phenyleneterephthalamide), known in fiber form as Kevlar, and fibers with an extremely high tensile strength have been obtained from this polymer by a wet-spinning method by virtue of high orientations of the polymer molecules owing to the lyotropic liquid-crystalline properties of the polymer solutions.

It was of interest to investigate the lyotropic properties of a rigid polymer having polar groups within the backbone of the polymer chain since these polar groups may interact with each other in concentrated solutions. Therefore, fully aromatic polyamides containing pyridine moieties have been synthesized and the lyotropic liquid-crystalline properties of the concentrated solutions in sulfuric acid have been investigated [1].

An aqueous solution of p-phenylenediamine containing an equimolar equal amount of potassium carbonate was mixed with a solution of various solvents containing 2,5-pyridinedicarboxylic acid chloride (PyCC). Polymers formed at the interface were isolated by an ultracentrifuge and washed repeatedly with water, followed by washing with methanol and dried.

Lyotropic properties of the polymers having pyridine moieties were investigated by dissolving the polymers in concentrated sulfuric acid or in a mixed solvent (1:1) of N-methylpyrrolidone/hexamethylphosphorictriamide containing 3 wt% of LiCl. The lyotropic behavior of the solution was observed under a polarized microscope at various temperatures using a heating stage.

An interfacial polycondensation of 2,5-PyCC with p-phenylenediamine in an aqueous system was carried out and it was found that water-immiscible solvents such as methylene chloride, chloroform, benzene, toluene, and m-xylene gave a quantitative yield of the corresponding polyamide. Among these solvents the methylene chloride/water system attained the highest solution viscosity (over 1.2) for the resulting polymer. Since 2,5-PyCC is more reactive than terephthaloyl chloride, such hydrophilic solvents as cyclohexanone or tetrahydrofuran caused a partial hydrolysis so that both yields and solution viscosities of the polyamides were not satisfactory.

The polymer and copolymer derived from 2,5-PyCC were quite stable up to 400°C, although increasing the units of incorporated 2,5-PyCC reduced the thermal stability in comparison with poly(p-phenylene terephthalamide). Neither distinct melting points nor glass transition temperatures were observed in these polymers by means of DTA or DSC.

Sulfuric acid dissolved the polymers to a high concentration and a lyotropic behavior of the concentrated solutions was observed from isotropic to anisotropic solutions with increasing concentrations of the polymers. The color of the concentrated solutions, which showed an anisotropic property, was dark brown, and an opaline luster appeared upon stirring or allowing the solutions to flow, owing to a high orientation of the rigid polymer chains.

The phase diagram of the solution of poly(p-phenyleneterephthalamide), PT, exhibited a different behavior from that of the polyamide from PyCC in that the PT solution showed an optically anisotropic behavior at much lower concentrations and higher temperatures in comparison with the solution of the polyamide from PyCC. These phenomena might be ascribed to the better solubility of the later polyamide owing to the basic pyridine moieties in the rigid polymer chains in comparison with PT molecules in concentrated solutions.

## Interfacial Polycondensation in Non-Aqueous Systems

An interfacial polycondensation of terephthaloyl chloride with bisphenol A or resorcinol was carried out by using such immiscible binary solvents as adiponitrile/carbon tetrachloride, adiponitrile/dibutyl ether, ethylene glycol/carbon tetrachloride, and ethylene glycol/benzene systems. Bisphenol A or resorcinol was dissolved in a polar solvent at a desired concentration in the presence of a two fold number of moles of triethylamine as acid acceptor, while terephthaloyl chloride was dissolved in a non-polar solvent. The two immiscible solutions were mixed with vigorous stirring at room temperature for 5 minutes and the solutions were poured into excess acetone and the precipitated polymers were recovered by filtration.

It was found that the interfacial polycondensation using binary solvents was useful for preparing aromatic polyesters in terms of yields and solution viscosities of the resulting polyesters, when compared with the interfacial polycondensation in aqueous systems.

Among the combinations of immiscible binary solvent systems, the adiponitrile/carbon tetrachloride system was superior to other solvent systems and aromatic polyesters having inherent viscosities as high as 0.9 were obtained in quantitative yield.

Copolycondensation of bisphenol A and resorcinol with terephthaloyl chloride was carried out by the interfacial polycondensation using non-aqueous systems. Generally speaking, the compositions of the resulting copolyesters were almost the same as those of the monomer feed ratios, while solution viscosities reached a maximum value of 0.8 in the range of feed ratio of 50/50 of bisphenol A and resorcinol.

Partition coefficients of bisphenols between two immiscible solvents were determined [3] in binary solvents at room temperature and found to be very different according to the binary systems. Therefore, sequence distribution of the copolyesters could be controlled from random to block types by the selection of the binary solvents.

## REFERENCES


1. N. Ogata, K. Sanui, and T. Koyama, accepted in Journal of Polymer Science, Polymer Chemistry Edition.

2. N. Ogata, K. Sanui, and S. Imanishi, accepted in Polymer Journal.

3. H. Nakamura, S. Imanishim, K. Sanui, and N. Ogata, Polymer Journal, 11, 661 (1979).

[illegible]

[illegible] quantity of [illegible] binary solvent system [illegible] the [illegible] each of [illegible] other solvent was [illegible] and aromatic polyesters [illegible] which [illegible] were obtained in [illegible] concentration of [illegible] and [illegible] specifically speaking [illegible] Those of [illegible] while [illegible] reached a maximum value of [illegible] the range of [illegible] respectively.

[illegible] between [illegible] different [illegible] to the [illegible] distribution of the [illegible] from random to block types [illegible] binary solvents.

REFERENCES

1. [illegible]

2. [illegible]

3. [illegible]

# Index

F



G



H



I



K



L



M

R